Revised edition

Knowing and recognizing

The biology of glasshouse pests and their natural enemies

M.H. Malais - W.J. Ravensberg

Knowing and recognizing, the biology of glasshouse pests and their natural enemies

is published by:

Koppert B.V.
P.O. Box 155
2650 AD Berkel en Rodenrijs, The Netherlands
Tel. +31 (0)10 514 04 44
Fax +31 (0)10 511 52 03
E-mail: info@koppert.nl
www.koppert.com

Reed Business Information
P.O. Box 4
7000 BA Doetinchem, The Netherlands
Tel. +31 (0)314 34 99 11
Fax +31 (0)314 34 38 39
E-mail: agriboek@reedbusiness.nl
www.agriboek.nl

This edition is a completely revised and expanded version of the first edition published in 1992 under the same title.

Authors
   M.H. Malais, W.J. Ravensberg
   With contributions of J.N. Klapwijk, J. van Schelt,
   S. Mulder, J. Douma, R.K. van der Pas, A.E. van Baal

Working-group (Koppert B.V.)
   A.E. van Baal (R&D Entomology), R.G. Boer (Marketing),
   S.C. Bol (Sales), J.B. Douma (Marketing), Y.M. van Houten
   (R&D Entomology), J.N. Klapwijk (R&D Entomology),
   M.H. Malais (Marketing), Y.T.A. van Markus (Public Relations),
   P.A. Moerman (Marketing), S. Mulder (R&D Entomology),
   R.K. van der Pas (R&D Microbials), W.J. Ravensberg (R&D
   Microbials), J. van Schelt (R&D Entomology),
   R. Timmer (R&D Entomology)

Art direction:
   G. Massop

Digital realisation:
   Grafisch ontwerp bureau Ap Timmers

Illustrations:
   Drawings: R. Slinger
   Aquarelles: I. van Noortwijk

Photography:
   See page 288

Project management:
   R. Koopmans

Translation:
   M.J. Pearson
   M. Whittacker

ISBN 90 5439 126 X

All rights reserved. No part of this book may be reproduced, transmitted and / or made public for consultation by third parties by means of print, photocopy, microfilm, video or magnetic disc or tape, or in any system, or in any manner, whether electronic, mechanical or otherwise without the express prior permission of the publishers. This prohibition applies both to the whole book and to any part thereof. The sole exception is that which is defined in copyright relating to reproduction rights.
The publishers accept no liability for any damage, of whatever nature, that may result from actions or decisions taken on the basis of information in this book.

Preface

PREFACE to the 1992 edition

For a long time Koppert B.V. has been confronted with questions about the behaviour of pest organisms in glasshouse crops and of the natural enemies used against them. As a leading company in the area of biological control, it has tried to answer these questions by producing this publication. The information in this book has been kept accessible to various groups of users. Most important are growers, who are facing disease and pest control daily. It also applies to suppliers and advisors involved in horticulture and those in horticultural education. In fact this book is a guideline for everyone who is now, or in the future, concerned with growing a healthy crop.

In compiling the book, several scientific publications were used as references, and that information is supplemented with the specialist knowledge of Koppert B.V. Illustrations and colour photographs help to explain the descriptions.

Thanks are due to many people who have given their co-operation in preparing this book. First the board of Koppert B.V., which offered the opportunity to compile the book and have given it a well presented appearance. For supplying data and constructive criticisms thanks are due to: Prof. J.C. van Lenteren and Dr. O.P.J.M. Minkenberg (both of the Faculty of Entomology of the Agricultural University of Wageningen), Dr. P. van Rijn (Faculty of Fundamental and Applied Ecology of the University of Amsterdam), Dr. S. Steinberg (Biological Control Insectaries, Sde Eliyahu, Israel) and to everyone else who has contributed to this book in their own specific field.

We hope that the information in this book will contribute to the success of biological control. A knowledge of pest organisms and their natural enemies is essential to the correct application of this technique. Only with this knowledge can there be an adequate response to pest situations.

PREFACE to the 2003 edition

In this new edition the number of organisms dealt with – both pests and their enemies - has been enormously expanded. More attention has also been given to ornamental crops. The book has also been more richly illustrated and the latest scientific information incorporated.

We owe a debt of gratitude to the many who contributed to the preparation of this new edition. Various colleagues supplied information and critically commented on the text. The critical comments on chapter 9 provided by J.P. Kaas (Bio Pré) have been particularly useful. We are grateful to a number of individuals from the *Praktijkonderzoek Plant & Omgeving (PPO)* [Plant and Environmental Practice Research Unit], and the *Plantenziektenkundige Dienst* [the Plant Pathology Service] who supplied answers to various questions, colour slides and critical comments on parts of the text. We especially wish to thank G. Vierbergen and H. Stigter of the *Plantenziektenkundige Dienst* for their contribution. All those who responded to an appeal on the internet by supplying illustrative material are also very warmly thanked. Finally we are most grateful to Professor J.C. van Lenteren (Department of Entomology at the University of Wageningen) for writing the Foreword to this edition.

Marleen Malais
Willem Ravensberg

Knowing and recognizing :
the basis for responsible and successful biological control

"Biological control does not stand still"

When in 1967 the project of controlling pests in glasshouses was tentatively embarked on, few expected that this form of crop protection would take off quite so spectacularly. In those days each grower had to be carefully instructed in the role that useful insects can play in the glasshouse, whereas nowadays the point of discussion is more likely to concern which of several natural enemies can best be used. In Europe, the biological control of pests is now applied on a large acreage of glasshouse crops, and increasingly so among ornamental plants. This is not so much a consequence of environmental awareness on the part of horticulturalists - although in the last thirty years this has increasingly become part of the equation - but more especially the fact that biological control strategies allow pests to be controlled effectively and with economic benefits. But biological control also offers a series of other advantages over conventional chemical pesticides. With biological control, neither the persons putting it into practice and harvesting the crop nor the consumer are exposed to chemical pesticides with their associated health risks; nor does the plant suffer any of the effects of phytotoxicity, and the environment is also spared. Further, the release of natural enemies usually takes less time and is far more enjoyable work than spraying crops in protective clothing. Once biological agents have been introduced, their work of controlling the pests continues for several months instead of requiring weekly spraying. Indeed, some pests are much more effectively controlled by natural enemies than by pesticides. Another significant advantage is that there is no harvest interval, as is the case with chemical sprays, and if an efficient natural enemy has been discovered this species can be used for hundreds of years without fear that the pest will develop resistance, as in the case of chemical pesticides. And finally, both consumers and government set great store by this form of pest control.

Are there then no disadvantages attached to this method? Yes, of course there are. Living organisms cannot be patented, so anyone with green fingers can culture these 'bugs' with the result that there is always the risk that organisms of lesser quality sometimes come onto the market. However, major companies like Koppert Biological Systems employ rigorous quality controls before any biological agents are sent out. Another disadvantage is that the biological control requires considerable knowledge on the part of the grower, knowledge that needs to be regularly supplemented and brought up to date. The demand for knowledge, in fact, was the original reason for writing this book.

What role does *Knowing and recognizing* play in the biological control of diseases and pests in glasshouses? The philosophy of Koppert Biological Systems has always been that biological control agents should not be treated like chemical agents. The starting point had to be proper guidance on all problems relating to crop protection in the glasshouse. The horticulturalist was thus not simply buying a control agent but was paying for a service which includes the biological control agents. In this way, Koppert's advisors could provide expert advice on what chemical agents might also be used, where necessary, alongside the natural enemies without destroying those natural enemies. This book is not designed to provide all this information for the grower, but that was never its intention.

The first edition of this book, which appeared in 1992, fulfilled an important role in satisfying the curiosity of growers and all those interested in this form of pest control by presenting the biology of the more important pests and their natural enemies. On the basis of a good description of the different organisms and their interactions, growers and others could get a better understanding of how biological control worked as an effective method, and also of the knowledge demanded of the user. That first edition turned out to have fulfilled an enormous need. It was used much more widely than merely by those working in the horticultural industry and it appeared in several languages.

Biological control in glasshouses, however, did not stand still. For a start, new pests appeared, but also, following research and exhaustive tests in the field, a range of new natural enemies appeared on the market. For this reason, after 11 years there is now a wholly new edition in which, in each chapter, a group of pests is described together with their natural enemies. In view of the success of the first edition, this new, much expanded edition should have no trouble in finding its place on the bookshelves of all those who work in horticulture, or have an interest in the biological control of pests.

J.C. van Lenteren
Professor of Entomology
Wageningen University

Introduction

In the control of pests through the use of natural enemies, knowledge of both kinds of organisms is of the utmost importance if good results are to be obtained. In this book you will find descriptions of the life habits of the most important invertebrate pests in glasshouse horticulture (which in this book refers to al forms of protected cultivation) and of the natural enemies used to control them. The description of each organism includes an account of its life-cycle with illustrated descriptions of each stage, information on the organisms reproduction, how it spreads in the crop, the disease and damage that it can cause, and its specific behavioural characteristics. Where similar pest species or natural enemies are encountered, those characteristics that allow them to be distinguished are clearly indicated.

The accounts of these life-cycles and life habits have been compiled from numerous sources. Much of the information comes from the scientific literature, especially where more specialized and detailed subjects are dealt with. References to the most important sources in the literature, containing further information on the pests and their natural enemies, can be found toward the end of the book. In the text itself, only the tables refer to the original sources of the information displayed.

Data on development times and rate of reproduction are normally obtained under controlled laboratory conditions, and can therefore by no means be directly translated into the conditions one finds in practice. This kind of information must therefore be interpreted with a degree of caution. Nevertheless, the data give a clear guideline and contain much useful, comparative information.

In the text, each chapter places the organisms to be discussed in the groups to which they belong. The animal and plant kingdoms are divided into a number of hierarchical groups that provide for a clear classification - a taxonomy. A general overview of the groups that are dealt with in this book, and their position within this system of biological classification, is given in chapter 1.

We shall not spend time on the actual methodology of pest control using natural enemies; these methods of control vary enormously from one pest to another, and between different crops. Moreover, conditions change rapidly, such as culture techniques, the number of chemical agents that can be integrated with the natural enemies, the quantities of the natural enemies to release, new pests that appear in crops, and so forth. It must also be borne in mind that any integrated control scheme must be crop-oriented; this calls for a totally different approach from that used in this book. In chapter 14, however, we do briefly look at the question of observing and monitoring pests, their natural enemies and crop damage.

At the back of the book can be found an index with English and Latin names of the pest organisms and their natural enemies that are discussed. This book is a new edition of a book published earlier under the same title. In this new edition, the number of both pest organisms and their enemies has been enormously expanded. More attention has also been given to ornamental crops and, in addition, the text has been more richly illustrated, and incorporates the latest scientific information. Please, keep in mind that this is an international publication. Some of the products described may not be approved for use in certain countries.

Contents

Preface 3

Biological control does not stand still 4

Introduction 5

1. Classification of insects and other organisms 9

Taxonomic classification of organisms: main outline 14
Taxonomic classification of organisms: exopterygote insects 16
Taxonomic classification of organisms: endopterygote insects 18
Taxonomic classification of organisms: mites 20

2. Spider mites and their natural enemies 21

Spider mites 22
  - *Tetranychus urticae* / two-spotted spider mite 23
Natural enemies of spider mites 27
  - *Phytoseiulus persimilis* 28
  - *Amblyseius californicus* 31
  - *Feltiella acarisuga* 34
  - *Stethorus punctillum* 36
Biological control of spider mites 37

3. Other mites 39

Bryobia spp. 40
Tenuipalpidae / false spider mites 42
Eriophyidae / gall mites 44
  - *Aculops lycopersici* / tomato russet mite 44
Tyrophagus spp. / storage mites 46
Tarsonemidae / tarsonemid mites 48
  - *Polyphagotarsonemus latus* / broad mite 48
  - *Tarsonemus pallidus* / cyclamen mite 51
Oribatida / moss mites 53

4. Whiteflies and their natural enemies 55

Whiteflies 56
  - *Trialeurodes vaporariorum* / glasshouse whitefly 59
  - *Bemisia tabaci* / tobacco whitefly 61
  - Differences between the various developmental stages of *Trialeurodes vaporariorum* & *Bemisia tabaci* 65
  - *Aleyrodes proletella* & *A. lonicerae* / cabbage whitefly & strawberry whitefly 66
Natural enemies of whiteflies 67
  - *Encarsia formosa* 68
  - *Eretmocerus eremicus* 72
  - *Eretmocerus mundus* 75
  - *Macrolophus caliginosus* 76
  - *Verticillium lecanii* 79
Biological control of whiteflies 81

5. Thrips and their natural enemies — 83

Thrips — 85
- *Thrips tabaci* / onion thrips — 88
- *Frankliniella occidentalis* / western flower thrips — 89
- *Thrips fuscipennis* / rose thrips — 91
- *Echinothrips americanus* — 92
- Overview of important thrips species — 94

Natural enemies of thrips — 96
- *Amblyseius cucumeris* — 97
- *Amblyseius degenerans* — 100
- *Hypoaspis* spp. — 101
- Overview of important species of predatory mite — 102
- *Orius* spp. — 103
  - *Orius laevigatus* & *Orius majusculus* — 105
  - Overview of important *Orius* species — 107
- *Verticillium lecanii* — 108
- Entomophthorales — 108

Biological control of thrips — 109

6. Leaf miners and their natural enemies — 111

Leaf miners — 112
- *Liriomyza bryoniae* / tomato leaf miner — 115
- *Liriomyza trifolii* / american serpentine leaf miner — 117
- *Liriomyza huidobrensis* / pea leaf miner — 118
- *Chromatomyia syngenesiae* / chrysanthemum leaf miner — 119
- Differences between various leaf miner species — 120

Natural enemies of leaf miners — 121
- *Diglyphus isaea* — 122
- *Dacnusa sibirica* & *Opius pallipes* — 124
- Differences between the various stages of *Dacnusa sibirica*, *Opius pallipes* & *Diglyphus isaea* — 127

Biological control of leaf miners — 128

7. Aphids and their natural enemies — 129

Aphids — 130
- *Myzus persicae* subsp. *persicae* & *Myzus persicae* subsp. *nicotianae* / peach potato aphid & tobacco aphid — 135
- *Aphis gossypii* / cotton aphid — 136
- *Macrosiphum euphorbiae* / potato aphid — 137
- *Aulacorthum solani* / glasshouse potato aphid — 138
- Overview of important aphid species — 140

Natural enemies of aphids — 144
- *Aphidoletes aphidimyza* — 145
- Coccinellidae / ladybirds — 148
  - *Adalia bipunctata* / two spotted ladybird — 150
  - Overview of important ladybird species — 152
- Lacewings — 154
  - Chrysopidae / green lacewings — 154
  - Hemerobiidae / brown lacewings — 156
- Hoverflies — 156
  - *Episyrphus balteatus* / marmalade hoverfly — 157
- Parasitic wasps — 159
  - *Aphidius colemani* — 159
  - *Aphidius ervi* — 160
  - *Aphelinus abdominalis* — 162
  - Hyperparasites of aphid-parasitic wasps — 163
    - *Dendrocerus carpenteri* — 164
    - *Alloxysta* spp. — 164
  - Overview of important parasitic wasp species — 166
- Entomopathogenic fungi — 165

Biological control of aphids — 168

8. Butterflies and moths and their natural enemies — 171

Butterflies and moths — 173
- *Chrysodeixis chalcites* / tomato looper — 177
- *Lacanobia oleracea* / tomato moth — 178
- *Mamestra brassicae* / cabbage moth — 179
- *Spodoptera exigua* / beet armyworm — 180
- *Autographa gamma* / silver-Y moth — 182
- *Spodoptera littoralis* / Egyptian cotton leafworm — 183
- *Helicoverpa armigera* / tomato fruitworm — 184
- *Clepsis spectrana* / cabbage leafroller — 185
- *Duponchelia fovealis* — 186
- *Opogona sacchari* / banana moth — 188
- *Cacoecimorpha pronubana* / carnation leafroller — 189
- Overview of important lepidopteran species — 191

Natural enemies of butterflies and moths — 198
- *Bacillus thuringiensis* — 199
- *Trichogramma brassicae* — 201

Biological control of butterflies and moths — 203

9. Mealy bugs and scale insects and their natural enemies — 205

Overview of mealy bugs and scale insects — 206
Mealy bugs — 207
- *Planococcus citri* / citrus mealy bug — 210
- *Pseudococcus affinis* / obscure mealy bug — 211
- *Pseudococcus longispinus* / long-tailed mealy bug — 212

Soft scale insects — 213
- *Coccus hesperidum* / brown soft scale — 215
- *Saissetia coffeae* / hemispherical scale — 216
- *Parthenolecanium corni* / brown scale — 217

Armoured scale insects	218
Natural enemies of mealy bugs and scale insects	220
- *Cryptolaemus montrouzieri*	221
- *Leptomastix dactylopii*	223
Biological control of mealy bugs and scale insects	224

10. Beetles and their natural enemies — 225

Beetles	226
- *Leptinotarsa decemlineata* / Colorado beetle	227
- *Otiorhynchus sulcatus* / black vine weevil	228
- Elateridae / wireworms, click beetles	230
- *Coccotrypes dactyliperda* / palm seedborer	231
Natural enemies of the black vine weevil	232
- Entomopathogenic nematodes	232
- *Heterorhabditis bacteriophora*	232
Biological control of beetles	234

11. Flies and their natural enemies — 235

Flies	236
- Sciaridae / sciarid flies	237
- Ephydridae / shore flies	239
- *Scatella stagnalis*	239
- Overview of different flies	240
Natural enemies of flies	242
- *Steinernema feltiae*	243
- *Hypoaspis* spp.	244
- *Hypoaspis miles*	244
- *Hypoaspis aculiefer*	246
Biological control of flies	247

12. Other bugs — 248

Heteropteran bugs	249
- *Lygocoris pabulinus* / common green capsid	251
- *Liocoris tripustulatis* / common nettle capsid	252
- *Lygus rugulipennis* / (European) tarnished plant bug	254
- *Nezara viridula* / southern green stink bug	255
- Differences between harmful bugs	257
Leafhoppers	258
- *Empoasca vitis* / grape leafhopper	260
- *Philaenus spumarius* / common froghopper, spittle bug	261
Natural enemies of leafhoppers	262
- *Anagrus atomus*	262

13. Other harmful organisms — 263

Collembola / springtails	264
Formicidae / ants	265
Dermaptera / earwigs	266
Gryllidae / crickets	267
Psocoptera / dust lice	268
Diplopoda / millipedes	269
Scutigerella immaculata / garden symphylan	270
Gastropoda / slugs and snails	272
Isopoda / woodlice	274

14. Observing and monitoring — 275

Index of Latin and English names — 277

References — 284

Photography — 288

1. Classification of insects and other organisms

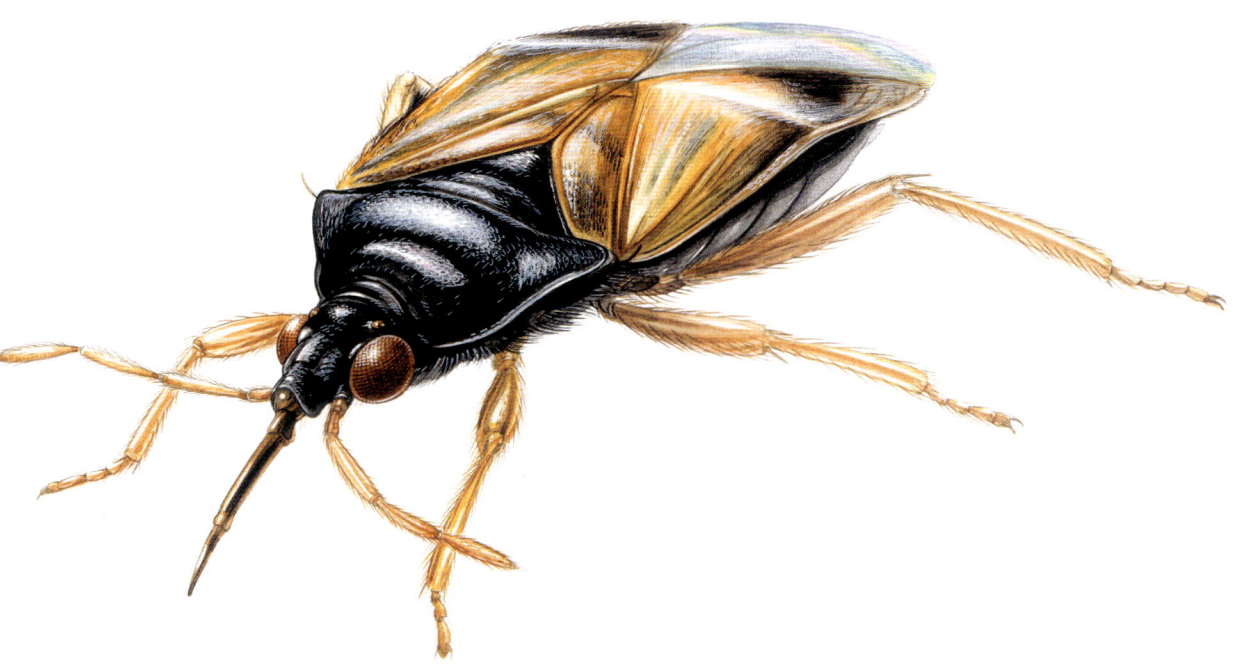

In this book, all organisms are referred to by their species name, with a common English name also being given where such exists. Every species is given a Latin name that consists of two parts, *e.g. Tetranychus urticae*. The first part names the genus, and always begins with a capital letter. The second part identifies the particular species of that genus. This system leaves no room for ambiguity when referring to any species. Latin names of genera and species are conventionally printed in italics. The names have not been assigned at random, but contain information about the species. For example, *Coccinella* means "like a red berry" and *septempunctata* "with seven spots". The name *Coccinella septempunctata* thus describes the appearance of the 7-spot ladybird.

All organisms are grouped according to a hierarchical classification. This classification, based on the relationship of common characteristics, systematically arranges the diversity of species into conveniently ordered groupings, or divisions. A genus thus contains a number of species, with a number of related genera forming a family. Many related families in turn constitute an order, and so on. In order to indicate clearly the different levels of this system of nomenclature, a different suffix, or ending, is used for the name of each level; family names, for example, end with "idae" (*e.g.* Thripidae), and sub-families end in "inae" (*e.g.* Thripinae). As well as the species names, division names also say something about the characteristics of the individuals within it. Most of the organisms described in this book, for example, belong to the phylum Arthropoda, meaning "jointed legs" – a reference to the segmentation of limbs. Table 1.1 gives an overview of the divisions to which organisms are assigned, each with its suffix indicated. A division can often be sub-divided into different categories; a super-family thus consists of different families which in turn may consist of different sub-families.

Table 1.1. Overview of the hierarchical classification of organisms	
Name of division	**Characteristics**
Kingdom	
Phylum	
Sub-phylum	
Class	
Sub-class	
Order	-a
Sub-order	-a
Super-family	-oidea
Family	-idae
Sub-family	-inae
Genus	with capital letter
Species	genus name with capital first letter, species name lower case.
Sub-species	as for species, with sub-species also in lower case.

Table 1.2. Classification of modern man		
Kingdom	Animalia	animals
Phylum	Chordata	chordates
Sub-phylum	Vertebrata	vertebrate animals
Class	Mammalia	mammals
Order	Primates	prosimians, monkeys, apes and humans
Family	Hominidae	hominids (only modern humans are extant in this group)
Genus	*Homo*	man
Species	*Homo sapiens*	"intelligent" man
Sub-species	*Homo sapiens sapiens*	modern man

As an example, table 1.2 shows the classification of humans. No classification is absolute. Different biologists can sometimes classify an organism differently to reflect their own view of the most logical ordering of relations. To complicate matters further, confusion can also sometimes arise through names being changed. Nevertheless, taxonomy, the science of classification, remains the basis for any biological research. Only when we are certain that we are discussing the same organism can data and results be compared. An understanding of classification is therefore a necessity for anyone who works with organisms.

In the description of species, several names are often in use for the same species. These are known as synonyms. Occasionally, the same name may be attached to different species. In both cases a choice has to be made. As far as possible, the first of the named species retains its original name, and a new name is given to the second species. This can become very confusing, especially when searching the literature for information on a particular species.

It is estimated that the number of animal species (organisms belonging to the kingdom Animalia) is somewhere between 20 and 50 million. Of these, at present only 1.7 million have been classified. Insects form by far the largest group, comprising 80% of all animal species.

Characteristics of the groups discussed

In the schematic diagram on pages 14 to 20, the organisms discussed in this book are assigned to their taxonomic position. As noted, opinions can differ, and not all biologists will agree with the finer details of this classification. Nor is the classification by any means complete. However, it should be clear from these diagrams how organisms are related to each other.

In the past, organisms were divided between an animal kingdom and a plant kingdom. Today, however, five different organic kingdoms are recognized:
- Monera (unicellular organisms with no nucleus, such as bacteria)
- Protoctista (unicellular nucleate organisms, such as amoebae and unicellular algae)
- Fungi (including moulds, mushrooms and toadstools)
- Plantae (higher algae, mosses, ferns and other spore-bearing plants and seed plants)
- Animalia (animals)

Viruses do not belong to any of these kingdoms because they do not consist of cells, and can only exist within a living host organism.
This book is mainly concerned with insects (kingdom Animalia, phylum Arthropoda, class Insecta), but not exclusively so: it also deals with other arthropods and even several organisms from other kingdoms. There follows a brief description of the groups that we shall meet in the subsequent chapters, with emphasis on their distinguishing characteristics.

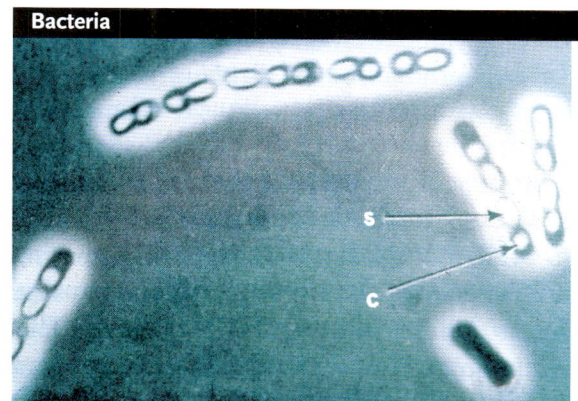

Bacteria are small unicellular organisms belonging to the kingdom Monera. They are minute organisms, usually visible with the aid of a light microscope. They possess no nucleus, but have a circular chromosome and reproduce by division. Under favourable conditions bacteria can multiply very rapidly. Several genera form spores by which they survive unfavourable conditions.
Bacteria are difficult to distinguish from each other on the basis of external appearances, and are mainly classified by metabolic characteristics, many of which show extraordinary differences. For example, there are species that are able to convert ammonia to nitrate to obtain their energy (nitrifying bacteria). Bacteria are highly flexible in their nutritional requirements and, as a result, species can be found living in the most hostile places.
Bacillus thuringiensis is a bacterium discussed in this book which is a well-known insect pathogen.

Fungi, in contrast to higher plants, are not differentiated into roots, stems and leaves, and possess no chloroplasts. They constitute a separate kingdom, and are mostly multicellular organisms composed of threads, known as hyphae. The food source is usually surrounded by a branching network of these hyphae, known as a mycelium, through which the cytoplasm streams, constantly transporting food and enzymes.

Reproduction and dispersal are mainly achieved by means of single-celled or multicellular spores known as conidia (which may be produced either sexually or asexually) on specialized hyphae or mycelial fruiting bodies. These are sometimes highly visible - the familiar mushrooms and toadstools. Spores may be dispersed by means of water, air or physical contact.

Most fungi are parasitic (living at the expense of a host organism) or saprophytic (living on dead organic material), but there are also many species that live in symbiosis with plants. This means that they both benefit from each other's growth, such as the mycorrhizal association between fungi and the roots of plants, and in the case of lichens, between fungi and algae.

The fungal kingdom contains a huge number of different species, whose classification often depends on the sexual stage and/or form of the fruiting body in which spores are produced. In those species where this reproductive phase is unknown, identification and classification can only be based on morphological characteristics.

In this book, several entomopathogenic fungi (fungi that parasitize insects) will be discussed.

Mites

Mites are not insects, but instead form the class Arachnida together with spiders and other spider-like organisms. Both groups belong to the largest of all animal phyla, the Arthropoda (animals with jointed limbs). The class Arachnida contains 11 sub-classes, of which the mites fall under the sub-class Acarina, or Acari as they are also known. They are distinguished from insects by the fact that, with a few exceptions, adult mites possess four pairs of legs whereas adult insects have three. Also, unlike insects, mites have no antennae. They are distinguished from spiders by their fused thorax and abdomen, which forms a single unit, whereas spiders have a fused head and thorax.

Predatory mites are important in biological control, feeding on harmful insects and/or other mites, whilst several other mite species are harmful. These are also discussed.

Nematodes

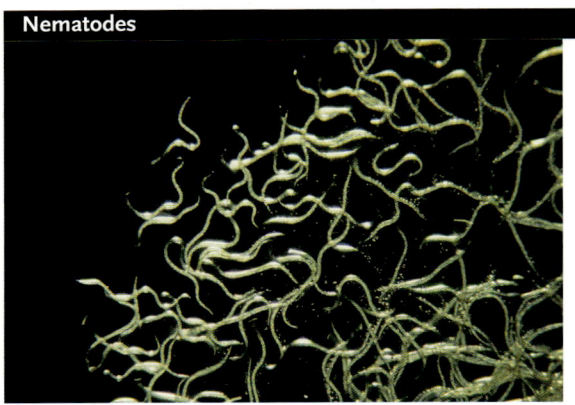

Nematodes or roundworms, sometimes also called eelworms or threadworms, belong to the phylum Nemathelminthes and the class Nematoda. There are at least 12,000 species adapted to highly diverse, and sometimes extreme environments. There are free-living nematodes as well as many species that spend their lives within hosts. Humans can also be parasitized by nematodes, although these are mainly species that live in the gut and cause little more than digestive disturbances. Although nematodes are generally very small, there are a few species that can be several metres long. The microscopic eggs can sometimes survive for long periods, even under extreme conditions.

There are many harmful species. For example, there are cyst nematodes, root gall nematodes, stem nematodes and leaf nematodes, many of which are poorly controlled by chemical pesticides.

However, the nematodes dealt with in this book, *Steinernema feltiae* and *Heterorhabditis bacteriophora* are useful organisms that can parasitize and kill insect larvae.

Centipedes and millipedes

Centipedes and millipedes form the class Myriapoda, also within the phylum Arthropoda. These are terrestrial arthropods with a flattened body. Some are carnivorous, others are plant-eaters. Millipedes can sometimes be harmful, and belong to the sub-class Diplopoda. The predatory centipedes belong to the sub-class Chilopoda, and the garden symphylan, *Scutigerella immaculata*, belongs to the sub-class Symphyla. The latter is discussed in this book along with millipedes.

Slugs and snails

Slugs and snails belong to the phylum Mollusca, and constitute the class Gastropoda, which will be discussed in chapter 13. They have a soft skin and a pulpy un-segmented body without a skeleton. Unlike slugs, snails have a spiral shell. Snails rarely constitute a problem in glasshouses, although slugs may sometimes cause damage to crops.

Woodlice

Woodlice belong to the order Isopoda (phylum Arthropoda, class Crustacea). They are thus more closely related to shrimps and crabs than to insects. Woodlice that appear in the glasshouse belong to the sub-order Oniscoidea, the terrestrial woodlice. Under normal circumstances they rarely cause problems, although they can cause considerable damage in organic crops. This group of organisms is also discussed in chapter 13.

Insects

Most of the organisms dealt with in this book belong to the class Insecta. This is an enormous class of organisms whose members outnumber any other class. They have managed to colonize every environment on Earth apart from salt water.

Insects are constructed according to a general body plan on which there exist a wide range of variations. The body consists of a head, a thorax and an abdomen. On the head are a pair of antennae. The thorax consists of three segments, each of which is equipped with a pair of legs. In many insects the second and third thoracic segments also carry a pair of wings. All insects at some stage of their life-cycle have 6 legs. Because the cuticle of insects is relatively rigid, development and growth are only possible by periodically moulting the old cuticle and forming a new one. At each moult, the old cuticle is shed to allow the underlying epidermis to expand with its new cuticle until that, in turn, has to be shed. The number of moults varies between different kinds of insect, with each stage between moults known as an 'instar'.

When each moult results only an increase in size and no change of form, development is referred to as 'ametabolous'. The adult differs from earlier stages (instars) only in possessing developed sex organs. An example of an ametabolous insect would be the springtail.

When the young instars do not differ radically from the adults in either form or behaviour but undergo minor, stepwise changes with successive moults, development is called 'hemimetabolous' or incomplete. In this case, the wings for example, become visibly more developed at each instar. The juvenile insects are then referred to as 'nymphs', as in the case of the true bugs.

In complete metamorphosis, or 'holometabolous' development, clear and radical differences are evident between juvenile and adult stages. In this case the juvenile insects often bear no resemblance to the adult form, and are referred to as 'larvae'. The larva must then pass through a special instar, called the pupa, before becoming an adult. A well known example of holometabolous development is that of caterpillars that pupate to become butterflies or moths.

Many orders of insects derive their names from the structure of their wings. The names of the winged orders therefore end in "–ptera", from the Greek *pteron*, meaning wing. Thus, we have the Coleoptera (shield-winged), the Hymenoptera (membrane-winged), the Diptera (two-winged) and the Lepidoptera (scale-winged).

Knowing and recognizing **Classification**

Taxonomic classification of organisms: main outline

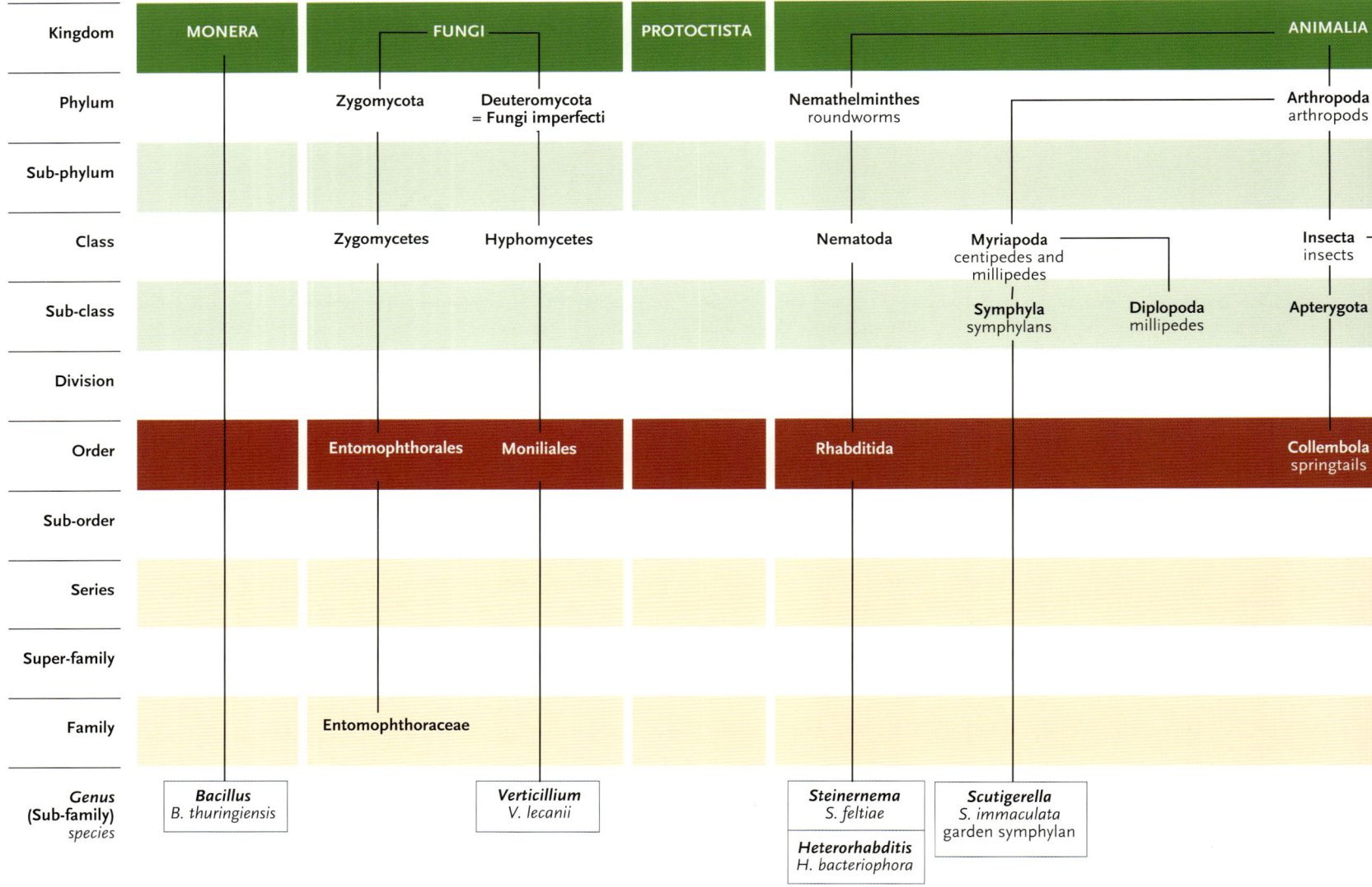

Knowing and recognizing **Classification**

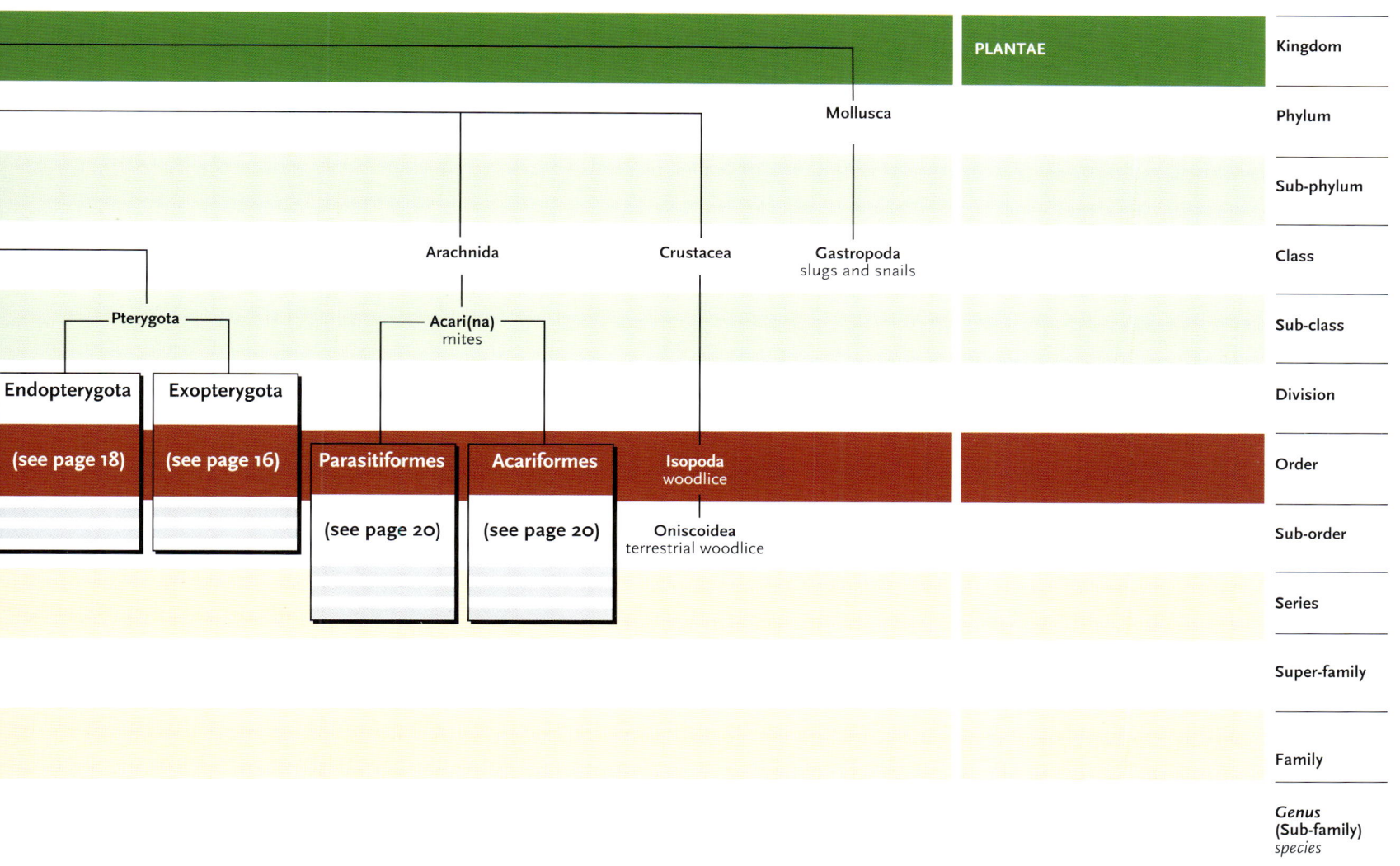

Knowing and recognizing **Classification**

Taxonomic classification of organisms: exopterygote insects

EXOPTERYGOTA (Hemimetabola)

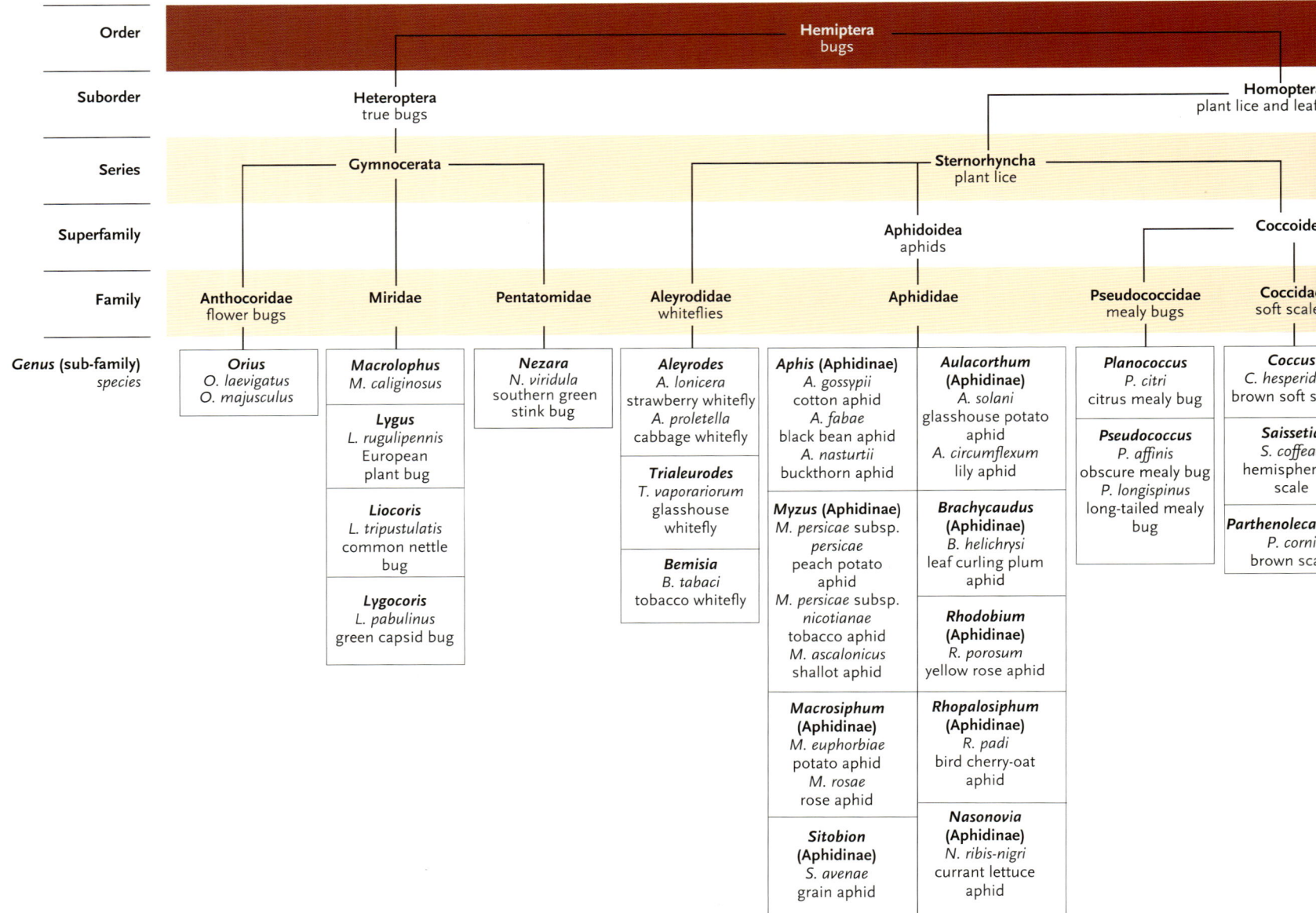

Knowing and recognizing **Classification**

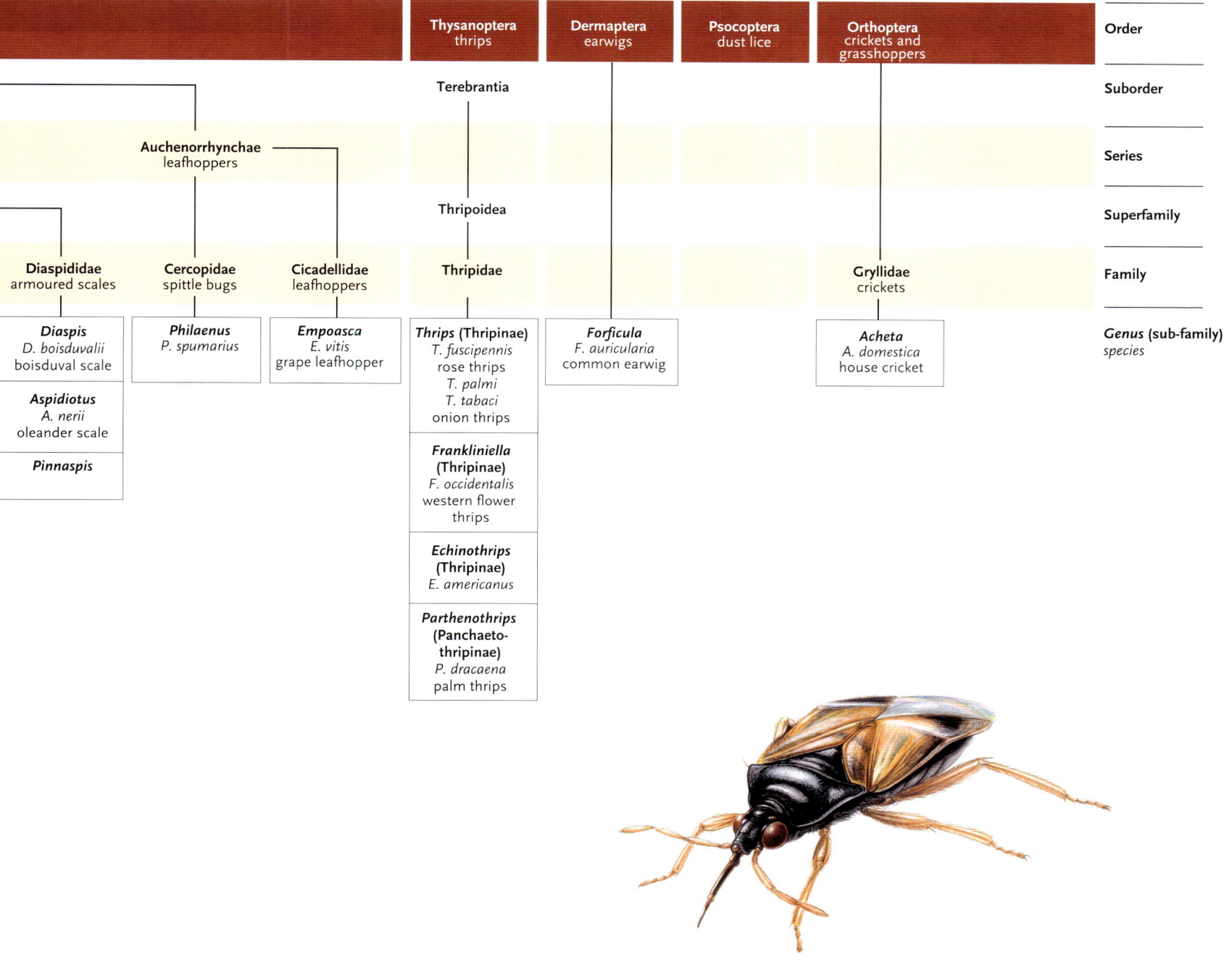

			Thysanoptera thrips	**Dermaptera** earwigs	**Psocoptera** dust lice	**Orthoptera** crickets and grasshoppers	Order
			Terebrantia				Suborder
	Auchenorrhynchae leafhoppers						Series
			Thripoidea				Superfamily
Diaspididae armoured scales	**Cercopidae** spittle bugs	**Cicadellidae** leafhoppers	**Thripidae**			**Gryllidae** crickets	Family
Diaspis D. boisduvalii boisduval scale *Aspidiotus* A. nerii oleander scale *Pinnaspis*	*Philaenus* P. spumarius	*Empoasca* E. vitis grape leafhopper	*Thrips* (Thripinae) T. fuscipennis rose thrips T. palmi T. tabaci onion thrips *Frankliniella* (Thripinae) F. occidentalis western flower thrips *Echinothrips* (Thripinae) E. americanus *Parthenothrips* (Panchaetothripinae) P. dracaena palm thrips	*Forficula* F. auricularia common earwig		*Acheta* A. domestica house cricket	***Genus*** (sub-family) *species*

Knowing and recognizing Classification

Taxonomic classification of organisms: endopterygote insects
ENDOPTERYGOTA (Holometabola)

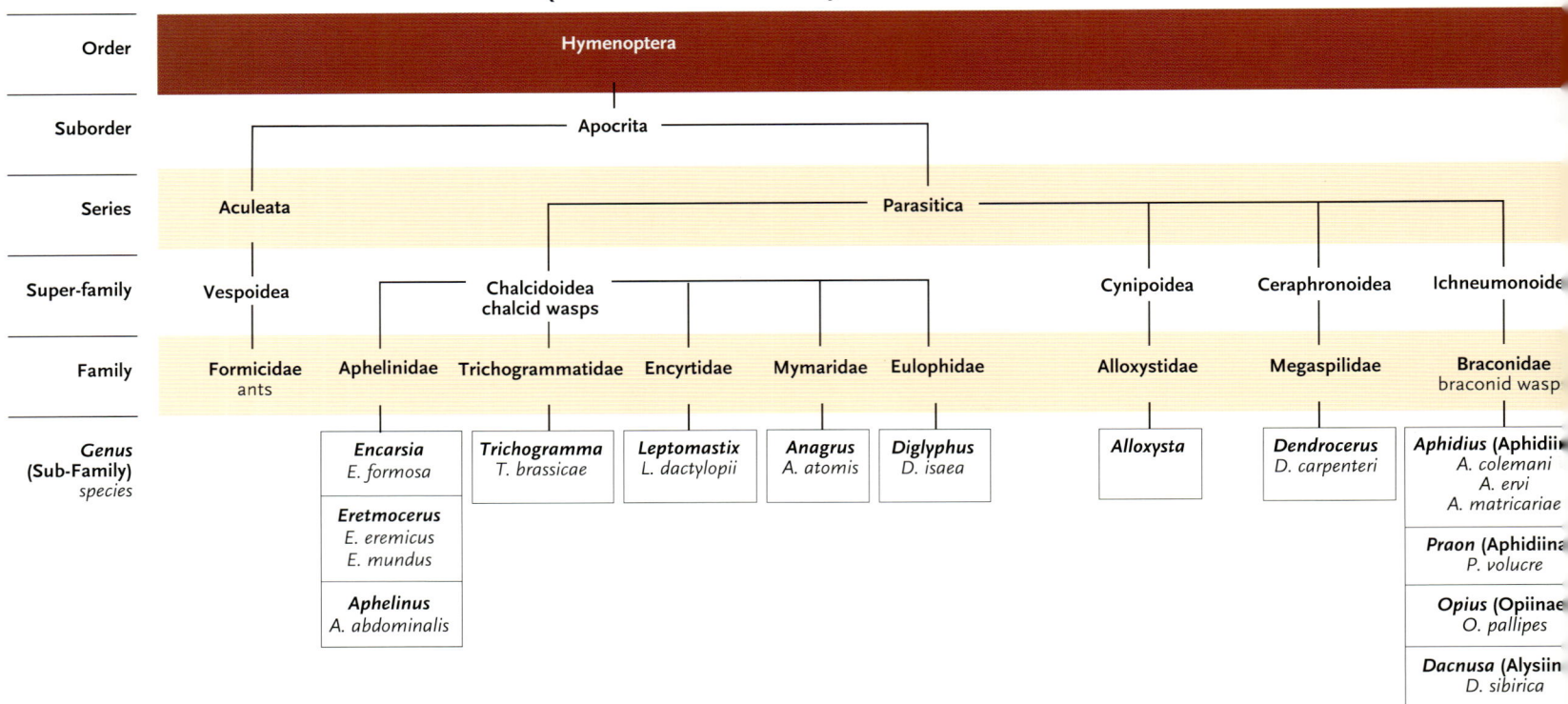

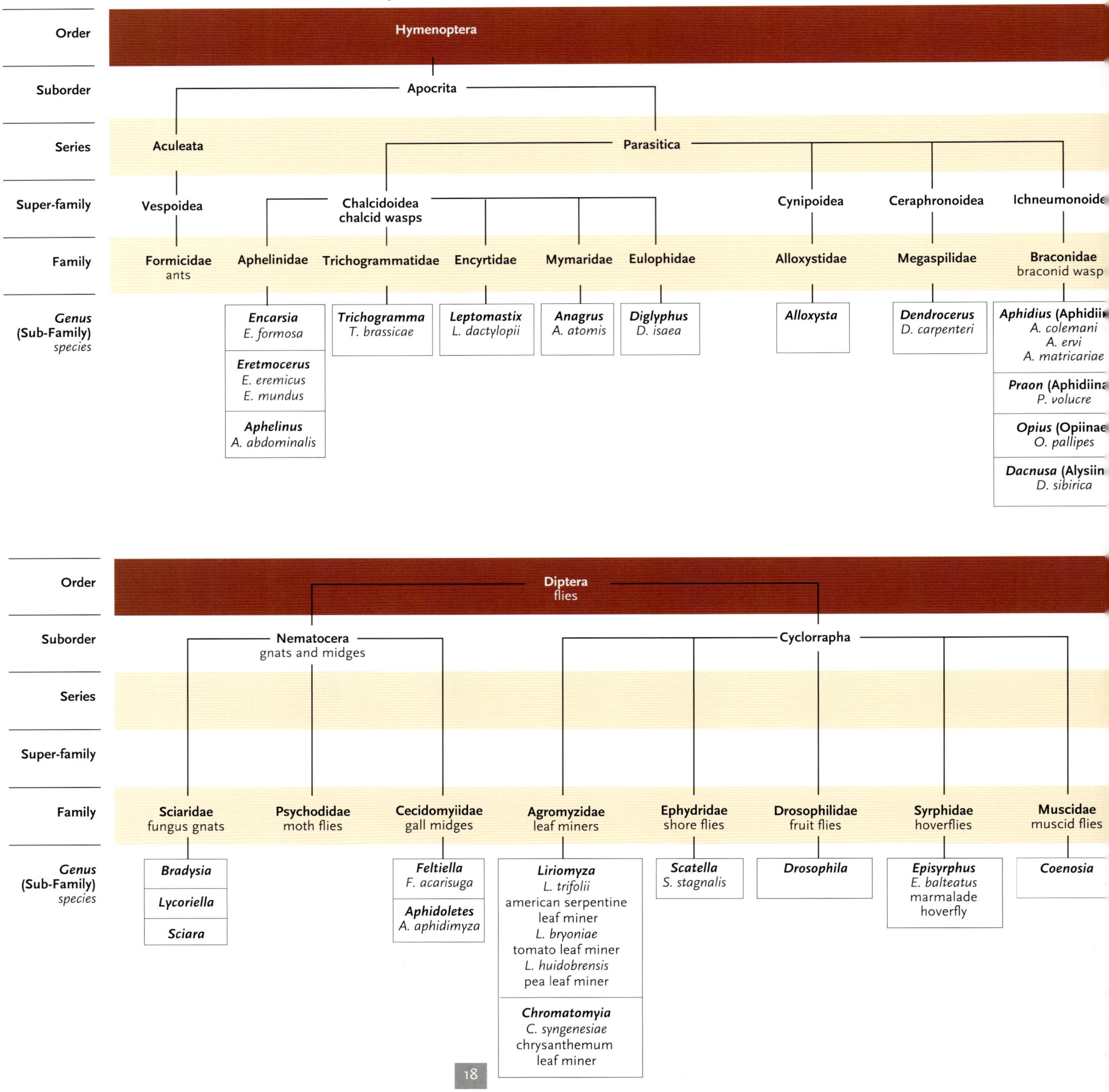

Knowing and recognizing **Classification**

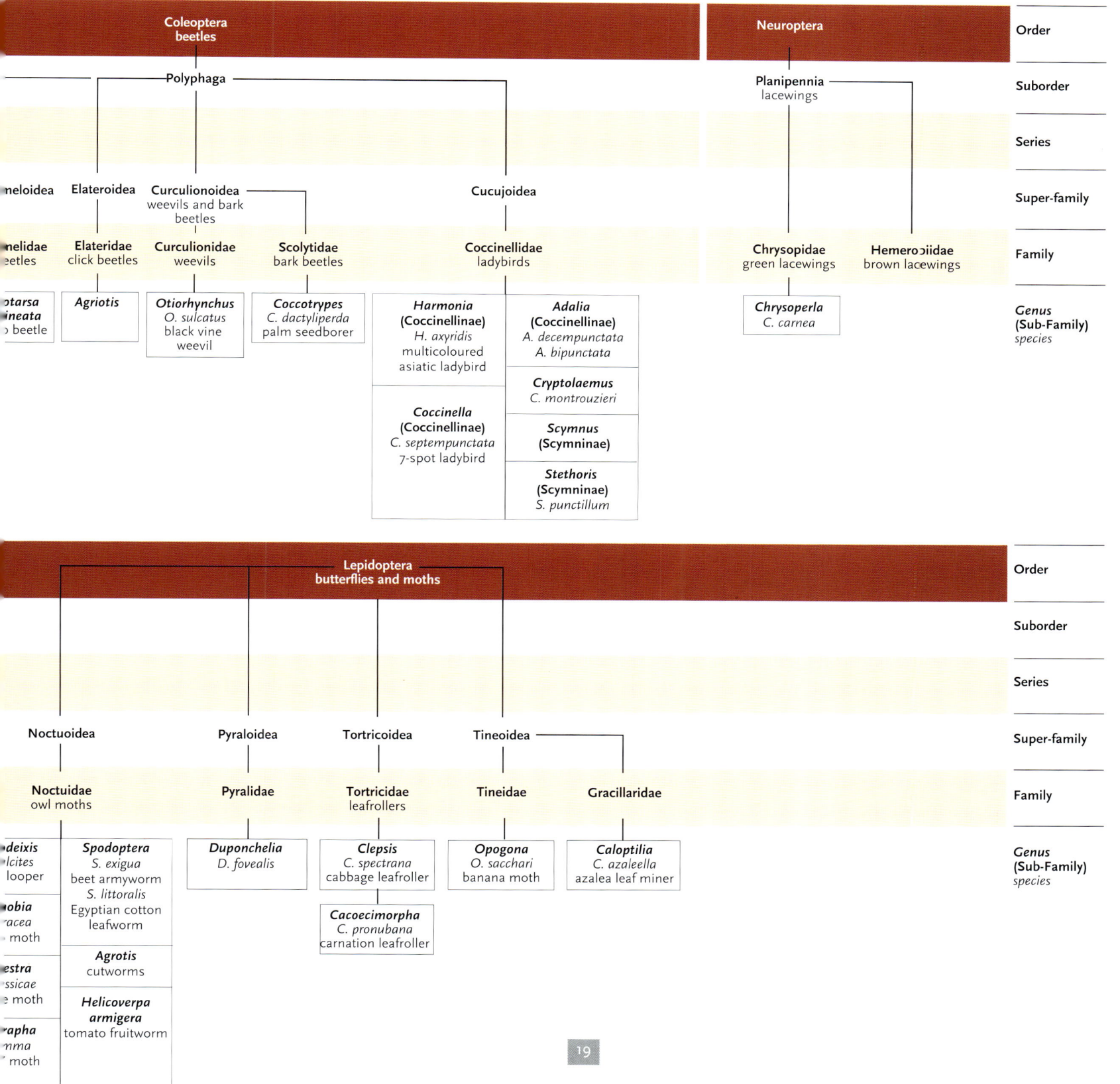

Taxonomic classification of organisms: mites

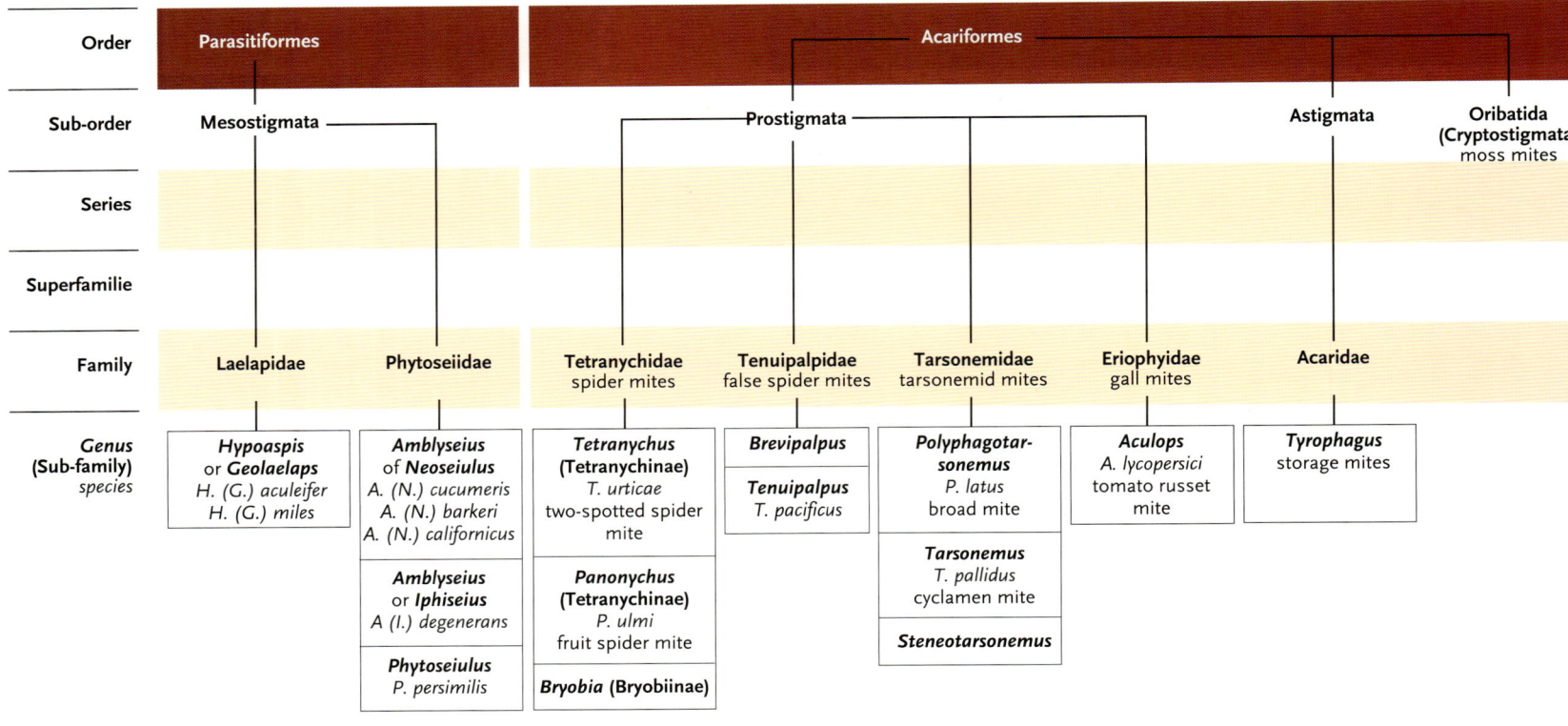

Knowing and recognizing **Spider mites**

2. Spider mites and their natural enemies

Spider mites are pests of many crops throughout the world. Their great reproductive capacity means that they are capable of destroying plants very rapidly. By far the most important species in glasshouses is the two-spotted spider mite, *Tetranychus urticae*, the two-spotted or glasshouse spider mite.

The widespread use of chemical control agents has led to many two-spotted spider mite populations developing resistance to a range of pesticides. For this reason, biological control of this pest is now widely practiced. In fact the first commercial application of natural enemies against pests in glasshouse horticulture was against the two-spotted spider mite; in the mid 1960's Koppert released the first predatory *Phytoseiulus persimilis* mites on cucumbers grown under glass. After an erratic start the experiment proved a success, and laid the foundations for the biological control of fruiting vegetable pests. This predatory mite is now successfully used to control spider mites in many crops throughout the world. In certain situations however, *P. persimilis* is not wholly effective. In such situations, another predatory mite, *Amblyseius californicus*, can provide a useful back-up. This predatory mite has been used for the biological control of spider mites since 1994.

The gall midge *Feltiella acarisuga* has been used in the same capacity since about 1996, as the larvae are capable of rapidly controlling concentrated spider mite colonies.

In this chapter we shall look at the characteristics of the species mentioned above, in addition to another spider mite predator - the ladybird *Stethorus punctillum* – which sometimes appears in the crop spontaneously.

Some of the natural enemies discussed in later chapters can also contribute to the biological control of spider mites, such as the predatory bug *Macrolophus caliginosus* in tomato, described in chapter 4.

The predatory mites *Amblyseius cucumeris* and *Amblyseius degenerans*, normally used for controlling thrips, will prey to a lesser extent on spider mites, although on their own they do not provide effective control. Finally, the brown lacewings discussed in chapter 7 can also predate spider mites.

Spider mites

Different colour forms of *Tetranychus urticae* ranging from green to red.

Spider mites belong to the class Arachnida (along with spiders), the subclass Acarina or Acari (mites), the order Acariformes and the family Tetranychidae. This family contains many harmful, plant-eating species. The two-spotted spider mite, known also as the red or glasshouse spider mite (*Tetranychus urticae*) belongs to the sub-family Tetranychinae and is by far the most important species in glasshouse horticulture. This species will be discussed in thorough detail in this chapter.
Another common species belonging to the same family is the fruit spider mite, *Panonychus ulmi*. Although this species is responsible for occasional infestations in small glasshouse fruit, it is mainly found in orchards, and as such is not dealt with in this book.
Finally, species of *Bryobia*, which also belong to the family Tetranychidae, can also cause slight damage in glasshouse crops from time to time. Unlike the two previous species, *Bryobia* spp. belong to the sub-family Bryobiinae, and are dealt with in chapter 3.

- **A major pest in many glasshouse crops**
- **Can cause problems throughout the year**
- **Extremely rapid population growth**

Tetranychus urticae
Two-spotted spider mite

Of the family Tetranychidae, *T. urticae* is the most polyphagous species (*i.e.* it feeds on a wide range of different plants), and is thus an important pest in both ornamental and food crops throughout the world. A similar species often encountered in the literature is *T. cinnabarinus*, the carnation spider mite. However, many researchers are convinced that this is not a separate species, but rather one of a complex of different races or strains of the two-spotted spider mite which vary in one or more of the following:
- host plant preference: some strains perform better on a particular host plant,
- colour: green and red spider mite populations are known,
- form: there can be minor morphological differences between different populations,
- hybridisation: some populations cross-breed, others do not,
- damage: sometimes only a few spider mites on tomatoes are enough to cause extensive damage.

Because there is no correlation between the features listed above (for example, it is not always the red coloured spider mite that causes the most serious damage to tomatoes) the different populations are now all regarded as a *Tetranychus urticae* complex and not as separate species.

Population growth

Population growth in two-spotted spider mite depends on temperature and relative humidity, as well as the variety, age and quality of the crop. Of these, temperature is certainly the most important. Different strains of spider mite may also show differing growth rates.
No development occurs at temperatures below 12°C, and temperatures above 40°C are deleterious. The rate of population growth increases with temperature up to 30°C, but declines again at higher temperatures. Atmospheric humidity also has an effect on population growth, with *T. urticae* laying more eggs and developing faster at lower humidities.
The spider mite inhabits a microhabitat provided by the boundary layer of air at the surface of the leaf. Temperature and humidity within this boundary layer can differ significantly from the conditions prevailing elsewhere in the glasshouse. In the event of a severe attack of spider mite the leaf becomes necrotic and transpiration is reduced, resulting in a higher temperature and a lower humidity at the boundary layer. This change in the microclimate at the leaf surface suits the spider mite and promotes rapid population growth. Under summer temperatures, a spider mite population can develop extremely rapidly under glasshouse conditions. As a result of this rapid population growth and the frequent use of chemical pesticides, resistance can develop very rapidly.
Table 2.1 shows data relating to the population growth of *T. urticae* under different conditions. The table shows that the duration of the egg stage is long compared to subsequent stages. It is also clear that under identical conditions, the development of populations can differ on different varieties of the same plant. For example, on the mite-sensitive cultivar of the gerbera Sirtaki, the generation time is shorter, the number of eggs laid by a single female per day is higher and the mortality among eggs is lower than on the less susceptible cultivar Bianca.

Males are often found close to females that are just about to become adults. As soon as the female develops into an adult mite, a male will mate with her, and if several males are waiting for the same female, they will fight, with the largest male usually winning. A single mating is sufficient to fertilize all her eggs.

Fertilized females produce both males and female progeny whereas unfertilized females produce only males. The sex ratio in a population of *T. urticae* is usually 1 ♂ : 3 ♀.

Overwintering

When environmental conditions change adversely, female spider mites enter diapause. Such adverse conditions include:
- decreasing daylength,
- falling temperature,
- decline or deterioration in food supply.

Females entering diapause become an orange-red colour within three to five days of becoming adult. After being fertilized they overwinter in a dormant condition, hidden within the structure or contents of the glasshouse. During this period they eat nothing, lay no eggs and are less susceptible to chemical pesticides. They are also more difficult for predatory mites to find, not only because they hide in concealed places, but because the cessation of feeding activity means they no longer cause the release of attractive volatile compounds from damaged plant tissue. As soon as conditions become more favourable in the spring, the females become active again and resume egg laying.

2.4. Overwintering spider mite

Table 2.1. The population growth of *Tetranychus urticae*

2.1.1. The population growth of *Tetranychus urticae* at different temperatures on rose at a relative humidity of 55-85% (Sabelis, 1981)

temperature (°C)	15	20	25	30	35
development time (days)					
egg	14.3	6.7	4.3	2.8	2.4
larva	6.7	2.8	1.8	1.3	1.0
protonymph	5.3	2.3	1.5	1.2	1.0
deutonymph	6.6	3.1	2.0	1.4	1.3
total egg-adult	32.9	14.9	9.6	6.7	5.7
po-period ★	3.5	1.7	0.9	0.6	0.6
total egg-egg	36.4	16.6	10.5	7.3	6.3
mortality (%)					
egg	6.1	6.3	4.3	6.6	10.1
larva	2.4	1.1	1.9	2.6	4.1
protonymph	1.0	1.3	0.0	1.0	3.9
deutonymph	2.2	1.0	2.0	0.0	5.1
po-period★	0.0	0.0	1.5	1.0	4.0
total generation	11.3	9.6	9.3	10.8	24.4

★ po-period = pre-oviposition period, *i.e.* period from becoming adult to first egg-laying

2.1.2. The population growth of *Tetranychus urticae* at 25°C in 2 gerbera cultivars at a relative humidity of 65% (Krips et al., 1998)

		cultivar
development time egg-egg (days)	11.4	Sirtaki
	17.0	Bianca
eggs/♀/day	9.1	Sirtaki
	1.6	Bianca

2.1.3. Total eggs laid per *Tetranychus urticae* female at 25°C in different crops

		temperature (°C)					crop	source
15	16	20	25	28	30	35		
84.2		129.9	129.7		119.5	120.6	rose	Sabelis, 1981
		37.9					strawberry	Laing, 1969
	94			170			cucumber	Popov *et al.*, 1992

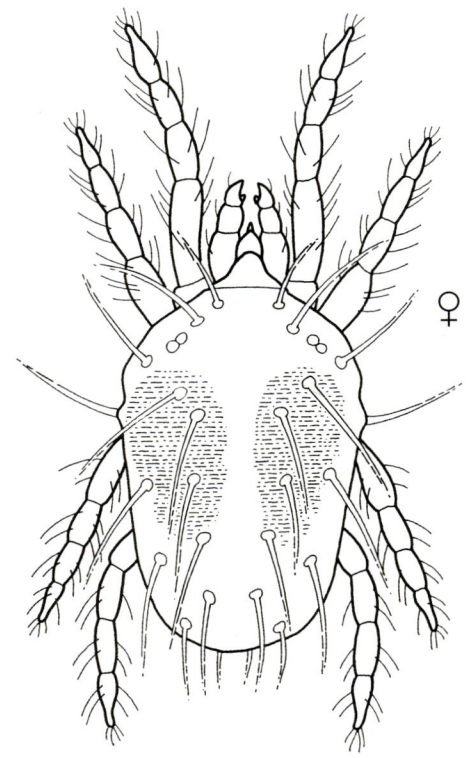

♀

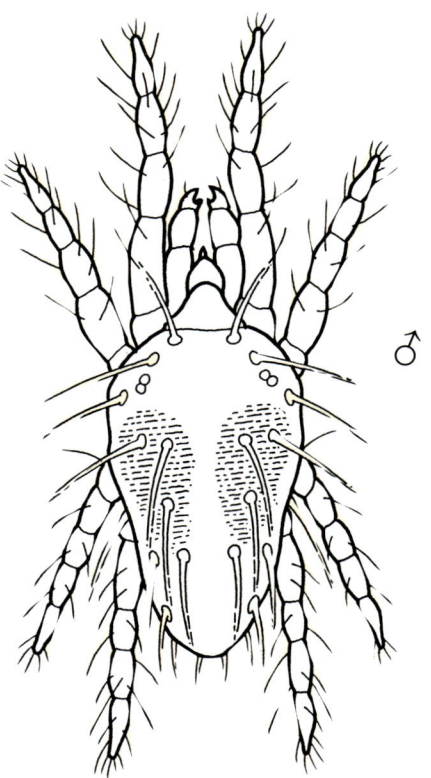

♂

Life-cycle and appearance
Tetranychus urticae

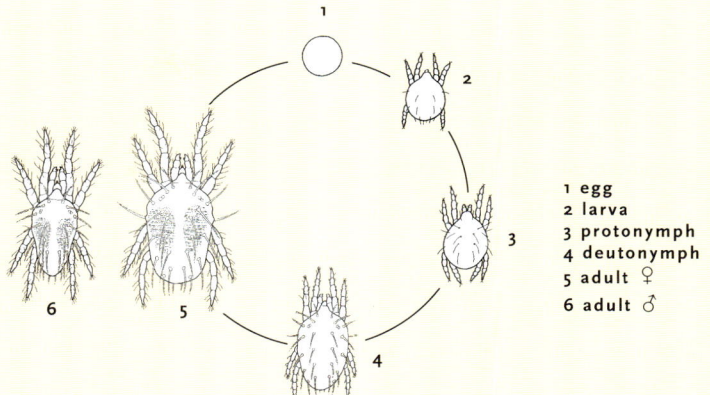

1 egg
2 larva
3 protonymph
4 deutonymph
5 adult ♀
6 adult ♂

Two-spotted spider mites pass through five developmental stages: egg, larva, first stage nymph (protonymph), second stage nymph (deutonymph), and adult mite. In the larval and nymphal stages, an active period and a resting period of roughly equal duration can be distinguished.

Two-spotted spider mite eggs are usually found on the underside of leaves. They are round, with an average diameter of 0.14 mm, and are transparent when freshly laid, later becoming opaque. By the time of hatching they are straw-coloured.

The larvae have three pairs of legs, and on hatching are virtually colourless with two dark red eyes. Once they begin feeding their colour changes to a light green, a brownish yellow or even dark green. Two dark spots also develop in the middle of the body. Once sufficient food has been consumed, they become inactive on the leaf surface and withdraw their legs until they develop into a protonymph.

Protonymphs have four pairs of legs and are somewhat larger than the larvae. Their colour varies from light to dark green. The two body spots are larger and clearer than in the larval stage. Here too, a period of feeding is succeeded by a resting period that terminates in a moult, out of which develops the deutonymph. This second stage nymph is larger than the protonymph but has the same colouration. Male and female individuals can usually be distinguished at the deutonymph stage, the male being more elongate and slightly smaller, with the female rounder in form and slightly larger than the male. Again, there is a feeding period followed by a resting period, and a moult out of which develops the adult mite. At each moult the mite leaves behind an empty grey-coloured cuticle.

The adult female *T. urticae* has an oval body that is rounded at the posterior. The colour can vary from orange, light yellow or light green to dark green, red, brown, or almost black. The male is smaller and more active than the female, with a body form that is narrower and more pointed. The colour varies from light yellow or orange to dark yellow or brown. The colour of the adults often depends upon the crop on which they occur: on cucumbers two-spotted spider mites are often yellow-brown, whilst red-brown specimens are found on tomatoes. Strains also occur that differ in colour regardless of the host plant. Both males and females usually have two large black spots, giving rise to the common name "two-spotted spider mite". The spots can vary in both form and size.

2.5. Eggs of spider mite

2.6. Different stages of *Tetranychus urticae*

2.7. Red coloured *Tetranychus urticae*

2.8. Dark coloured *Tetranychus urticae*

In regions where conditions do not become harsh enough to initiate diapause (e.g. in Florida) T. urticae reproduces throughout the whole year, although population growth may be slowed during winter. Plate 2.4. (page 23) shows a dormant spider mite.

Damage

Both larvae, nymphs and adults cause damage to the host plant by feeding on plant tissue and plant sap. They mainly occur on the underside of leaves where they pierce the cells and suck out the contents. These dead cells become yellow, and in many plants the damage can also be seen on the upper surface of leaves as small yellow dots. As damage increases whole leaves turn yellow, and as the chlorophyll is removed, the leaf, and eventually the whole plant, may die. The nymphs and adults also produce webs, and plants can be completely covered with such webs in which the mites swarm.

For the plant, this has the following consequences:
- the chlorophyll is destroyed or disappears, resulting in a reduction of photosynthesis and plant growth. In tomatoes and cucumbers, 30% damage of the leaf surface can lead to loss of the crop,
- the webbing and spotting on the leaves affects the appearance of the crop. This is of particular concern in ornamental crops,
- toxic substances can be introduced into the plant by the mites, although little is known about this at present.

In tomatoes, a small population of some strains of T. urticae can cause enormous damage. The damage begins as small spots on leaves that subsequently become necrotic. These spread until the whole leaf shrivels and dies. Dark stripes can be observed in the leaf stalks, though the main stems are not affected. Feeding spots may or may not be visible. It is unclear why such a violent reaction can occur to a spider mite infestation. Plate 2.9 shows several examples of damage.

Distribution and dispersal

The two-spotted spider mite can spread through a crop in various ways. If plants are heavily infested, mites fall to the ground and migrate to other plants, or they can migrate to new plants along crop wires. They also secrete silk threads on which they are dispersed by air currents. Mites can also be dispersed mechanically, either by the movement of infected plant material, or by transport on clothing and other objects. Despite the ease of dispersal, spider mites often appear locally in the glasshouse in particular places. Such places are often difficult to reach with pesticides, and may have a more favourable (drier, warmer) climate.

2.9. Damage caused by attack of *Tetranychus urticae*

Two-spotted spider mite webbing

Leaf damage in cucumber

Leaf damage in tomato

Natural enemies of spider mites

2.10. *Phytoseiulus persimilis*

2.11. *Amblyseius californicus*

The most important natural enemies of *Tetranychus urticae* are:

Phytoseiulus persimilis
Amblyseius californicus
Feltiella acarisuga
Stethorus punctillum

2.12. *Feltiella acarisuga*

2.13. *Stethorus punctillum*

The predatory mite *Phytoseiulus persimilis*, by far the most important natural enemy of *Tetranychus urticae*, has been introduced into many crops worldwide. This predatory mite was the first biological control agent used in glasshouses and is still extremely effective. However, under dry, warm conditions *P. persimilis* has difficulty keeping spider mites under control. Under such conditions, the predatory mite *Amblyseius californicus* can be used, as it is more tolerant of higher temperatures, lower humidities and pesticides than *P. persimilis*, and will combat small spider mite populations more effectively over the longer term.

Feltiella acarisuga, a gall midge that often establishes naturally, can also be introduced into glasshouses to control large, concentrated spider mite populations.
The ladybird *Stethorus punctillum*, another spontaneously occurring species, can also be used under glass to combat spider mites. All four species will be described in this chapter. The predatory bug *Macrolophus caliginosus* (see chapter 4), the thrips predators *Amblyseius cucumeris* and *Amblyseius degenerans* (chapter 5) and the brown lacewing, which also sometimes occurs naturally in glasshouses (chapter 7) are also capable of controlling spider mites to a greater or lesser extent.

- Predatory mite
- Nymphs and adults feed on all stages of spider mites
- Capable of rapid and highly effective control of spider mites

Phytoseiulus persimilis

The predatory mite *Phytoseiulus persimilis* belongs to the class Arachnida (spiders and spider-like animals), the subclass Acarina, the order Parasitiformes and the family Phytoseiidae. All the species of this family are predators of spider mites and/or other plant-dwelling mites and small insects.

In 1958, *Phytoseiulus persimilis* was accidentally imported into Germany on orchids from Chile. Once discovered by a researcher, it was distributed to research institutions throughout the world. *P. persimilis* is now used effectively in many different glasshouse crops and outdoor cultures.

Population growth

Comparison of the development data in table 2.2 shows why *P. persimilis* is such an effective predator of *T. urticae*, particularly when temperatures are not too high.

At 20°C a female predatory mite lays more eggs than a female spider mite. A population of predatory mites also contains more females (the sex ratio is usually approximately 1♂ : 4♀), and since the average generation time of the predatory mite is shorter than that of the spider mite, the predator population will increase faster than that of the pest. Mating in *P. persimilis* usually takes place a few hours after they become adults, with a female often mating with several males. Once a female has mated she can lay eggs throughout her life, whereas an unmated female will lay none. A fertilized female will continue to lay eggs until she has laid the maximum number possible under the environmental conditions, or until she dies.

However, *P. persimilis* is more susceptible than *T. urticae* to temperatures above 30°C, and thus struggles to control spider mite at higher temperatures. A relative humidity lower than 60% also has an adverse effect on the hatching of eggs and, to a lesser extent, on the duration of development. Although both the activity of the predator and the number of prey consumed increases in proportion to the decrease in relative humidity, this does not offset the adverse effects on population growth. Therefore *P. persimilis* does not perform well under dry, warm conditions (see table 2.3).

Feeding behaviour

An adult predatory mite will feed on all stages of spider mite, and although the larva does not eat, the nymphs will eat the eggs, larvae and protonymphs. The quantity consumed depends on the predator and prey populations, the age of the predator, and the temperature and relative humidity. In general, if the temperature increases, the consumption of prey also increases. However, *P. persimilis* is sensitive to temperatures above 30°C, and at 35°C it will stop feeding altogether. The optimal temperature for controlling spider mite is between 15 and 25°C. At roughly 20°C and with a plentiful supply of prey, an adult predatory mite will

Table 2.2. The population growth of *Phytoseiulus persimilis* and for comparison (in parentheses) of *Tetranychus urticae*.
- on rose with *Tetranychus urticae* (T.u.) as prey (Sabelis, 1981)
- on bean with *Tetranychus pacificus* (T.p.) as prey (Takafuji and Chant, 1976)
- on strawberry (Laing, 1968, 1969)

	\multicolumn{6}{c}{temperature (°C)}	crop	prey
	12	15	20	25	30	33		
development time (days)								
egg		8.6	3.1		1.7		rose	T.u.
				2.3			bean	T.p.
larva		3.0	1.1		0.6		rose	T.u.
				0.7			bean	T.p.
protonymph		3.9	1.4		0.8		rose	T.u.
				1.1			bean	T.p.
deutonymph		4.1	1.6		0.8		rose	T.u.
				1.3			bean	T.p.
total egg-adult		19.6	7.2		3.9		rose	T.u.
				5.4			bean	T.p.
po-period*		5.6	1.9		1.1		rose	T.u.
				1.5			bean	T.p.
total egg-egg		25.2	9.1		5.0		rose	T.u.
		(36.4)	(16.6)		(7.3)		rose	
				6.9			bean	T.p.
eggs/♀								
			53.5				strawberry	T.u.
			(37.9)				strawberry	
					79.5		bean	T.p.
	20.2	40.9		62.8		60.8	rose	T.u.

*: po-period = pre-oviposition period, *i.e.* period from becoming adult to first egg-laying

Table 2.3. Effect of relative humidity (RH) and temperature (T) on the development of immature instars of *Phytoseiulus persimilis* (on *Tetranychus urticae* on bean) (Stenseth, 1979)

T (°C)	RH (%)	eggs hatched (%)	eggs developing to adults (%)	nymphs developing to adults (%)
21	40	84.5	73.1	86.1
21	80	95.1	78.7	84.0
27	40	7.5	7.2	95.0
27	80	99.7	79.3	79.8

Life cycle and appearance
Phytoseiulus persimilis

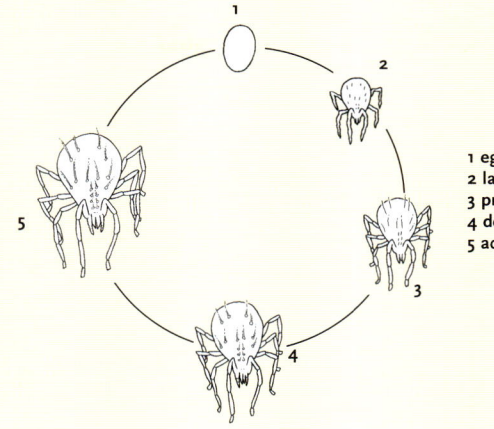

1 egg
2 larva
3 protonymph
4 deutonymph
5 adult

The developmental stages of *P. persimilis* are the same as those of the spider mite; egg, larva, first stage nymph (protonymph), second stage nymph (deutonymph) and adult mite. Unlike *T. urticae*, however, the larval stage and the two nymphal stages do not pass through a resting period. The oval eggs are deposited close to a food source. When freshly laid they are a transparent light pink colour, turning darker later on. They thus differ from the eggs of the spider mite in colour and form, and are also roughly twice the size.

The larva has three pairs of legs. It does not feed, and remains inactive unless disturbed. Once the larva has moulted to the first stage nymph it begins feeding immediately, and during the first and second nymphal stages the mite feeds almost incessantly.

The adult mite develops from the second nymphal stage, and is light red with long legs. The adult is highly active, particularly at higher temperatures. Females are approximately 0.6 mm in length, and although males are slightly smaller, flatter and more elongated, they are very difficult to distinguish from the females.

2.14. Eggs of *Tetranychus urticae* (l) and *Phytoseiulus persimilis* (r)

2.15. Egg and larva of *Phytoseiulus persimilis*

2.16. Different stages of *Phytoseiulus persimilis*

2.17. Preying *Phytoseiulus persimilis*

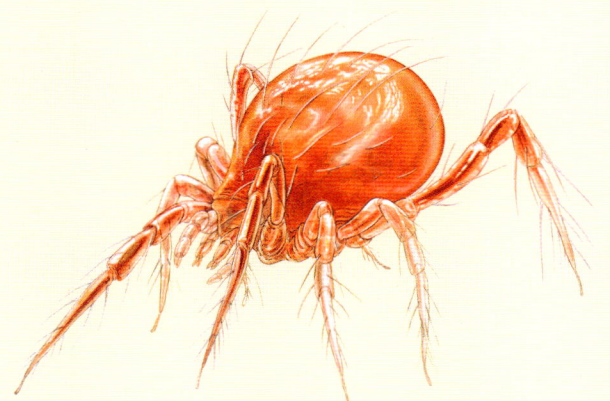

consume five adult spider mites or twenty young larvae or eggs per day. Because the predatory mite's development rate is roughly twice as fast as that of the spider mite, the number of predatory mites increases rapidly, and will quickly reduce the spider mite population by predation. Female *P. persimilis* are generally more effective at controlling spider mites than the males. In addition to utilizing consumed prey for maintaining metabolism, the majority is used for egg production; an adult female can lay five eggs per day, with a total weight equal to that of the mite itself.

P. persimilis is almost entirely dependent on spider mites as food. However, when prey is in short supply they will become cannibalistic. In the total absence of food, they can survive for some time on water and nectar, although reproduction will cease under these conditions. As a result of cannibalism and the lack of an adequate food supply, the predatory mite population can decline dramatically in the absence of spider mites. Thus, with the appearance of new concentrations of spider mite it may be necessary to reintroduce predatory mites.

P. persimilis is highly specific and is only effective against spider mites. This has the advantage that an outbreak is quickly and effectively dealt with, but the disadvantage that in the absence of suitable prey, it disappears rapidly. As spider mite density increases, *P. persimilis* kills more mites and spends correspondingly less time consuming each individual prey item.

Searching behaviour and dispersal

The ability of *P. persimilis* to spread and find new colonies of prey depends on the environment, the density and distribution of prey and the presence of spider mite webs. The density of the plants is also very important, as where infested plants are in contact with adjacent plants the predatory mite can spread with ease.

The damage and the webbing produced by spider mites helps the predator to locate its prey. If webs or leaf damage are detected, the predatory mite will extend its search in the vicinity. Although the predatory mite cannot see, it detects webs by touch, and the leaf damage by scent; the plant reacts to the effects of spider mites by producing chemical substances that act as signals to the predatory mites. This three way interaction, between the herbivorous spider mite, the plant on which it feeds and the predatory mite, is known as a tritrophic interaction. The chemical stimulants are present most strongly where spider mites are actually present, but will also be produced by the rest of the affected plant. It is these stimulants that attract the predatory mites. If the prey density is relatively low compared to predator density, adult *P. persimilis* will spread out in search of new food sources. The nymphs often remain on the plant and eat everything they can find. The latter behaviour is partly the reason the prey dies out. Also, the adult predatory mite is generally more densely distributed through the crop than its prey. Tomatoes provide an exception to this rule, as the predatory mite has difficulty negotiating the glandular hairs, and finds it more difficult to migrate to adjacent plants. However, whenever *P. persimilis* is reared on

2.18. Adult *Phytoseiulus persimilis* attacking *T. urticae*

tomatoes, the walking behaviour changes such that it has fewer problems negotiating the hairs. Predatory mites can also adapt to their prey, as tomato-dwelling spider mites often contain substances from the plant that make them unpalatable.

Under low relative humidities, *P. persimilis* move down to the bottom of the crop, whereas spider mites seek the upper, drier parts of the plant. As a consequence, the pest and its natural enemy can become separated. Further, at high temperatures and low humidities the predatory mite population may fail to develop sufficiently.

- Predatory mite
- Larvae, nymphs and adults all eat spider mites
- Can survive for some time in the absence of prey

Amblyseius californicus

Amblyseius californicus (also often known as *Neoseiulus californicus*), is a predatory mite that occurs naturally in tropical and subtropical areas of North and South America, and around the Mediterranean Sea. Like *Phytoseiulus persimilis*, it belongs to the order Acarina and the family Phytoseiidae. It is a good predator of various species of spider mite, including *T. urticae*.

Population growth

In common with *P. persimilis*, *A. californicus* develops faster at higher temperatures. In fact, *A. californicus* can still develop healthily at 33°C, a temperature at which *P. persimilis* struggles. The lower limit for development is around 10°C, and a relative humidity of less than 60% has a negative effect on population growth as among other things, egg mortality increases with decreasing humidity. However, the adverse effect of low humidities is weaker than in *P. persimilis*. Population growth under warm dry conditions is therefore better in *A. californicus* than in *P. persimilis*.

Table 2.4 shows data relating to population growth in *A. californicus*. Mating is essential for egg-laying, although the frequently observed multiple-mating is not essential. Egg-laying also depends on temperature (as shown in table 2.4) and the availability of prey; with fewer prey available, both the numbers of eggs laid and the percentage of females that hatch declines. However, a low prey density has less of an effect on population growth than in *P. persimilis*.

At 23°C the average number of eggs laid per day by *A. californicus* is 2.2 when provided with spider mite eggs as food, and 0.7 when given pollen. *P. persimilis*, conversely, will not reproduce on a diet of pollen alone.

Table 2.4. The population growth of *Amblyseius californicus* at different temperatures with *Tetranychus urticae* as food supply and a relative humidity of 75% (Castagnoli & Simoni, 1991)

temperature (°C)	13	17	21	25	29	33
development time (days)						
egg	8.0	4.4	3.1	2.2	1.7	1.6
larva	3.6	1.6	1.0	0.8	0.7	0.3
protonymph	5.7	2.9	2.0	1.7	1.1	1.0
deutonymph	5.1	2.7	1.4	1.2	0.9	0.8
total egg-adult	22.4	11.6	7.5	5.9	4.4	3.7
po-period*	7.3	5.1	2.6	2.0	0.8	0.9
total egg/♀	48	50	64	60	67	65
eggs/♀/day	0.7	1.1	1.9	2.9	3.6	3.5
egg-laying period (days)	74	45	35	21	19	19
sex ratio (%♀♀)	59	57	66	64	61	54

* po-period = pre-oviposition period, *i.e.* period from becoming adult to first egg-laying

2.19. Predation of *Tetranychus urticae* by *Amblyseius californicus*

Life-cycle and appearance
Amblyseius californicus

The developmental stages are the same as other members of the Phytoseiidae; egg, larva, proto- and deutonymphs and the adult mite. The species closely resembles the thrips predator *Amblyseius cucumeris*, described in chapter 5. The eggs are oval, transparent to white in colour, and are deposited on the underside of the leaves (see plate 2.20). They are smaller than those of *P. persimilis*.

The larva has three pairs of legs and, unlike the larvae of *P. persimilis*, feeds on spider mite eggs. When nymphs and adults feed on *T. urticae* they are a transparent white in colour with an X-shaped orange marking on the back. If there has been no food supply for some time they become thinner and a uniformly light colour. Some of the predatory mites will then disperse in search of new prey, migrating both over the crop and along the ground. However, the majority remain in the crop and await the arrival of new prey.

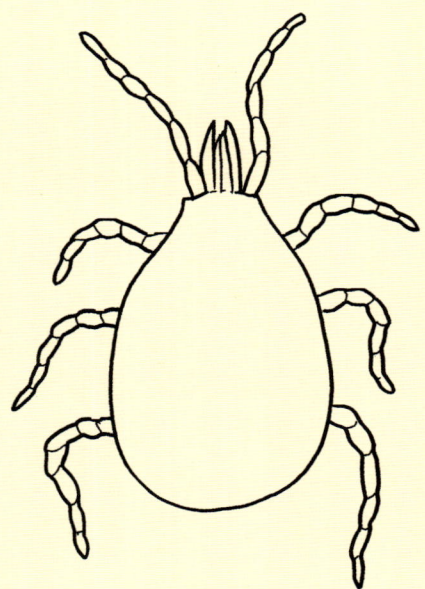

2.20. Egg of *Tetranychus urticae* (l) and of *Amblyseius californicus* (r)

2.21. Larva of *Amblyseius californicus*

2.22. Adult *Amblyseius californicus*

2.23. Preying *Amblyseius californicus*

Feeding behaviour

A. californicus shows a preference for the larval and nymphal stages of two-spotted spider mite even when these stages are present only at low densities. However, the female will eat all stages of *T. urticae*, the larvae eat mainly the eggs, and the nymphs will eat eggs, larvae and nymphs. Consumption depends on the density of predator and prey, as well as the temperature and relative humidity. An egg-laying female consumes by far the most food. *A. californicus* will not consume as much prey as *P. persimilis* (see table 2.5), however, at lower spider mite densities the effect of *A. californicus* is greater than that of *P. persimilis*. This is because at low prey densities *A. californicus* populations will decline less than those of *P. persimilis*. In the absence of spider mites, *A. californicus* can survive on other mites, thrips, moulds and nectar. They can also survive for some time in the complete absence of food, remaining in the crop to await the arrival of new prey.

Searching behaviour and dispersal

A. californicus is unable to detect its prey from the same sort of distances as *P. persimilis* and thus disperses less widely through the crop. They can also spend time on the ground and then return to the crop later

Comparison with *Phytoseiulus persimilis*

Table 2.6 compares *A. californicus* with *P. persimilis*. A comparison with other predatory mites can be found in chapter 5.

2.24. Predation of *Tetranychus urticae* by *Amblyseius californicus*

Table 2.5. Total number of spider mite eaten by different instars of *Phytoseiulus persimilis* and *Amblyseius californicus* (on *Tetranychus urticae*, relative humidity 50%, temperature 26°C) (Gilstrap & Friese, 1985)

instar	A. californicus	P. persimilis
larva	1.1	0.0
protonymph	4.7	6.1
deutonymph	5.6	7.3
adult ♀	156.2	503.1

Table 2.6. Comparison of characteristics of *Amblyseius californicus* and *Phytoseiulus persimilis* of significance in biological control

characteristics	P. persimilis	A. californicus
predation on spider mite	+++	++
speed of development	++	+
reproductive capacity	++	+
effectiveness at high temperatures	+	++
effectiveness at low humidity	-	+
survival without food supply	-	+
survival on alternative food sources	-	+

- **Gall midge**
- **Larvae eat all stages of spider mites**
- **Particularly good at combating concentrations of spider mite**

Feltiella acarisuga

The gall midge *Feltiella acarisuga* (formerly *Therodiplosis persicae*) belongs to the order Diptera and the family Cecidomyiidae. Gall midges are found almost world-wide. The larvae of all species of the genus *Feltiella* feed exclusively on spider mites, and although they are often introduced, from May to September they occur spontaneously in numerous crops including tomato, cucumber, rose and sweet pepper.

Population growth

The speed of gall midge development depends on temperature, relative humidity and the density of prey. The total period of development varies from 30 days at 15°C to less than 10 days at 27°C. Temperatures in excess of 30°C are lethal, and development ceases below 8°C.

Adults mate within a day of emergence. After a further one or two days, the female lays roughly 100 eggs. The eggs are deposited separately on the underside of spider mite infested leaves. Adults only live for 2-3 days, with survival and fertility being enhanced by the presence of a sugar source (*e.g.* in the form of the honeydew exuded by aphids).

The optimal conditions for *F. acarisuga* are 80% relative humidity and a temperature of 20-27°C. *F. acarisuga* is able to overwinter in temperate regions.

Feeding behaviour

Immediately after emerging from the egg, the midge larva creeps toward an egg, nymph or adult spider mite, and uses its mouthparts to first immobilise it, and then suck out the contents. The larvae will eat all stages of spider mites, including those in diapause, and can consume roughly 30 mites or 80 mite eggs per day. They are aggressive, and adapt the amount of food they consume to the density of the spider mite population. Where there are large, localised concentrations of spider mite, *F. acarisuga* is a good control agent. The adults are not predatory, and need only water and sugar.

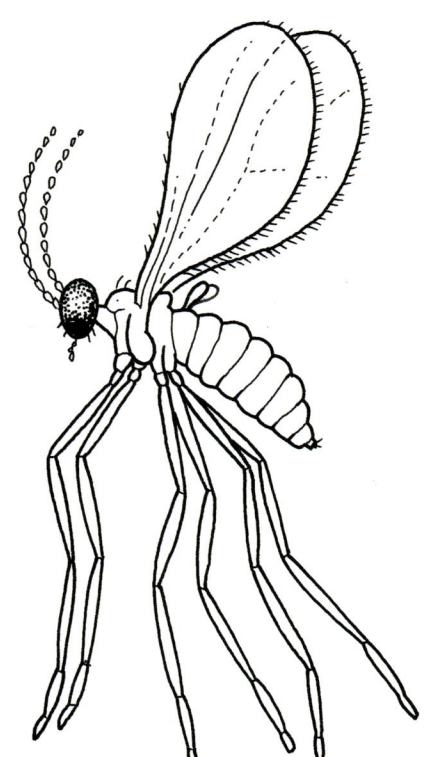
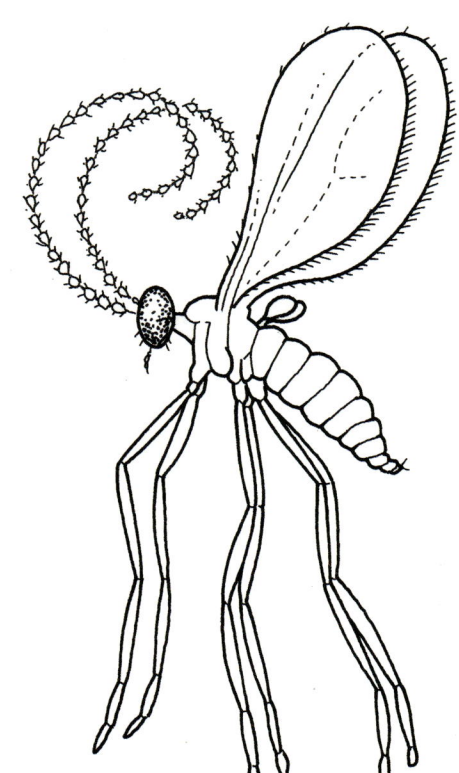

Life-cycle and appearance
Feltiella acarisuga

1 egg
2 larva
3 pupa
4 adult ♀
5 adult ♂

The life-cycle of *F. acarisuga* consists of the egg stage, several larval stages (probably 3), a pupal stage and adult stage or imago. *F. acarisuga* adults are almost incapable of locating individual spider mites, and so the female deposits her eggs on leaves and webs in and around spider mite colonies. The eggs are often laid near leaf veins. They are transparent to light yellow, slightly longer than the diameter of spider mite eggs and almost banana-shaped.

The colour of the larva is usually orange, but varies according to the body contents and may sometimes be yellow or red. Immediately prior to hatching the larva is yellowish white. The eggs and the first larval stages are very difficult to find in the crop on account of their small size and colour (a lens with at least x20 magnification is needed). As development progresses, a red spot appears anterior to the midline, after which the entire larva becomes darker. Finally, white spots appear on the body. A fully grown larva is about 2 mm long. At the end of the last larval stage, the larva spins a white cocoon near a leaf vein and pupates within it.

If the food supply is inadequate, the larva enters a resting phase during which, although still alive, it neither moves nor pupates. The resting phase is broken once food is present again. It is also possible that a shortage of food may induce pupation, although this is only possible if the larva has already reached a sufficient size.

The adult midges are small, rosy-brown in colour with a light covering of hairs on wings and body. They are difficult to distinguish from other midges. Males and females can be distinguished from each other by their antennae: the male antennal segments are longer and hairier than those of the female, making the male antennae longer than those of the female. Adults are nocturnal, and rest during the day on the undersides of leaves.

2.25. Egg

2.26. Preying larva

2.27. Pupa

2.28. Female adult

- A small black ladybird
- Larvae and adults feed on all stages of spider mite
- Can occur naturally under glass from May to September

Stethorus punctillum

Stethorus punctillum belongs to the family Coccinellidae, or ladybirds, which are discussed in chapter 7. This small beetle sometimes appears in a crop naturally and can contribute significantly to the control of spider mites.

Population growth

At 21 - 23°C, development from egg to adult takes 17 - 20 days. The lower temperature threshold for development is 15°C, while the maximum temperature is 35°C. Once an adult female has had sufficient food (usually by around early May) she is capable of laying eggs. For eggs to be viable, mating is essential. However, virgin females may lay unfertilized eggs, which they frequently eat. The sex ratio is 1:1.
In temperate regions there are normally two to three generations per year, although under glass there can be more, with the beetle remaining active for longer. The adults overwinter.
Once food supplies dwindle, the females cease laying eggs. If food supplies are replenished, egg production resumes immediately.

Feeding behaviour

Larvae and adults of *S. punctillum* can prey on all stages of spider mite. As their feeding requirements are rather specialized, they feed almost exclusively on spider mites.
Young larvae feed solely on eggs, larvae or young nymphs and seem to find spider mite webs a major obstacle; older larvae appear to have no trouble with them. The third and fourth larval stages are very active, feeding on all stages of spider mite. Once the larva finds its prey, it injects salivary juices to pre-digest the prey before it is consumed. Adults actively search for mite colonies and consume the whole mite, with females being particularly good predators. However, they are less aggressive than the late larval stages. The beetle can eat around 9 mite eggs per hour, consuming 75 to 100 eggs per day. The larvae eat an average of 10 eggs per hour. At higher temperatures, more prey are eaten and more eggs are produced. Just as with other ladybirds, there is a positive correlation between efficacy and prey density.

Searching behaviour and dispersal

S. punctillum adults are highly active and are therefore very efficient at locating concentrations of spider mite. When disturbed, they frequently drop from the plant to the ground. They are good fliers, and often congregate where there are signs of spider mite colonies. As soon as the spider mite population has been reduced, they leave. The beetle struggles to negotiate the glandular hairs on tomatoes, preferring plants with smooth leaves, and for this reason it establishes poorly on tomatoes.

Life-cycle and appearance
Stethorus punctillum

S. punctillum undergoes the following developmental stages: egg, 4 larval stages, pupa and adult. The adults are small oval shaped beetles (1.2 - 1.5 mm in length), and are good fliers. They are a shiny black colour and covered with fine, yellowish-white hairs. The antennae, jaws and most of the legs are a reddish yellow. All species of the genus *Stethorus* are very similar in appearance and can only be distinguished by examination of the male genitalia.

The eggs are very small, oval and cream-coloured. They are deposited separately, mainly in or close to a colony of spider mites, and usually close to the main leaf vein. Depending on the density of spider mites, a female will lay 1 to 10 eggs per leaf. The eggs appear to be attached by their lateral surface and adhere strongly to the surface of the leaf, with most eggs found on the underside of leaves. Shortly before emergence, the larva becomes visible through the egg cuticle, with takes on a grey colour.

The larvae are a light grey colour at first, becoming darker grey to black as soon as they begin to feed. They have numerous long, branching hairs and black spots.

As the larvae develop, their colour changes, beginning laterally, to a light brown or reddish colour. Shortly before pupation, the whole larva becomes light brown. The larva attaches itself to the underside of the leaf and remains there, immobile, for one to two days before pupation occurs.

The pupa is small and flattened, and the initial light brown to orange colour rapidly turns to black. The whole pupal body is covered with yellow hairs, and the wing buds are clearly visible. The adult insect emerges from the pupa.

2.29. *Stethorus punctillum*

2.30. Larva of *Stethorus punctillum*

Knowing and recognizing *Spider mites*

Biological control of spider mites

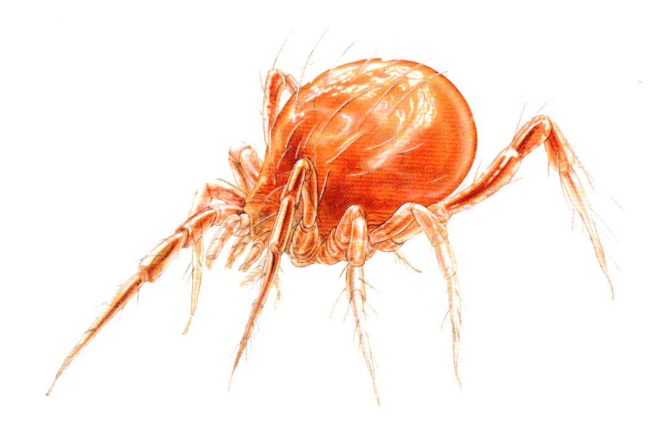

OVERVIEW OF THE BIOLOGICAL CONTROL OF *TETRANYCHUS URTICAE*				
product name	natural enemy	controlling stage	crop	remarks
SPIDEX	*Phytoseiulus persimilis*	nymph and adult	all crops	efficient predator in many crops
SPIDEX-CPR	*Tetranychus urticae*	-	sweet pepper particularly	followed by SPIDEX for rapid achievement of balance between pest and control
SPICAL	*Amblyseius californicus*	larva, nymph and adult	all crops	mainly in tall crops
SPIDEND	*Feltiella acarisuga*	larva	all crops	at high prey densities
MIRICAL	*Macrolophus caliginosus*	predominantly nymph	tomato	also effective in other crops

The predatory mite *Phytoseiulus persimilis* can be used for the control of spider mites in many crops under glass or plastic, and in several outdoor cultures. They should be introduced as soon as the first spider mites are noticed. For effective control the relative humidity must not be too low and the temperature regularly above 20°C. If these conditions are maintained, *P. persimilis* is capable of reducing spider mite numbers quite quickly, and can keep them under control throughout the growing season.

This predatory mite is supplied in a handy shaker together with carrier material, allowing the predator to be distributed throughout the crop evenly and rapidly. The product is known internationally by the trade name SPIDEX.

Controlling spider mites on tomatoes is often more of a problem for the following reasons:
- a great many *P. persimilis* get stuck in the glandular epidermal hairs of the tomato plant,
- tomatoes secrete substances that are toxic for both insects and mites. In *P. persimilis*, these substances cause high mortality among the nymphs, and reduced fertility and longevity in the adults.

Knowing and recognizing **Spider mites**

Where *P. persimilis* is able to adapt, it has less difficulty in combating spider mites. *P. persimilis* can adapt to the particular strain of spider mite found on tomatoes, and adopts a different walking action. SPIDEX contains this specific strain of predatory mite. Research has shown that this adaptation does not reduce efficacy in other crops.
In some situations it may be desirable to introduce the pest mite first by means of SPIDEX-CPR (Controlled Pest Release) followed by the application of *P. persimilis*. This allows the rapid and controlled build-up of a population of spider mite and their predators in the crop. In this way, a balance between pest and predator can be achieved early on.
Amblyseius californicus is best suited to crops of long duration or as a preventive measure, giving long term control of small spider mite populations. *A. californicus* is also more tolerant of higher temperatures, lower relative humidities and of the use of chemical pesticides. In situations where *P. persimilis* struggles, *A. californicus* can provide a useful alternative, and can be combined with *P. persimilis* in cases of heavy of spider mite infestation. *A. californicus* is produced in shakers under the trade name SPICAL.
Feltiella acarisuga is not capable of locating individual spider mites and is therefore specifically recommended where spider mites have already established colonies, particularly in susceptible crops. It is an especially welcome back-up in warm conditions. It is recommended that
F. acarisuga be used together with *P. persimilis*. *F. acarisuga* is supplied as pupae in small plastic tubs under the international trade name SPIDEND.
In salad crops, the predatory bug *Macrolophus caliginosus* can be used, and is available under the trade name MIRICAL. Although predominantly employed to combat whitefly, the nymphs will consume spider mite until they reach adulthood, at which point they disperse in search of whitefly. This bug is supplied in the nymphal stage, in shaker bottles with carrier material.

Knowing and recognizing Other mites

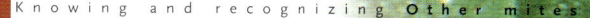

3. Other mites

3.1. *Bryobia* sp.

3.2. *Brevipalpus* sp.

3.3. *Tyrophagus* sp.

This chapter deals with the following:
Bryobia spp.
Tenuipalpidae
Eriophyidae
Tyrophagus spp.
Tarsonemidae
Oribatida

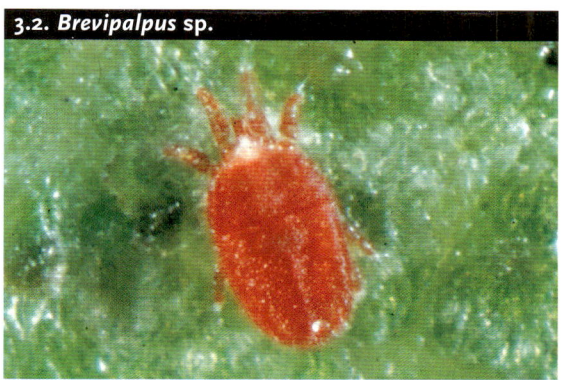

Mites form the subclass Acari (or Acarina) within the class Arachnida (spiders and their relatives). Chapter 2 dealt with an important group, the spider mites. However, there are other mites that can cause varying degrees of damage in glasshouse crops, and all are extremely small. As a result, it is sometimes difficult to establish that damage is being caused by mites, and which species are involved.
In this chapter, the following mites are discussed in order:

- *Bryobia* spp. (plate 3.1),
- false spider mites (Tenuipalpidae) (plate 3.2),
- gall mites (Eriophyidae), specifically the tomato russet mite *Aculops lycopersici*,
- storage mites (*Tyrophagus* spp.) (plate 3.3),
- broad mites (Tarsonemidae) particularly the broad mite *Polyphagotarsonemus latus* and the cyclamen mite *Tarsonemus pallidus*,
- moss mites (Oribatida).

- **Generally occur outdoors**
- **Can occur in glasshouses in spring, and to a lesser extent in the autumn**
- **Rarely causes damage**

Bryobia spp.

In common with the two-spotted spider mite, members of the genus *Bryobia* belong to the family Tetranychidae. However, *Bryobia* species belong to a different subfamily, the Bryobiinae. These mites occur mainly in temperate regions, but also in some subtropical areas.

There has been considerable confusion over the genus *Bryobia*, but it is currently accepted that there are a number of separate, though highly similar, (sub)species, differing in their life cycles, habits and host plants. Most *Bryobia* species are pests of outdoor crops. They can sometimes occur in glasshouses (particularly in the spring and to a lesser extent in the autumn) where slight damage may be caused, particularly by certain *Bouvardia* species. The main culprits are *B. cristata*, a species found on outdoor grasses, and *B. praetiosa*, the clover mite. In general, these mites are easily controlled with selective acaricides.

Population growth
Under glasshouse conditions, the life cycle lasts about 4 weeks and can give rise to 5 or 6 generations per year. The influence of temperature on the development of *Bryobia praetiosa* is shown in table 3.1. The mites overwinter either as eggs or active stages. They are resistant to cold conditions, and development does not stop entirely during winter. They are thus active again early in spring, before most other plant-eating mites.

Damage
The damage caused by these mites is characterised by a meandering trail of small yellow dots on the upper surface of the leaves (see plate 3.4). This pattern of damage can be confused with that caused by leaf miners, spider mites or thrips. In flowering plants both leaves and blooms may be damaged, with heavily affected leaves turning yellow or brown and subsequently dying.
Unlike *Tetranychus* species, *Bryobia* mites do not produce webs.

Biological control
Almost nothing is known about the natural enemies of *Bryobia* spp. It is possible that predatory bugs may prey on them when released against other pests. It is also possible that predatory mites can feed on eggs and larvae of *Bryobia* species, but whether this contributes to control or not has never been investigated.

Table 3.1. Influence of temperature on development time of *Bryobia praetiosa* (Kremer, 1956)

temperature (°C)	15	20	25	30
development time (days)				
egg	25.2	17.6	11.4	7.9
larva	12	5.7	4.4	3.0
protonymph	12.4	5.3	4.5	2.9
deutonymph	11.7	5.9	4.6	3.3
egg-adult	61.3	34.5	24.9	17.1
po-period *	4.2	2.4	1.5	1.1

*: po-period = pre-oviposition period, *i.e.* period from becoming adult to first egg-laying

3.4. Damage caused by *Bryobia*

Life-cycle and appearance
Bryobia spp.

Bryobia spp. have five developmental stages: egg, larva, protonymph, deutonymph and adult mite. In common with spider mites, larval and nymphal stages can be divided in to an active and a resting period. During the resting period, the outer skin, or cuticle, is shed. The newly moulted mite then emerges from the old cuticle.

In most *Bryobia* species very few males are found, and often none. Reproduction is parthenogenetic, and produces only females.

The egg is spherical, smooth, dark red and around 0.15 - 0.2 mm in diameter.

The active stages are slightly larger than those of the spider mite. On hatching, the larva is orange-red and disc-shaped. Once the larva begins feeding the colour becomes greener and the shape rounder. The larva is approximately 0.25 mm long and only has three pairs of legs.

The protonymph can be distinguished from the larva by its larger size (approximately 0.3 mm) and the fact that it has four pairs of legs. Immediately after the moult they are brown, becoming green-brown as soon as more food has been consumed.

The deutonymph is 0.40 – 0.45 mm long. The colour of the deutonymph also changes from brown to green depending on how much food has been consumed.

The adult is approximately 0.6 – 0.7 mm long, and has a flattened oval form. The colour is brown to green, while the anterior part of the body and the foremost pair of legs are red. The adult is further distinguished by a very strongly developed first pair of legs, which are held forward.

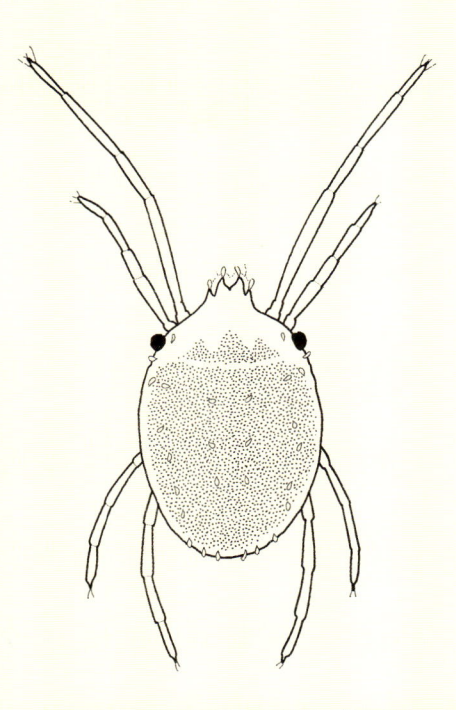

The mite seeks out sheltered places in which to rest, moult, and for the adult to lay eggs.

The mites react immediately to disturbance by dropping from the plant, retracting their legs and remaining still.

3.5. *Bryobia* **sp.**

3.6. *Bryobia* **sp.**

- Introduced on imported material
- Can occur on various ornamental crops
- Population growth is much slower than that of spider mite

Tenuipalpidae
False spider mites

Members of the family Tenuipalpidae can occur on various ornamental crops, and several species have been reported in glasshouse cultures. The family belongs to the same suborder as the spider mite (Prostigmata). Members of the family Tenuipalpidae are known as false spider mites because they closely resemble spider mites (family Tetranychidae). Common species include *Brevipalpus russulus*, which can cause damage to cacti, *Tenuipalpus pacificus*, which causes damage to orchids, *B. phoenicis*, which can be found on several ornamental plants including Dizygotheca and Aeschynanthus, and *B. obovatus*, which occurs on Chamaedorea, among others. *B. obovatus* and *B. phoenicis* are species that occur almost world-wide on a wide range of host plants.

Population growth
The development time of false spider mites is longer than that of spider mites. Reproduction is for the most part parthenogenetic, with females producing almost only female progeny, and males occurring only in very small numbers. At 18 - 21°C the eggs hatch in roughly 2 - 3 weeks, with development to the adult stage taking another 3 - 4 weeks.

Damage
False spider mites are found on the underside of the leaves where they cause a brown, scabby discolouration spreading from both sides of the main vein into the leaf blade. The damage often leads to premature ageing of the plant. In some crops, plants can become misshapen, presumably because the mites secrete a substance while feeding that is poisonous to the plant. In places where mites have been feeding, sunken patches can often be found. Unlike most true spider mites, false spider mites do not make webs. Plate 3.7 shows the damage caused by *B. russulus* and *T. pacificus*.

Biological control
Hardly anything is known about the natural enemies of false spider mites. It is possible that predatory bugs, when introduced against other pests, can prey on them. It is also possible that predatory mites may prey on the eggs and larvae, but it is not known whether this makes a significant contribution to their control.

3.7. Damage caused by false spider mites

Brevipalpus russulus damage on cactus

Tenuipalpus pacificus damage on Phalaenopsis

Life-cycle and appearance
Tenuipalpidae

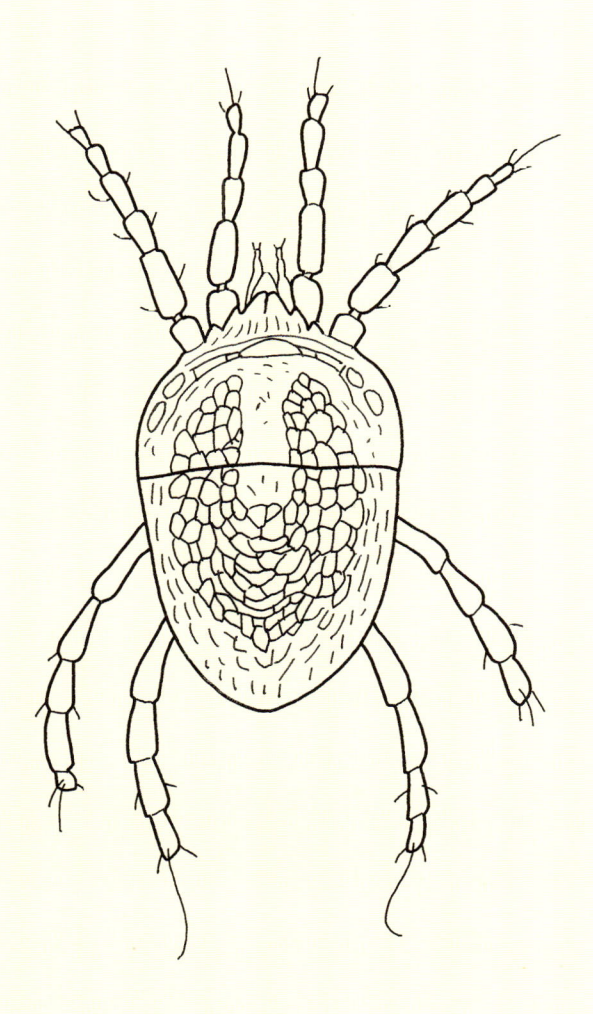

False spider mites are brick-red to orange in colour. They have the same body form as spider mites but are somewhat smaller (around 0.25 – 0.3 mm in length). The different species are all very similar. Plates 3.8 and 3.9 show an adult *Brevipalpus*, and the line drawing shows an adult *Brevipalpus obovatus*.

False spider mites pass through the same stages as spider mites: egg, larva, protonymph, deutonymph and adult. They are rather sluggish and are mostly found along the veins on the underside of leaves. In form they are flattened and egg-shaped from above, with their dorsal surface showing a net-like pattern. The legs appear crumpled.

The eggs are a clear red colour and elliptical. They are laid in a fold in the leaf or along the mid-vein, often in dense clusters of several hundred.

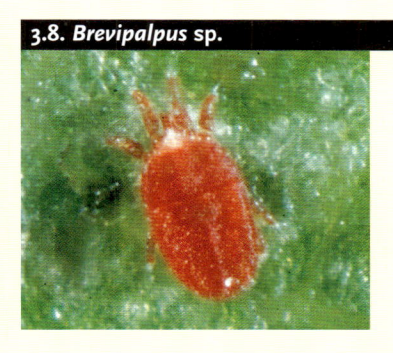

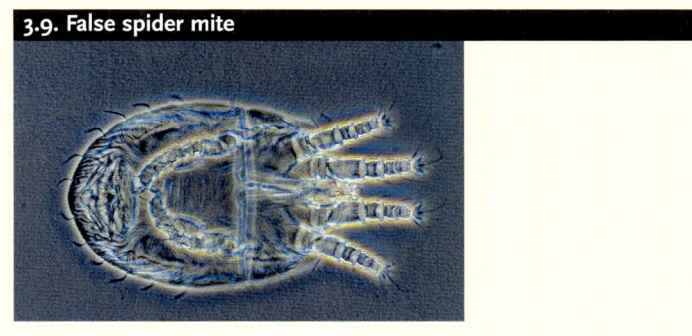

3.8. *Brevipalpus* sp.

3.9. False spider mite

- Very small mites with only two pairs of legs
- *Aculops lycopersici* is the most important species in glasshouses
- Can cause problems for IPM, as there are no known means of biological control

Eriophyidae
Gall mites

Mite species within the large family Eriophyidae live exclusively on plants. Despite the generic name 'gall mites', the majority of species do not form galls, although many species do cause great economic damage. They are generally host-specific, and whilst some species cause damage to only one kind of plant, others have a wider host range including related plant species or even related genera. A particular characteristic of some gall mites is that they can transmit viruses. To date, however, not much is known about this phenomenon due to the difficulties of conducting research in this area.

Gall mites pass through an egg stage, a larval stage, two nymphal stages and an adult stage. In some species an overwintering form of female develops in the autumn. They often overwinter in the buds of host plants or in other sheltered places.

Gall mites are elongated, soft and segmented, and have a few hairs that are important diagnostic features. They are small and almost invisible to the naked eye. The body appears to be constructed from two parts; the head with its mouthparts, and the rest of the body. The key feature that distinguishes gall mites from other mites is that all stages have only 2 pairs of legs, whereas all the other mites have 4 pairs. They are so small that they have no respiratory tubes, and no circulatory system or heart. Although they do not have any eyes, they do have several other sensory organs. They have tiny mouthparts that are short and dagger-shaped for piercing. Non-gall forming gall mites prefer a high temperature and a low relative humidity. They often live in dense colonies, laying minute eggs (0.04 – 0.06 mm) by leaf hairs, in grooves on the leaf surface or along leaf veins. A gall mite population normally contains far more females than males. The total development time from egg to adult is around one to two weeks.

The damage caused by these mites varies enormously depending on the species, and can include curling of leaf edges, blistering of leaves, swollen buds, galls, and the browning of leaves and stems.

Aculops lycopersici
Tomato russet mite

The tomato russet mite (*Aculops lycopersici*) belongs to the gall mite family Eriophyidae, and the subfamily Eriophinae. Like other species that belong to this family the tomato russet mite does not produce galls. There is a certain amount of confusion over the correct Latin name for the tomato russet mite, which is variously referred to as *Vasates lycopersici*, *Vasates destructor* and *Phyllocoptes destructor*. *Aculops lycopersici* originally comes from Australia, and is a pest in tomatoes in all regions where they are grown. Even wild tomatoes, and to a lesser extent other members of the Solanaceae can be affected.

Population growth
Table 3.2 shows information on the population growth of *Aculops lycopersici*.
The optimal climate for population growth is a relative humidity of 30% and a temperature of 27°C. At higher temperatures, a high relative humidity has an adverse effect on survival.
Fertilized eggs give rise to both males and females, whereas unfertilized eggs produce only males. A female can lay 10 – 55 eggs at 21°C.

Table 3.2. The population growth of *Aculops lycopersici* at 25°C (Abou Awad, 1979)

development time (days)	
egg	2.3
larva	0.9
nymph	1.7
total egg-adult ♂	4.6
total egg-adult ♀	5.2
life span (days)	
♀	22.1
♂	16.5

Damage
Damage is caused by the mites sucking out the contents of plant cells. Affected leaves are lightly curled and acquire a silvery appearance on the underside. Later they become brown and brittle. Affected stems look rusty brown, and in serious cases they may snap. The fruit can also be affected, and when this happens in tomatoes, the skin becomes coarse and turns a reddish brown, and the fruit itself is sometimes deformed. Considerable damage can occur, especially under high temperatures when population growth is most rapid and affected leaves quickly dry out. The damage is evident first lower down on the plant, later moving upwards.

Although an entire tomato crop can be ruined, other members of the Solanaceae are generally less severely affected. Plate 3.10 shows damage caused by tomato russet mites in tomato.

Distribution and dispersal
The tomato russet mite spreads through a crop mainly by means of wind-blown threads which they produce. However, they can just as easily walk from one plant to the next, or be spread by mechanical means such as on clothing or other materials.

3.10. Damage in tomato caused by *Aculops lycopersici*

Overview

Damage to fruit

Life-cycle and appearance
Aculops lycopersici

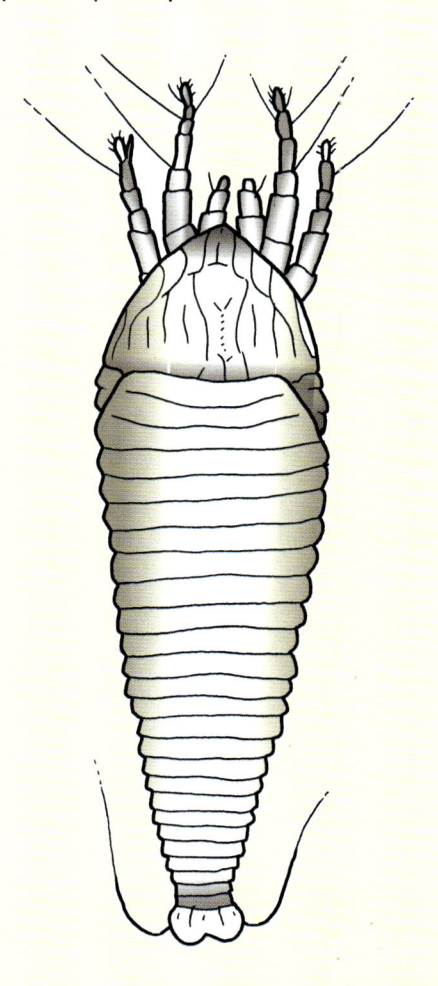

The eggs of the tomato russet mite are relatively large (roughly 0.05 mm in diameter). When newly laid they are creamy white in colour but turn a patchy yellow as they age. From these eggs, transparent white larvae (0.09 – 0.1 mm in length) develop. The larval stage is followed by two nymphal stages. The yellow nymphs are very similar to adult mites in appearance and are 0.14 – 0.16 mm in length. After a short resting period the adults develop. They are reddish brown to orange in colour, wedge-shaped and extremely small (roughly 0.17 mm in length), with males being slightly smaller than the females.

- Occur commonly, but are not always harmful
- Sometimes cause damage on ornamental crops
- Naturally combated by *Hypoaspis* spp., *Amblyseius cucumeris* and *Amblyseius barkeri*

Tyrophagus spp.
Storage mites

Storage mites, also known as fungal mites, belong to the family Acaridae and are the sole members of the genus *Tyrophagus*. These mites occur world-wide in various environments, such as grain or hay stores, in the earth, in moss, and in the litter and nests of a wide range of animals. The different species can only be clearly distinguished from each other by examining them under a microscope.

Storage mites mainly inhabit dark, humid places, and with favourable temperature, humidity and food supply they can suddenly and rapidly multiply. The majority of *Tyrophagus* species constitute part of the soil fauna where they feed on rotting organic material or on moulds. The mites often occur in the soil of potted plants. Although under normal conditions they cause no problems, they sometimes appear in a crop, and if conditions are favourable (particularly high relative humidity) they can establish and cause damage. This applies to many different species, such as *T. longior, T. palmarum, T. perniciosus, T. neiswanderi* and *T. putrescentiae*. The most prevalent species are *T. neiswanderi* and the mould mite *T. putrescentiae*. To date, storage mites are known to cause damage in a limited number of ornamental crops, including species of kalanchoë, begonia, Bromeliaceae, gerbera and anthurium.

Population growth

Storage mites can only survive under high relative humidity. The lowest relative humidity at which *T. putrescentiae* is capable of developing is 65% with a temperature between 15 and 25°C. Beyond these limits a higher relative humidity is required (80% at 10°C, 70% at 30°C and 80% at 35°C). At temperatures below 10°C development stops, and temperatures above 35°C are lethal. Optimal conditions are between 15 and 27°C with a relative humidity of between 85 and 95%.

Apart from relative humidity and temperature, the condition of the substrate is also important for population growth. When there is plenty of organic material present and little competition from other organisms (perhaps as a result of sterilization of the substrate) and few natural enemies, a population can multiply very rapidly. The adult females live in glasshouses for roughly 30 days and can lay about 50 eggs in that time. The eggs hatch after about 6 days.

Table 3.3 shows data concerning population growth.

Damage

Storage mites often occur in large numbers in humus-rich soil, where they help to break down organic material. These populations may infest parts of growing plants where they can multiply in shoots, flower and leaf buds and in rolled leaves. Damage depends on the plant and its growth stage, as well as the density of mites.

In cucumbers, damage is caused by mites piercing the young expanding plant cells. When the leaves with these pierced cells expand, small holes or irregular-shaped cavities appear (see plate 3.11). Furthermore, malformation and discolouration can arise in several crops from the mites piercing soft plant parts. In kalanchoë, storage mites sometimes cause the growing tips to die off and damage the stalks. In cyclamen they can damage the flowers so severely that they impair seed-setting. In begonia, the damage is inflicted on flower and leaf buds. In gerbera, the ray florets are damaged. This damage is difficult to distinguish from the damage caused by thrips. One distinction may be that thrips affect only the ray florets, whereas an attack of storage mite affects both the ray florets and the receptacle. Tapping out the flowers or inspection under a microscope will show which pest is responsible

Table 3.3. The population growth of *Tyrophagus putrescentiae* and *Tyrophagus neiswanderi* (temperature = 25°C; relative humidity = 85%) (Czajkowska & Kropczynska, 1983)

host plant	T. putrescentiae		T. neiswanderi	
	development time (days)	mortality %	development time (days)	mortality %
freesia scales	12.5 (11-16)	16	13.4 (12-15)	29
crocus scales	13.0 (11-17)	8	17.0 (12-20)	29
tulip bulbs	17.5 (12-22)	56	18.0 (10-23)	55
hyacinth bulbs	21.0 (12-26)	28	22.0 (19-22)	50

Knowing and recognizing **Other mites**

Biological control
Hypoaspis spp. can prey on storage mites (see plate 3.12). Once storage mites are present in the crop *Amblyseius* spp. can also contribute to control.

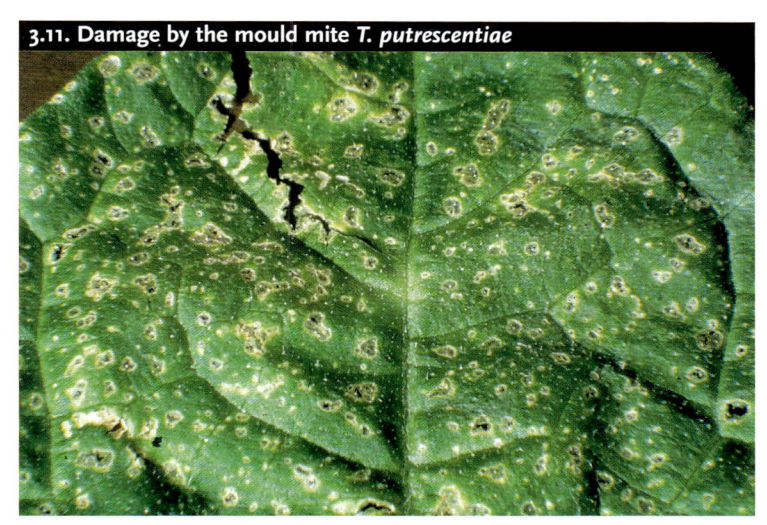

3.11. Damage by the mould mite *T. putrescentiae*

3.12. *Hypoaspis aculeifer* preying on *Tyrophagus putrescentiae*

Life-cycle and appearance
Tyrophagus spp.

Storage mites are minute creatures, scarcely visible to the naked eye. These mites pass through the same five stages as the two-spotted spider mite and *Bryobia*: egg, larval stage, protonymph, deutonymph and adult stage. They are spherical, glassy white in colour, and avoid light. They can be recognized by their long hairs which are clearly visible, particularly at the posterior end of the body. Hairs can be used to distinguish different species under the microscope. They move very slowly.

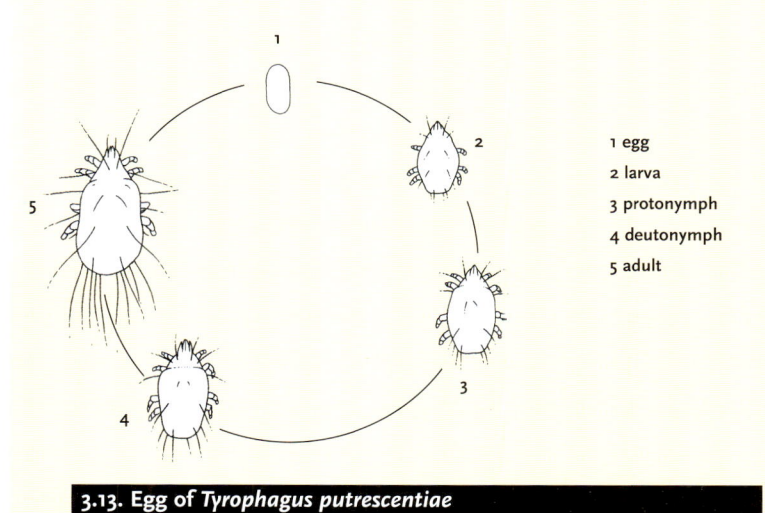

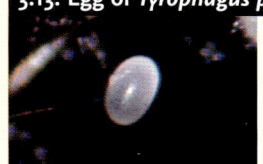

1 egg
2 larva
3 protonymph
4 deutonymph
5 adult

3.13. Egg of *Tyrophagus putrescentiae*

3.14. Developmental stages of *Tyrophagus putrescentiae*

3.15. *Tyrophagus putrescentiae* adult

- ***Polyphagotarsonemus latus* and *Tarsonemus pallidus* are the most important species**
- **Prefer high relative humidities**
- **Damage resembles the effects of viral attack**

Tarsonemidae
Tarsonemid mites

Mites belonging to the family Tarsonemidae (broad mites or tarsonemids) display a greater diversity of feeding habits than any other mite family. There are species that eat moulds, algae, plants, as well as insect predators and parasites. Those living on plants can cause considerable damage to their host.

Tarsonemids are only 0.1 to 0.3 mm long and are invisible to the naked eye; a good magnifying glass or a microscope is needed to see them. Unlike spider mites and gall mites, they need a high relative humidity (at least 70% and preferably above 90%) in order to survive.

Adult tarsonemids have a broad appearance. The body and first pair of legs are sparsely covered with hairs, while the other legs have more hairs, some of which are sensory. The females are larger than the males and more oval in form. Eggs are large in comparison to the size of the adult. Unfertilized eggs give rise to males, and fertilized eggs to females. The larvae have three pairs of legs, adults possess four pairs. The last pair of legs in both males and females are different to the others. Tarsonemid mites have no eyes. They also have a soft skin because the cuticle covering the body contains very little chitin.

Tarsonemids can occur both in vegetable crops under glass or in ornamental crops. The most important species in glasshouse crops are the broad mite (*Polyphagotarsonemus latus*) and the cyclamen mite (*Tarsonemus pallidus*). Other tarsonemids can also sometimes cause damage, such as *Steneotarsonemus ananas* which can occur in bromelia, and the bulb scale mite *Steneotarsonemus laticeps*, which can cause red discolouration in amaryllis (see plate 3.16).

Biological control

The control of tarsonemids is often difficult because of their concealed habit. Removal of affected plants or plant parts and lowering the relative humidity are effective measures, as is the selective use of locally applied acaricides.

The thrips predators *Amblyseius cucumeris*, *A. barkeri* and *A. degenerans* also prey on tarsonemids and thus contribute positively to their control. *A. californicus* can also survive on *P. latus*, and although the population growth is slower than on *Tetranychus urticae*, this predatory mite can complete its development on broad mite and is thus a promising candidate for biological control.

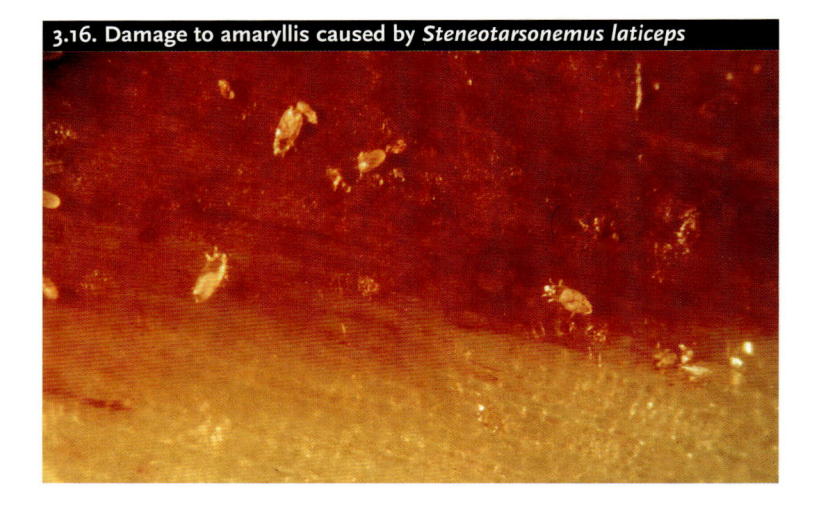

3.16. Damage to amaryllis caused by *Steneotarsonemus laticeps*

Polyphagotarsonemus latus
Broad mite

The broad mite (*Polyphagotarsonemus latus*) occurs in the tropics and in glasshouses in temperate regions. The species is also known by the following names: *Hemitarsonemus latus*, *Tarsonemus latus* and *Steneotarsonemus latus*. The broad mite has a wide range of host plants, including agricultural crops, ornamental crops and wild plants. It was known that this mite was responsible for damage to cotton crops in Africa as early as 1890. The economic damage caused in important crops such as cotton, tea, rubber, citrus fruits, tobacco and potatoes can be considerable. In the glasshouse, peppers are particularly susceptible to damage from this mite, but aubergine, tomato and cucumber, together with ornamental crops such as begonia, gerbera, cyclamen and bouvardia are also affected. It appears that there are several different strains, and it is not entirely clear whether all belong to the same species; so far they cannot be distinguished from each other with any certainty.

In a temperate climate, broad mites are not a serious problem outside since they are unable to overwinter.

Population growth

The development time from egg to adult is extremely short; under favourable conditions it may be 3 to 5 days, and in the winter approximately 7 to 10 days depending on temperature. A large part of the generation time is taken up by the egg stage. *P. latus* prefers high atmospheric humidities; development requires a relative humidity of 75 - 90%, while population growth is fastest at a temperature of 20°C and a relative humidity of 90%. Temperatures above 30°C coupled with a low relative humidity are lethal for egg and nymphal stages. Irrespective of temperature, population growth becomes faster as relative humidity increases. The female mates only once, and begins to lay eggs the following day.

Under summer conditions the life span of a female is about 10 days, although in the winter they may survive for up to 30 days. Males live a few days longer. According to the literature, the number of eggs laid varies from 15 to 50, laid at a rate of 1 to 7 eggs per day depending on the season. Eggs are deposited separately in well-hidden, damp and shady places. Fertilized eggs produce females, whilst unfertilized eggs produce males. The ratio of females to males is roughly 3:1.

In the winter, both the activity of the mites and the rate of reproduction declines. The broad mite depends for its survival on living plant material. Broad mites are not cold hardy and do not enter diapause, and unlike two-spotted spider mites, they are unable to overwinter by hiding in the glasshouse framework. They will remain active in the crop until cultivation ends and the glasshouse is left unheated.

Damage

The damage caused by *P. latus* is similar to that caused by viruses. The typical pattern of damage, depending on the host plant, consists of malformation and distortion of the above-ground growth of the plant. The mites preferentially attack young, developing plant tissue, the growing tips, young leaves and flower buds. Whilst sucking out the contents of plant cells they secrete substances that disturb local growth and lead to deformations, although damage only becomes visible later on.

Leaf feeding is mainly concentrated on the underside near the leaf stalk, which tends to cause the leaf to turn brown and curl up. A typical indication of an attack of broad mite, as with cyclamen mite, is the appearance of dark brown edges to the base of young leaves. If the attack is light, one can often see brown, often collapsed spots, or brown stripes forming a fine network above the green mesophyll beneath. With more serious infestation, this network becomes so dense that there is no more green tissue visible. In most cases, the main veins are untouched, making them stand out as a green pattern against the browned leaf mesophyll.

3.17. Damage caused by *Polyphagotarsonemus latus*

Damaged to the growing head of pepper

Damage to flower of Saintpaulia

Damage to fruit of aubergine

Life-cycle and appearance
Polyphagotarsonemus latus
Tarsonemus pallidus

There are four stages in the life-cycle of both begonia and cyclamen mites: egg, larval stage, false pupa or nymph stage and the adult stage. Female broad mites lay their eggs in unobtrusive hollows of the leaf surface (mainly on the underside of the leaf) or on the fruit surface. The elongated, oval eggs are strongly attached to the surface and are rather large (about 0.07 mm) compared with the subsequent, active stages. The side that attaches to the leaf surface is flattened. They are transparent, and speckled with white (see plate 3.20).

The larva of the broad mite resembles the adult, but is slightly smaller and has only three pairs of legs. Young larvae are white, with heavy cuticle folding to allow for considerable growth without the need to moult.

Without further moult, the larva enters a resting period, called the 'false pupa' or 'resting nymph'. This stage is fixed to the underside of the leaf and the adult develops within. The emerging adult mite is roughly 0.15 mm in size. The female is relatively large, oval and broad, and has a shiny yellow-green or dark green colour, depending on the type and quantity of food consumed. Female mites have a striking white dorsal stripe.

The fourth pair of legs is unsuitable for locomotion and is used as a touch sense organ. The male is shorter and broader than the female, with a body that narrows towards the posterior and is held upward. Males have long legs, are colourless at first but later become a shiny yellow-green. The male mites often drag female 'pupae' behind them, using their fourth pincer-like pair of legs. The male is thus assured of a partner, and mating takes place immediately after the adult female emerges from her moult. Males are more active than females.

Eggs of the cyclamen mite are half the size of the adult mite. They are oval in form, smooth, clearly transparent and twice as long as they are wide. Both ends are equally rounded. Because the egg shell is transparent the embryo is clearly visible shortly before hatching. Figure 3.18 shows an egg.

Larvae of the cyclamen mite are a clear white colour and have 3 pairs of legs. They develop directly to the resting stage without a moult. This resting stage is identical to the larval stage except for the fact that legs are absent; it is a transitional stage between larva and adult mite during which the mite remains immobile on the leaf surface and does not feed.

Adult cyclamen mites are transparent to light brown in colour and roughly 0.25 mm in size, and their colour depends on the host plant on which they are feeding. They are elongated, slightly oval in form and possess 4 pairs of legs. In males, as in the broad mite, the last pair of legs is modified to form thread-like structures used to transport the resting stages of females from which the adult female will emerge.

In general, the cyclamen mite is larger and slower than the broad mite.

3.18. Egg of *Polyphagotarsonemus latus* (left) and *Tarsonemus pallidus* (right)

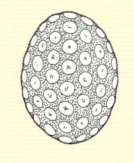

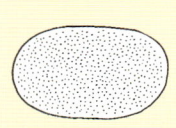

3.19. Female adult of *Polyphagotarsonemus latus*

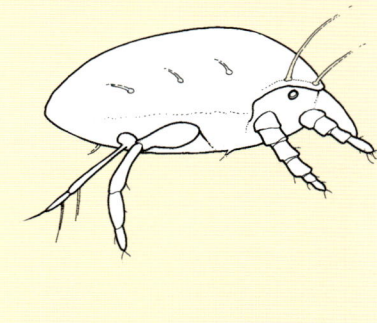

3.20. Egg of *Polyphagotarsonemus latus*

3.21. *Polyphagotarsonemus latus*

Brown, corky patches can arise on the leaf stalks and main stems. The heads of affected plants acquire a misshapen appearance, with contorted leaves and sporadic brown discolouration caused by cork formation. With a severe attack, the growing tip can be killed, plant growth stops and in time the whole plant dies off.

Cork tissue can also develop on fruit. Where pierced cells are killed, deformed corky patches frequently appear causing misshapen fruit that often cracks open at the site of deformation. Blooms are often discoloured, and with a severe attack the stamens are blackened and the flowers deformed. Extensive damage can be caused at very low population densities.

As a rule, the lower leaves of a plant will remain unaffected while the younger leaves are severely damaged. Mites are only occasionally found on the older leaves, even if they have been badly damaged. Most broad mites are found on the undersides of young, expanding leaves, where they conceal themselves between the surface hairs. An attack can be localized, and even on a seriously diseased plant healthy buds can still be found. Symptoms of an attack remain visible several weeks after the mites have been removed.

Distribution and dispersal

Broad mites like a high atmospheric humidity and therefore hide in growing shoots, flower buds and beneath the bracts of the fruit calyx. However, unlike the cyclamen mite, these mites can also be found on leaves. Although the colony as a whole tends to disappear from older leaves and move to younger leaves, the adult females and larvae appear not to leave the leaf on which they occur. It seems that the movement of the colony as a whole is effected by the males transporting the female 'pupae'. Dispersal over larger distances is the result of nursery workers moving through the crop, sudden gusts of wind, or by mites hitching a ride with other insects, particularly whiteflies.

Whenever the atmospheric humidity falls, mites can simply drop to the ground and search for a more humid habitat. Once temperature and humidity become more favourable, the mites migrate upward again. Predatory mites are often present on pepper and cucumber and prey on broad mites. As a result, attacks of broad mite are often localized in these crops, and only gradually spread outwards.

Tarsonemus pallidus
Cyclamen mite

The cyclamen mite (*Tarsonemus pallidus* or *T. pallidus* subsp. *pallidus*) mainly occurs on ornamental crops. It is an important pest in cyclamen and azalea cultivation, but can also cause serious damage in other ornamental crops.

The strawberry mite (*Tarsonemus fragariae* or *Tarsonemus pallidus* subsp. *fragariae*) and the cyclamen mite were once considered to be a single identical species. It is now accepted that they are different species, and they are thus accorded different names. The two species of mite are morphologically identical and can only be distinguished by their host plants. Both mites may be referred to by various names, particularly in the older literature, and their respective names are often used interchangeably which has led to considerable confusion. Even the generic name is a matter of dispute; it is assumed here that these mites should be called *Tarsonemus pallidus*, despite the fact that they are often called *Phytonemus pallidus*. The cyclamen mite occurs throughout the world.

Population growth

The development of the cyclamen mite takes longer than that of the broad mite. Under favourable conditions (20 - 25°C and a high relative humidity) the generation time from egg to egg is roughly ten days. Table 3.4 shows data on the development of the cyclamen mite. Under glasshouse conditions, all stages of the mite can be found

Table 3.4. Development time of *Tarsonemus pallidus* at different temperatures and a relative humidity of 90-100% (Karl, 1965)

temperature (°C)	15	25
development time (days)		
egg	10	3.5
larva	5	2
nymph	3	1.5
total egg-adult	18	7
po-period*		2

*po-period = pre-oviposition period, *i.e.* period from becoming adult to first egg-laying

throughout the year, and therefore damage may be caused year round. The mite can also survive and reproduce under outdoor conditions, albeit at a much slower rate. Outside, there are on average about seven generations per year, but in a glasshouse there are many more. At 4°C embryonic development stops, and larval development ceases at 8°C. Only adult female mites are capable of overwintering, usually in the crown of the plant between furled leaves, or in buds. Some 95% of the population fail to survive the winter. Those females that do survive resume activity in spring, and the first males appear in the beginning of summer. The greatest numbers of mites appear in the summer. Both sexual and asexual reproduction occurs. In asexual reproduction both male and female progeny are produced. In the glasshouse a female lays 1 to 3 eggs per day, giving a total of 12 to 16 throughout her lifetime.

Damage
The cyclamen mite feeds on plant sap by penetrating the plant tissue with its piercing-sucking mouthparts. These mites are mostly found within the flower buds, growing tips and furled leaves where the humidity is highest and they are shielded from direct sunlight. In such places they are difficult to observe. Damage arises through the secretion of auxins (plant growth hormones) during feeding, which in very small concentrations can cause malformations and distortions in plant growth. Damage occurs in localized patches because the mite spreads very slowly.
The damage varies depending on the host plant and can resemble the damage caused by viruses. In cyclamen, the flowers are damaged and leaf growth is inhibited. In fatshedera the leaves remain very small, while in hedera a mite attack can cause baldness in parts of the stem. Gerbera flowers discolour and the leaf turns bronze. Even a mild infestation can cause enormous damage, and with heavier infestations plant growth is arrested and the flower buds are so badly affected they shrivel and die. The patterns of damage caused by *P. latus* and *T. pallidus* are indistinguishable, and so in order to identify the culprit, the mites themselves must be identified.
Plate 3.22 shows the damage caused by the cyclamen mite.

Distribution and dispersal
Cyclamen mites can be dispersed by various means, including walking from one plant to the next. This is a slow process and is affected by atmospheric humidity. Mites can also be dispersed by air currents, by human activity, or by certain insects. If the conditions in the plants become less favourable (drier, for example), the mites can migrate toward the base of the plant, returning when conditions improve. The males are able to transport the female 'pupae' with them. In response to a dramatic deterioration in conditions, mites may drop from the plant to the ground where they search for a more favourable environment. Cyclamen mites are found almost exclusively in flower buds and growing shoots and not, as with broad mites, on leaves.

3.22. Damage caused by the cyclamen mite

Damage to flower (cyclamen)

Damage to leaf (campanula)

- Many different species are known
- Occasionally cause damage in ornamental crops
- Slow population growth, but can survive a long time

Oribatida
Moss mites

Moss mites form the suborder Oribatida, also referred to as Cryptostigmata. They constitute a special group of some 7,000 known species of mites and can cause damage in ornamental crops.

Population growth

In general, moss mites develop slowly under natural conditions and often produce only a single generation per year, or sometimes every two to four years. Compared with other groups of mites, moss mites do not lay many eggs in their lifetime; a few dozen is usual, and sometimes less than 10.

When conditions are unfavourable the mites readily enter a resting state and can survive long periods of food shortage. Adults can sometimes survive several years.

Damage

Although moss mites are omnivorous, they mainly consume dead organic material, including various micro-organisms. They possess biting mouthparts, and are opportunists rather than hunters.

Moss mites are sometimes introduced into flower cultures along with the growth medium, and when this medium no longer provides sufficient food, the leaves or flowers may be attacked. Damage has been observed in anthurium (see plate 3.23) and phalaenopsis. In young leaves, holes appear and both leaves and flowers show deformities. However, because moss mites do not usually feed on living plant material, they seldom inflict serious damage in cultures.

Life-cycle and appearance
Oribatida

The life-cycle of moss mites consists of the following stages: egg, larva, protonymph, deutonymph, tritonymph (sometimes), and adult mite. Males do not come in to direct contact with females but produce spermatophores that are picked up by the females. Parthenogenesis is not uncommon, with unfertilized females producing female progeny. Plate 3.24 shows moss mites.

3.23. Damage in anthurium caused by moss mites

3.24. Moss mites

Knowing and recognizing **Whiteflies**

4. Whiteflies and their natural enemies

Although the glasshouse whitefly *Trialeurodes vaporariorum* has been quite common in glasshouses for some time, it became particularly important in around 1970, especially in tomato and cucumber. As the sudden increase of this pest was met by the use of non-selective pesticides, major problems were caused in the relatively new use of predatory mites to control spider mites in cucumber. Research into methods of biological control was begun, leading to the rediscovery of the use of the parasitic wasp *Encarsia formosa*. This wasp had already been used in England against whitefly in the early part of the 20th century, but with the shift of attention to the chemical insecticides developed as a consequence of two world wars, biological control faded from view. Following considerable success in tomato since 1972, this parasitic wasp is now widely used against the glasshouse whitefly in many crops.

In 1986 the tobacco whitefly (*Bemisia tabaci*) became established in Europe. This insect is now a serious pest in many parts of the world. This whitefly is also parasitized by *Encarsia formosa*.

Two other whiteflies that occur commonly outdoors and are sometimes also found in glasshouses are the cabbage whitefly, *Aleyrodes proletella*, and the strawberry whitefly, *A. lonicerae*. However, neither of these cause economic damage in glasshouses.

Because the use of *E. formosa* to combat whitefly is not always successful, alternatives have been sought. The parasitic wasp *Eretmocerus eremicus* has been used since 1994. Unlike *Encarsia formosa*, which prefers the glasshouse whitefly to the tobacco whitefly, *E. eremicus* has no preference, and will readily parasitize both species. The same year saw the arrival on the market of *Macrolophus caliginosus*, a polyphagous predatory bug with a strong preference for whitefly. Since 2002, *Eretmocerus mundus* has been used against *B. tabaci* in cases where neither *Encarsia formosa* nor *Eretmocerus eremicus* can maintain control. The fungus *Verticillium lecanii* can also infect and kill whitefly.

All these organisms are discussed below.

Knowing and recognizing **Whiteflies**

Whiteflies

The whitefly species *Trialeurodes vaporariorum* and *Bemisia tabaci* are commonly encountered in glasshouses. *Aleyrodes proletella* and *A. lonicerae* have been observed very occasionally.

4.1. *Trialeurodes vaporariorum*

4.2. *Bemisia tabaci*

4.3. *Aleyrodes proletella*

4.4. *Aleyrodes lonicerae*

Whiteflies are not true flies, but belong to the order Hemiptera (true bugs) and the sub-order Homoptera. Together with aphids and scale insects they belong to the division Sternorrhyncha. The whiteflies form the family Aleyrodidae, with those species encountered in European glasshouses all belonging to the sub-family Aleyrodinae. The most common species are the glasshouse whitefly, *Trialeurodes vaporariorum*, and the tobacco whitefly, *Bemisia tabaci*. Both are widespread, and are broadly comparable in outline.

Two other species of whitefly that are not so commonly encountered but sometimes occur in glasshouses are the cabbage whitefly, *Aleyrodes proletella*, and the strawberry, whitefly *A. lonicerae*. These will be discussed briefly.

Life-cycle and appearance
Whitefly

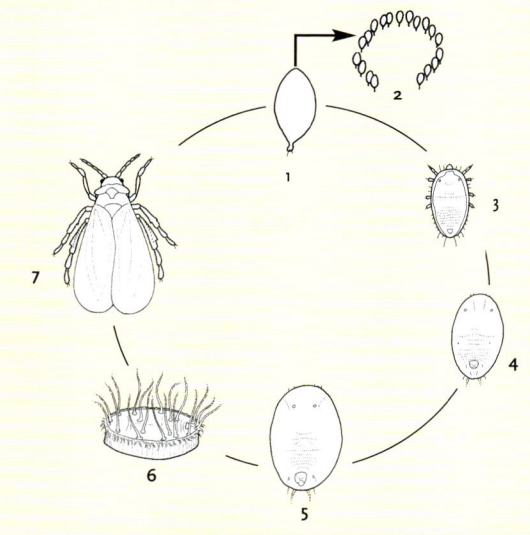

1 egg
2 egg circle
3 larva 1
4 larva 2
5 larva 3
6 larva 4 (= pupa)
7 adult

In whiteflies there are six distinct developmental stages: the egg, first, second, third and fourth larval instars and an adult insect. The late fourth larval stage is frequently referred to as a 'pupa', but strictly speaking it is not a true pupal instar. That is, there is no separate pupal moult, even though the external appearance of the insect changes markedly during this stage. Here, however, for the sake of convenience, we shall refer to four larval stages and a pupal stage.

The eggs are elliptical and often deposited in circles. The young larvae, which on hatching already have well-developed legs and antennae, are known as "crawlers". For several hours they are actively engaged in seeking a suitable place on the leaf where they can feed. Once they have found such a place they remain there throughout the period of larval development. In the second and third larval stages, both the legs and the antennae are reduced to one or two segments and are no longer visible. In the fourth larval stage the insect finally becomes flattened, then subsequently fatter, and the cuticle hardens.

The growth of hairs on the insect depends on that of the leaf; the longer the hairs on the leaf, the longer the hairs on the insect. In other words, the larva adapts itself to the structure of the leaf on which it lives.

On very hairy leaves the larva can sometimes become damaged if it obstructs the growth of stiff hairs.

As soon as the red eyes of the adult whitefly are visible in the fourth larval instar, the stage is referred to as a 'pupa'. The different species of whitefly can best be distinguished from each other at the pupal stage. The adult emerges from the hardened cuticle of the pupa through a characteristically shaped opening; the pupa splits on the upper surface along pre-existing seams to create a T-shaped opening.

On emergence, adult whiteflies immediately begin to penetrate and suck the leaf, and they will continue to do this for the rest of their life. Newly emerged whiteflies have two pairs of transparent wings; later these two pairs of wings and the body are covered with a white, waxy powder that gives the insect its characteristic appearance.

Adult whiteflies are mostly found on the underside of young leaves where they lay their eggs. If the plant is shaken, the adults fly up in a cloud, and return to the undersides of leaves.

4.5. Young egg of *Trialeurodes vaporariorum*

4.6. Larvae of *Trialeurodes vaporariorum*

4.7. Pupa of *Trialeurodes vaporariorum*

4.8. Adult of *Trialeurodes vaporariorum*

4.9. Damage caused by whitefly

Damage

The larvae need a lot of protein for growth, and thus consume a large quantity of plant sap. This contains a high proportion of sugar, and the excess is excreted as honeydew, with larger larvae expelling large quantities. The damage that whiteflies cause to a crop is the result of sucking out the sap from the plant leaves and secreting honeydew. This can have the following consequences:
- if the population is very large, feeding on plant sap can affect the physiology of the plant, as a result of which growth is retarded. In full sunlight, leaves can wilt and fall. Such leaf damage can in turn influence the development of fruit and lead to a reduction in yield,
- the honeydew deposited on fruit makes it sticky. Dirt adheres to the fruit, and the growth of sooty moulds (*Cladosporium* spp.) is encouraged. Such fruit is unsaleable. In serious cases the fruit will rot. Sooty moulds also develop on the leaves, reducing photosynthesis and transpiration,
- viruses can be transmitted. There are three species of whitefly that have been shown to transmit viruses, including *T. vaporariorum* and *B. tabaci*,
- through sucking plant sap and secreting honeydew, whitefly reduce the aesthetic value of crops. This is particularly important in ornamentals.

Damage in various different crops is shown in plate 4.9.

Leaf damage (aubergine)

Leaf damage (cucumber)

Fruit damage (tomato)

- One of the main pests in glasshouses in temperate areas
- Can be a problem throughout the year
- Highly polyphagous

Trialeurodes vaporariorum
Glasshouse whitefly

Since attention was first drawn to the glasshouse whitefly (*Trialeurodes vaporariorum*) as a pest in tomatoes in America in 1870, this species has been one of the major pests of vegetable and ornamental crops in glasshouses throughout the world. The insect originally comes from tropical and subtropical America, probably Brazil or Mexico. Several hundred plant species from different families serve as suitable hosts for this whitefly. In Europe, it occurs on many crops, including aubergine, beans, cucumber, sweet pepper, tomato, fuchsia, gerbera and poinsettia.

Life-cycle and appearance
Female *T. vaporariorum* lay eggs on the underside of young leaves, toward the top of the plant. These eggs are fixed to the leaf with the help of a short hook, by which food is taken up from the leaf. Females often deposit a number of eggs at once in the form of a circle, by turning around on the spot during deposition. This only happens if the leaf surface has a sparse covering of hairs. On hairy leaf surfaces the eggs are usually deposited singly. The eggs are white, elliptical and approximately 0.25 mm in size. Sometimes they may be covered with a kind of waxy powder from the wings of the female. One or two days after eggs are laid they turn brown to black (see plate 4.10).

The larvae that hatch from the eggs are oval in shape and approximately 0.3 mm long, with well developed legs and antennae. First instar larvae are capable of locomotion, whereas the second instar larvae remain flattened on the leaf. Because they are virtually transparent they are difficult to recognize. At this stage they are 0.3 - 0.4 mm long. During the third instar the larva is approximately 0.5 mm long but in all other respects appears the same as the second instar. Plate 4.11 shows larvae of
T. vaporariorum.

In the fourth larval stage the insects are at first flat but later change their form to become a white, oval case encircled by a ring of erect waxy rods. This is the 'pupa' (plate 4.12), which is approximately 0.7 mm in length. During this stage, important morphological developments occur, including the development of genitalia, and the renewed growth of legs and antennae. Finally the adult emerges from the pupa via a T-shaped fissure.

The adult insects are mostly found in the tops of the plant on the underside of the topmost leaves. They fly up when disturbed. Females are roughly 1.1 mm in length and the males 0.9 mm. The wings of newly emerged adults are still a transparent white, and there is no wax on the body at this stage. After a few hours, however, the insect is covered with a white waxy substance (plate 4.13).

The adult glasshouse whitefly has well-developed piercing-sucking mouthparts and begins to feed on plant sap very soon after emerging. The insect will then remain feeding for the rest of its life.
During the night they hardly move, and females stop laying eggs.

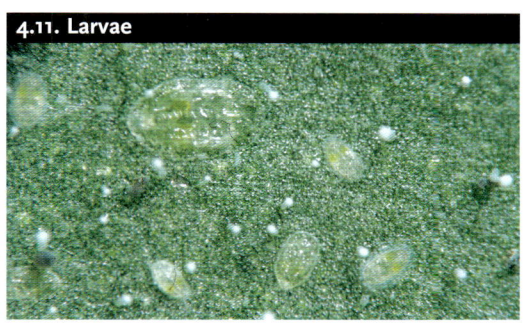

4.10. Eggs

4.11. Larvae

4.12. Pupa

4.13. Adult

Population growth

Population growth of the glasshouse whitefly depends mainly on temperature and host plant. Under normal conditions, atmospheric humidity has relatively little influence on population growth. The optimum relative humidity is 75 – 80 %.

The much greater influence of host plant and temperature on the development of a whitefly population is shown in tables 4.1.1 and 4.1.2. However, a population can adapt to its host plant to some extent. A population that originates on sweet peppers will develop much better on another sweet pepper crop than a population derived from a different crop. The data in the tables are taken from laboratory and semi-field trials in which the crop from which the whitefly originated was not always taken in to account.

Not only do the different species of plant differentially influence the population growth of whitefly, but so too do varieties of a single species. For example, population growth in beef tomatoes is faster than in round tomatoes.

Glasshouse whitefly, unlike most other insect species, does not show an explosive increase in population growth at higher temperatures. The reason for this is the high mortality rate under these conditions, as the species is adapted to somewhat lower temperatures.

One or two days after becoming adult, females of *T. vaporariorum* begin to lay eggs. If they have not yet mated, all their eggs develop into males. After mating, which usually occurs after several days, the fertilized eggs will hatch into either males or females. The sex ratio in *T. vaporariorum* is mostly 1:1. However, parthenogenetic strains of this whitefly are known that produce only females. The number of eggs laid by a single female varies enormously.

The life span of the adult whitefly can also vary from several days to more than 2 months.

Overwintering

The glasshouse whitefly has no special overwintering stage, and survival over winter depends on the host plant. At low temperatures the plant must have cold-hardy leaves. The eggs are able to survive best at low temperatures: at -3°C eggs can survive more than 15 days, but only 5 days at -6°C.

Damage

The damage caused by the glasshouse whitefly is similar to that caused by other whiteflies. *T. vaporariorum* transmits several viruses of vegetable and fruit crops, such as the beet pseudo yellows virus and the tomato chlorosis virus.

Distribution and dispersal

As only the first instar larvae are capable of movement, albeit only over short distances, there is a vertical distribution of the different larval stages on the plant. As the adults prefer to lay their eggs on the young leaves at the top of the plant, the eggs and young larvae are found on

Table 4.1.1. Development time of *Trialeurodes vaporariorum* on tomato at 20°C and 30°C (Tsueda & Tsuchida, 1998) and at 22°C (van Lenteren et al., 1977)

temperature (°C)	20	22	30
development time (days)			
egg	9.9	8	3.9
larva 1	4.4	6	4.9
larva 2	3.5	2	4.0
larva 3	4.3	3	3.8
larva 4	2.6	4	3.8
pupa	7.3	5	5.8
egg-adult	32	28	26.2

Table 4.1.2. Development time and survival at 24°C (van de Merendonk & van Lenteren 1978), 15°C, 20°C and 25°C (Dorsman & van de Vrie, 1987); lifespan, total number of eggs and number of eggs per day at 25°C (van Sas et al., 1978) and at 22°C (van Boxtel et al., 1978) of *T. vaporariorum* in various crops.

	temperature (°C)					crop
	15	20	22	24	25	
development time egg to adult (days)						
	58.4	33.7			21.7	gerbera
	50.9	33.4			21.0	tomato
				21.5		aubergine
				23		cucumber
				24		tomato
				27		sweet pepper
survival (%)						
				91.2		aubergine
				89.4		cucumber
				78.8		tomato
				7.5		sweet pepper
	94.9	85.8			89.7	gerbera
	97.1	98.2			93.5	tomato
lifespan (days) ♀ (♂)						
				50 (30)	74	aubergine
				38 (21)	75	cucumber
				29 (11)	49	tomato
				3 (2)		sweet pepper
					26	gerbera
					44	melon
number eggs per ♀						
				364	666	aubergine
				158	666	cucumber
				47	197	tomato
				2.3		sweet pepper
					130	gerbera
					306	melon
number eggs per ♀ per day						
				10.5	9.0	aubergine
				8.3	8.8	cucumber
				3.4	4.0	tomato
				0.6		sweet pepper
					5.0	gerbera
					6.9	melon

the upper parts of the plant, and the older instars are found further down.
Glasshouse whiteflies mostly stay close together until the population on an infected plant becomes too dense. The attack thus remains concentrated in a few places, especially during the early stages. Later in the season, they become more widely dispersed, probably because activity increases with increasing temperature.

The insects then spread over the entire crop. In addition, an attack can come from outside the glasshouse, or may originate from infected plant material that is brought in from elsewhere.

- **Very large host range**
- **Can cause problems at higher temperatures**
- **Notorious for transmitting viruses**

Bemisia tabaci
Tobacco whitefly

Like the glasshouse whitefly, the tobacco whitefly (*Bemisia tabaci*) belongs to the family Aleyrodidae and the sub-family Aleyrodinae. This insect was first described on tobacco in Greece in 1889, hence the name tabaci. It was later also discovered in Florida (1900), in Sri-Lanka (1926) and in Brazil (1928). It has subsequently been found in most tropical and subtropical countries of the world. The original habitat was probably a tropical or subtropical area, possibly in Pakistan.
B. tabaci has an enormous host range and has affected an extremely wide range of crops throughout the world. It causes particular damage in (sub)tropical areas.
The tobacco whitefly is feared because of its high level of resistance to many insecticides.
There is still much uncertainty over the different populations of *B. tabaci*. For example, in the mid 1980's in the United States there were huge problems with tobacco whitefly, even though it had been present there since 1897. It appeared that a new strain had arisen that caused far greater damage. In order to distinguish this new, serious strain from the old, less harmful strain, it was called the "B strain" (or poinsettia-strain) and the original strain was designated the "A strain" (or cotton strain).
In external appearance the two strains are identical and they share the same range of host plants, although they manifest different preferences for certain hosts. They are genetically different, however, and unable to hybridise, and for this reason some researchers insist that they are different species, naming the "B strain" *B. argentifolii* (meaning 'of silver leaf', a symptom caused solely by this strain in plants of the squash family). However, it was already clear by 1950 that there were different strains of *Bemisia tabaci* that did not differ appreciably in appearance but which did differ in host plant range, the ease with which they could adapt to a host plant, and in their powers of virus transmission. There are even populations that are restricted to a single host plant, or to a very restricted number of host plants. *B. tabaci* is therefore considered to be a complex of different biotypes.

The unstoppable spread of the B strain via the transport of ornamental plants since around 1985 has led to this strain becoming established throughout Europe, the Mediterranean area, Africa, Asia, north, middle and south America, (Argentina, Brazil, Colombia and Venezuela) and the Caribbean region. The A strain, long known in the United States, does probably not occur in Europe.

Life-cycle and appearance
The development of *B. tabaci* is the same as in *T. vaporariorum*: there is an egg stage, four larval instars and the adult insect. The adult insect closely resembles *T. vaporariorum*, but is slightly smaller and yellower. More distinctively, the wings of the tobacco whitefly are held vertical and parallel along the body, whereas the wings of the glasshouse whitefly are held horizontally, forming a triangle. However, even this difference is not always clear. The adult female is slightly larger than 1 mm in length, the males slightly less than 1 mm. On emergence, the adult body is yellow in colour and the wings transparent, but after several hours both body and wings are covered with a white wax. As a result, the body colour changes to a creamy white. This wax secretion is less intense than in the glasshouse whitefly and the body colour is therefore more yellow. Plate 4.16, page 62 shows an adult of *B. tabaci*.
B. tabaci lays its eggs spread widely over the whole plant with the result that, unlike *T. vaporariorum*, all stages of the life-cycle can be found together on the same leaf. The eggs are a yellowy green when newly laid and gradually colour to a light brown (see plate 4.14). They are smaller than the eggs of *T. vaporariorum* (average 0.18 mm). Eggs are preferentially laid on the undersides of leaves.
First instar larvae are transparent and approximately 0.25 mm in length. At this stage the larvae possess legs and antennae. After exploring the area for a while they settle in a suitable place, then begin to partially retract their legs and antennae and start to feed The body of a second instar larva is flat and around 0.3 mm in length. The third instar larva is

larger (around 0.5 mm), but unchanged in form. At the beginning of the fourth instar the larva is broad and flat, but during the course of this instar it changes shape to become almost circular (approximately 0.8 mm long and 0.6 mm wide), after which it is known as a pupa. Pupae can sometimes have an irregular, dented outline depending on the growth of hairs on the leaf, and the leaf structure of the host plant. In the pupal stage the red eyes are clearly visible. Because the pupal cuticle is transparent, the yellow colour of the whitefly is also visible. Laterally, the white of the developing wings can just be seen. This is the stage when the important morphological developments occur that lead to the adult form. Plate 4.15 shows larvae and pupae of *B. tabaci*.
T. vaporariorum and *B. tabaci* can be distinguished by several features during the pupal stage. In *T. vaporariorum* a wreath of short waxy threads can be seen around the body, projecting outwards. This is lacking in *B. tabaci*. In the tobacco whitefly the pupae are rather broad and flat and transparent or yellowish in colour, whereas those of the glasshouse whitefly are attached to the leaf like small, white, oval cases.

Population growth
B. tabaci is a naturally occurring pest in tropical and subtropical regions, and so development is optimal at fairly high temperatures (around 30°C). Above this temperature population growth declines rapidly, while below 16°C it comes to a standstill. At low atmospheric humidity and a temperature of 9°C the larvae die and the population is drastically reduced. Apart from temperature, the development time for *B. tabaci* appears to depend also on the crop and (to a lesser extent) on relative humidity. It is not only the species of host plant that matters, but also the nutritional quality of the crop. Stress conditions, such as low light intensity, (very) high temperatures and extremes of relative humidity therefore affect the development time both directly and indirectly. The optimal temperature for *B. tabaci* is higher than for *T. vaporariorum*, not least because *B. tabaci* suffers higher mortality at lower temperatures. Mating takes place one or two days after the appearance of the adult. After mating there is a pre-oviposition period of 1 - 5 days, after which egg-laying begins. The number of adult females always exceeds the number of males. Females that have not mated lay eggs that produce only male larvae.
Tables 4.2, 4.3 and 4.4 (p. 63) present data concerning population growth in *B. tabaci*.
The lifespan of adult *B. tabaci* is strongly dependant on temperature and the nature of the crop. At higher temperatures (28 - 30°C) an adult female will live for 10 to 15 days, whereas under winter conditions an adult may live for one or two months. Even without a host plant, *i.e.* in an empty glasshouse, an adult can survive several weeks at a low temperature. However, *B. tabaci* cannot survive below 0°C.

Damage
The effects of *B. tabaci* consuming plant sap and excreting honeydew are much the same as with an infestation of the glasshouse whitefly. There are, however, other pathological effects associated with the presence of this whitefly. The larvae inject enzymes into the plant that alter their normal physiological processes. In some host plants, this can cause

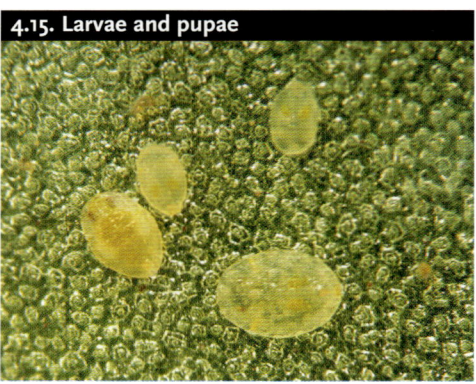

4.14. Eggs

4.15. Larvae and pupae

4.16. Adult

damage, including irregular ripening in tomatoes and sweet peppers, yellowing of the flower stalks in gerbera and serious yellowing of the leaves of French beans. Other symptoms include the appearance of chlorotic patches, yellowing, fruit- and leaf-fall and misshapen fruit. In various ornamental crops (including bouvardia and zinnia) the veins of young leaves turn yellow, giving rise to "mosaic vein". As the leaves age this symptom largely disappears.

As a result of the irregular ripening of tomatoes, fruit can show irregular colouring, with green or yellow patches (see plate 4.17, p. 64). Such fruit does not colour evenly, even several weeks after the onset of fruit colouration. The fruit remain hard and taste sour, and cutting into the fruit shows that internal colouring is also retarded.

In The Netherlands, no symptoms have been so far noted in squashes, but in France and the USA it is known that serious physiological damage can occur in these crops. These virus-like symptoms do not appear in cucumber and aubergine.

There are viruses that can cause very similar irregular colouring of fruit, but this is generally accompanied by leaf symptoms. When an attack is successfully controlled and subsequent fruit development is normal, it is obvious that the symptoms were the direct effect of whitefly on the plant. However, fruit that has been affected will not recover.

These physiological symptoms have been the main problem caused by *B. tabaci* since the mid 1980's in various parts of the world, and symptoms can appear at very low whitefly densities.

B. tabaci is a formidable and much feared carrier of viruses, transmitting more than 25 viruses and many other virus-like pathogens. These viruses can cause serious diseases, mainly in tropical crops. One of the most important of these is tomato yellow leaf curl virus (TYLCV). Both larvae

Table 4.2. The population growth of *Bemisia tabaci* in tomato (Tsueda & Tsuchida, 1998; Salas & Mendoza, 1995)

	temperature (°C)		
	20	25	30
development time (days)			
egg	12.4	7.3	6.1
larva 1	7.1	4.0	3.4
larva 2	3.5	2.7	2.0
larva 3	6.1	2.5	2.2
larva 4 to pupa	3.3	5.8	2.0
from pupa	6.7		2.2
total	39.1	22.3	17.9
po-period (days)*		1.4	
egg-laying period (days)		16.7	
mortality egg-adult (%)	34		21
sex ratio (% females)		73	
reproductive capacity			
av. number of eggs/♀		195	
number of eggs/♀/dag		11.7	

*po-period = pre-oviposition period, *i.e.* period from becoming adult to first egg-laying

Table 4.3. The population growth of *Bemisia tabaci* on poinsettia (Enkegaard, 1993)

	temperature (°C)				
	16	19	22	25	28
development time (days)					
egg	34.3	17.6	12.7	10.5	7.8
egg-adult	137.2	66.8	38.7	31.9	23.2
po-period (days)*	4.3		3.0		2.2
lifespan ♀	50.8		21.8		16.0
mortality (%)					
egg	19.3	9.2	2.1	1.0	2.8
egg-adult	95	60.4	60.6	39.3	6.1
sex ratio (% females)	60	63	69	76	
reproductive capacity (number of eggs per ♀)					
	60.2		90.9		96.3

*po-period = pre-oviposition period, *i.e.* period from becoming adult to first egg-laying

Table 4.4. The population growth of *Bemisia tabaci* in three crops at 25°C (Tsai & Wang, 1996)

	aubergine	tomato	cucumber
development time egg-adult (days)	17.3	18.0	19.3
lifespan ♀	24.0	20.6	9.9
mortality egg-adult (%)	11.3	39.8	53.6
vreproductive capacity (number of eggs per ♀)	224	168	66

and adults can pick up the virus, which is then transmitted by the adult insect.

Distribution and dispersal
Unlike the glasshouse whitefly, which is found mostly on upper parts of the plant, the tobacco whitefly can be found over the whole plant. Females deposit their eggs on the underside of leaves, and the winged adults may disperse over considerable distances by active flight or on the wind.

In the following table, the distinguishing features of the glasshouse whitefly and the tobacco whitefly are set out.

4.17. Irregular ripening of tomato caused by *Bemisia tabaci*

4.18

Differences between the various developmental stages of
Trialeurodes vaporariorum & *Bemisia tabaci*

	Trialeurodes vaporariorum glasshouse whitefly (T.v.)	*Bemisia tabaci* tobacco whitefly (B.t.)
appearance		
developmental stages		
egg	first 1-2 days white, later dark brown to black	light yellow-green, later light brown
larva	hardly distinguishable from B.t.	hardly distinguishable from T.v.
pupa	oval, white case with a wreath of projecting wax threads around the body	rather flat and transparent or yellow in colour; adult is visible (yellow with red eyes and white wing buds)
Encarsia formosa pupa	black, not transparent	brown, colour of parasite visible
Eretmocerus eremicus pupa	yellow, quite transparent	transparent yellow
adult	larger than B.t.; appears whiter as result of greater wax secretion; from above appears more triangular in form	smaller than T.v. appears yellower due to lesser wax secretion; from above appears more elongated
other characteristics		
optimal development temperature for population	20-25°C	25-30°C
lifespan	relatively short, particularly at higher temperatures	longer than T.v., good survival at high temperatures
distribution over the plant	in the top of the plant	distributed over the plant
damage	aesthetic damage and reduced yield due to honeydew and feeding damage	aesthetic damage and reduced yield due to honeydew and feeding damage, small population can cause major physiological changes leading to serious disease symptoms
resistance to pesticides	less resistant	highly resistant

- Clearly distinguishable from glasshouse whitefly and tobacco whitefly
- Moderate economic significance
- Difficult to combat with chemical pesticides because of copious wax secretions

Aleyrodes proletella & *A. lonicerae*
Cabbage whitefly & strawberry whitefly

The cabbage whitefly *Aleyrodes proletella* (previously known as *A. brassicae*) and the strawberry whitefly *A. lonicerae* (previously known as *A. fragariae*) are two common outdoor species that are only occasionally found in glasshouses. *A. proletella* can occasionally be a problem in gerbera and is sometimes spotted in cucumber, while *A. lonicerae* appears sporadically in strawberry crops. The economic damage caused by these two species, however, is slight. The cabbage whitefly occurs commonly in cabbage crops in Europe, north and east Africa, India, the United States, Brazil, Taiwan, Bermuda, New Zealand and Australia. The strawberry whitefly is commonly found on outdoor crops of strawberries, raspberries, currants and blackberries.

Eretmocerus eremicus and *E. mundus* are wholly ineffective parasites of either of these species, and although *Encarsia formosa* is capable of parasitizing them, it is probably not effective enough to control an infestation. Little is known about the effectiveness of any other natural enemies of these species, and chemical control is hampered by the thick waxy layer secreted by larvae and pupae.

Life-cycle and appearance

The cabbage whitefly and the strawberry whitefly have the same life-cycle as the other species discussed above: an egg stage, four larval instars, the last of which becomes the pupal stage, and the adult whitefly. Adults are much larger than those of the glasshouse and tobacco whiteflies and are further distinguished by the 2 (*A. lonicerae*) or 4 (*A. proletella*) grey patches on the wings. They also have a thicker powder covering than the glasshouse or tobacco whitefly. Plates 4.19 and 4.20 show adult cabbage and strawberry whiteflies.

Eggs are laid on the under-surface of leaves. The larvae and pupae are roughly 1.5 times bigger than the same stages of glasshouse and tobacco whiteflies and are covered with a thick layer (see plate 4.21). On outdoor crops the larvae and pupae are dark yellow, whereas under glass they are a lighter colour.

The strawberry whitefly produces four to five overlapping generations outdoors in a year in temperate climates, with the adult insect overwintering through the coldest months. More generations can be produced in glasshouses.

4.19. *Aleyrodes proletella*

4.20. *Aleyrodes lonicerae*

4.21. Parasitized and unparasitized larvae of *Aleyrodes lonicerae*

Natural enemies of whiteflies

The most important natural enemies of whiteflies in glasshouses are:
Encarsia formosa
Eretmocerus eremicus
Eretmocerus mundus
Macrolophus caliginosus
Verticillium lecanii

4.22. *Encarsia formosa*

4.23. *Eretmocerus eremicus*

4.24. *Macrolophus caliginosus*

4.25. *Verticillium lecanii*

Encarsia formosa has been used for decades to combat whitefly in glasshouses. For many years, this parasitic wasp was the only natural enemy used against whitefly. However, the parasitic wasp *Eretmocerus eremicus* performs better at higher temperatures, is more tolerant of pesticides, and parasitizes *Bemisia tabaci* better than *Encarsia formosa*. The parasitic wasp *Eretmocerus mundus* also parasitizes larvae of *B. tabaci*, and has been used for controlling *B. tabaci* in southern Europe since 2001. Another important control measure is *Macrolophus caliginosus*, a predatory bug with a preference for whitefly. Finally, the entomopathogenic fungus *Verticillium lecanii* can also be used. All these species mentioned above are discussed in detail in this chapter.
In soutern Europe *Nesidiocoris tenuis* is naturaly occuring. This predatory bug feeds on whitefly, but can also cause damage to the plant.

- Parasitic wasp
- Prefers to parasitize third and fourth instar larvae and to feed on second instar larvae
- Controls glasshouse whitefly particularly well

Encarsia formosa

Encarsia formosa is a well known and much used parasite of whitefly. This parasitic wasp probably originates from a tropical or subtropical region. Although its exact origin is unknown, it is assumed that *E. formosa* comes from the same areas as its host *Trialeurodes vaporariorum*, and can be found nowadays in Europe, Australia, New Zealand, Japan, Canada and the United States. *E. formosa* belongs to the family Aphelinidae, of the order Hymenoptera. In 1926 an English tomato grower discovered black pupae of *Trialeurodes vaporariorum* from which parasites emerged. These were subsequently identified as *Encarsia formosa*. This insect is capable of using various species of whitefly as hosts, such as *Trialeurodes vaporariorum* and *Bemisia tabaci*. The species has been used as a biological control for whitefly in glasshouses since 1972.

Population growth

Several investigations have been carried out into the population growth of *E. formosa* compared to that of *T. vaporariorum*. The development of *E. formosa* is mainly dependent on the age of the host attacked and on the prevailing temperature, but also on the species of host and the plant on which the host occurs. Atmospheric humidity is less important: between 50 and 85% is optimal for the growth of the population. *E. formosa* develops more rapidly than the whitefly. A comparison of the average development times from egg to adult in *T. vaporariorum* and *E. formosa* at different temperatures is shown in figure 4.26.
At temperatures of less than about 12°C, development of the parasitic wasp is arrested, whilst the maximum temperature for development is around 38°C; the wasp will not survive at higher temperatures. Development of the egg is of the same duration in any instar and takes roughly 4 days, but a parasitic wasp will not develop further than the first larval stage until the host insect itself reaches the fourth larval instar. The time taken for the development of the parasitic wasp thus depends on the age of the host.
A population of *E. formosa* consists almost entirely of females. Males are hardly ever encountered. Mating is thus unnecessary for reproduction, and females produce female progeny by parthenogenesis. The fact that many populations consist solely of females is due to the presence of certain bacteria (*Wolbachia* spp.) in the wasp. Because the relationship between the two organisms is beneficial for the growth of the population, it is considered a case of symbiosis (that is, both bacteria and parasitic wasps benefit from the relationship). The females of the

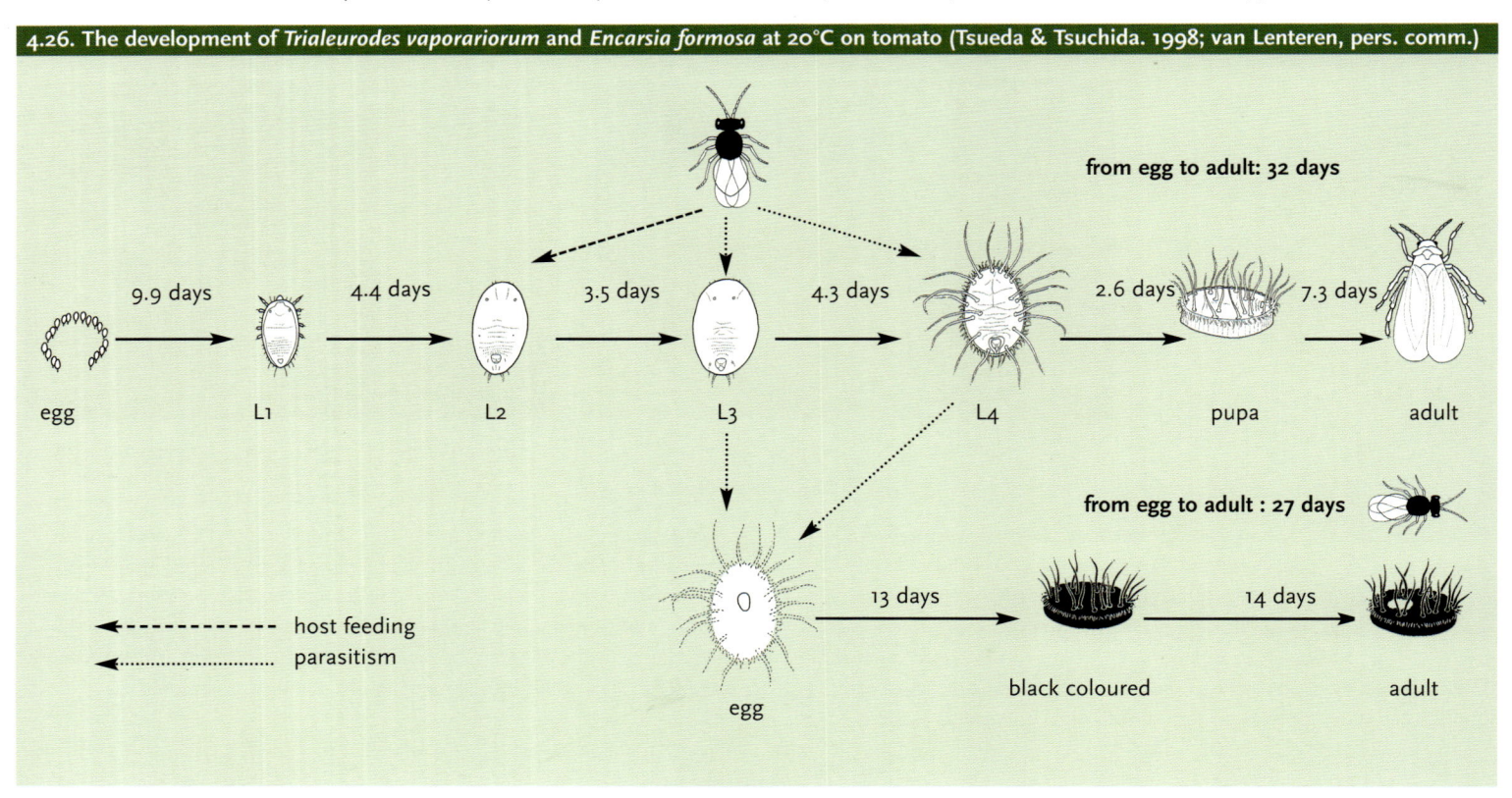

4.26. The development of *Trialeurodes vaporariorum* and *Encarsia formosa* at 20°C on tomato (Tsueda & Tsuchida. 1998; van Lenteren, pers. comm.)

Life-cycle and appearance
Encarsia formosa

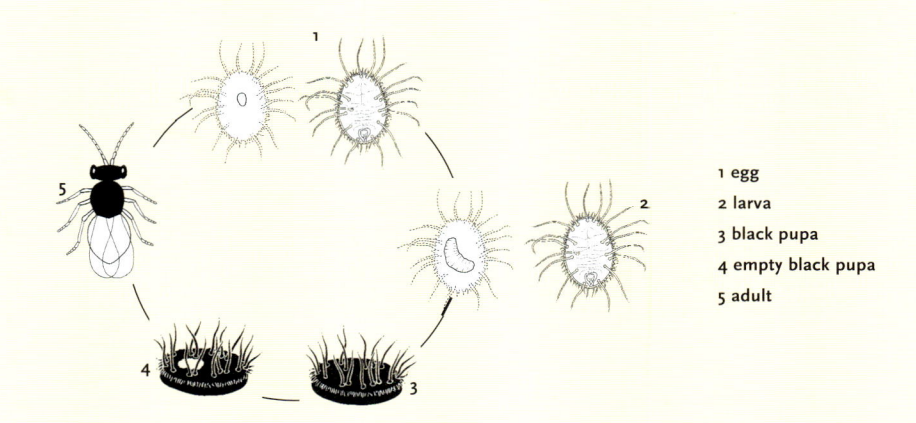

1 egg
2 larva
3 black pupa
4 empty black pupa
5 adult

The development of E. formosa consists of six stages: the egg, three larval instars, a pupal instar, and the adult wasp.

All these stages except the adult insect are found inside the host (i.e. larval and pupal whiteflies).

The female can deposit her egg in any larval stage of whitefly but preferably selects the third and young fourth larval instars. This offers the best chance of successful development. The egg is 0.08 mm long and 0.03 mm wide.

About halfway through the parasite's development, the host pupa turns black. Parasitized pupae are thus easily recognized. However, this blackening can be delayed, in which case the larva remains visible throughout its development.

When the tobacco whitefly has been parasitized, the pupa remains transparent to brown.

The pupa of E. formosa is unique among the species of Encarsia, in that it is formed immediately beneath and in contact with the whitefly pupal cuticle, whereas in other Encarsia species it can be clearly seen that the pupa develops freely within the host pupa.

When the parasitic wasp is full grown it eats a round hole in the pupal cuticle. This takes several hours before the wasp frees itself from its case.

The female parasitic wasp is approximately 0.6 mm long, with a head that is dark brown to black in colour, a dark brown to black thorax and a yellow abdomen. The male is entirely black, but a population consists of no more than 1 to 2% males. The full-grown insect feeds on honeydew and the body fluids of whitefly larvae (host feeding). For the latter, a female selects a larva (preferably a second instar, although other stages are also attacked), and uses her ovipositor to pierce it. The wasp then sucks out the whitefly body fluid from this opening, often killing the larva as a result.

4.27. Parasitized pupae of *Trialeurodes vaporariorum*

4.28. Parasitized pupa of *Bemisia tabaci*

4.29. Empty pupa of *T. vaporariorum* from which *E. formosa* has emerged

4.30. Parasitizing female

parasite can lay approximately 150 eggs under optimal conditions, at a rate of around 5 to 15 per day.

The lifespan of a female declines rapidly with increasing temperatures. At 30°C a female lives only a few days. The lifespan is always extended if there is a host to feed on.

At temperatures of 20 - 25°C the wasp operates most efficiently, but is still an effective biological control agent at lower temperatures. Females are very good at detecting whether or not a whitefly larva has already been parasitized, and always avoid parasitizing a larva that has already been parasitized. Table 4.5 presents data relating to the population growth of *E. formosa*.

Searching behaviour and dispersal

Female *E. formosa* will search the crop actively but randomly until they find a whitefly larva. Once this happens, they remain in the area until all host whitefly have been parasitized, after which they go in search of new hosts, often flying long distances. As a result of this highly active searching behaviour, small areas of whitefly in the glasshouse are located rapidly.

The density of hairs covering the plant plays an important role in determining the mobility of *E. formosa*; plants with dense hairs will restrict the wasp's movement. The honeydew produced by whiteflies also impedes the parasitization process by making it more difficult for the parasites to move about. In places with heavy whitefly infestation, *E. formosa* can often become less effective.

E. formosa searches a leaf using their antennae to detect whitefly larvae. Once located, the wasp probes the larva with the antennae and eventually also with the ovipositor to determine whether the larva is suitable either to parasitize or as food. *E. formosa* is highly skilled in identifying those larvae that are already parasitized and will not deposit an egg in such larvae. *E. formosa* do have more difficulty identifying larvae that have been parasitized by other species of parasitic wasps, particularly when those parasites in the larvae are still in the egg stage. But even then, in more than 90% of cases a female will not parasitize the host again.

Parasitic wasps are active in the daytime and hardly ever fly at night. Short daylength, low temperature and low light intensity all have a negative effect on activity. The searching behaviour of *E. formosa* is greatly influenced by the number of ripe eggs present: the more ripe eggs there are, the more active the searching behaviour. The wasps appear to continue searching longer on leaves infected with whitefly or whitefly excretions or contaminated with honeydew than on clean leaves.

Parasitization of whitefly by *E. formosa*

Parasitization of *T. vaporariorum* by *E. formosa*

The degree to which *T. vaporariorum* is parasitized by *E. formosa* is highly dependant on the crop concerned. Cucumber plants, for example, have several characteristics that create problems for the control of whitefly. The hairy leaves and the network of veins in the leaf can reduce the wasp's mobility. In addition, cucumber is an excellent host plant for whitefly, which further complicates control. The relation between the glasshouse whitefly and the parasitic wasp differs according to the crop. This needs to be taken into account when using biological control, for example by adapting the schedule for introducing the parasitic wasps. Because the glasshouse whitefly lays eggs on the youngest leaves, the larvae on any leaf are all at roughly the same stage of development and, as a result of parasitizing them, the population of *E. formosa* is fairly uniform. Non-parasitized whitefly larvae develop into adults some 7 days before the parasitic wasp. Defoliation can therefore affect the ratio of whitefly to parasitic wasps. For example, if one removes 5- to 6-week old leaves the whitefly is at an advantage, since these leaves will still carry many parasitized pupae. Before proceeding with such steps in a crop, this should be taken into account.

At low temperatures and light intensities the parasite is less active. Parasitization still occurs, but to a much lesser extent. The percentage of larvae parasitized increases with rising temperature, and thus when it is cooler in the glasshouse, biological control of whitefly is less effective than at higher temperatures.

Parasitization of *B. tabaci* by *E. formosa*

Bemisia tabaci is not the optimal host for *E. formosa*. In particular, its small size does not suit the normal development of the wasp. In a situation where there is a choice, *E. formosa* will always preferentially parasitize *T. vaporariorum* before *B. tabaci*.

Nevertheless, development on *B. tabaci* is possible.

With the exception of the first larval instar, all larval stages and the pupal stage of *B. tabaci* can be parasitized; the first larval instar of

Table 4.5. The development time (Arakawa, 1982), egg-laying and lifespan (van der Kaay & Jansen, 1993) of *Encarsia formosa* on *Trialeurodes vaporariorum* on tobacco

temperature (°C)	20	25	30	host stage*
development time (days)				
egg-adult		20.2		larva 1
		17.4		larva 2
		15.9		larva 3
		15.0		larva 4 early
		15.0		larva 4 mid
		14.4		larva 4 late
eggs laid/♀	181	143	68	-
eggs laid/♀ day	8.3	13.2	7.9	-
lifespan (days)	23.5	12.3	10.3	-

* at time of egg-laying

T. vaporariorum can be parasitized perfectly well as long as it is already fixed to the leaf. Older larval stages of both species of whitefly are preferred, and the wasp develops better in these stages. During development of the parasite from egg to adult, mortality in *B. tabaci* larvae can be 1.5 to 8 times greater than in *T. vaporariorum* larvae. The percentage of parasitic wasps that eventually emerge from the parasitized host is thus smaller for
B. tabaci than for *T. vaporariorum*. The development of *B. tabaci* also takes longer than that of *T. vaporariorum*. In parasitizing *B. tabaci*,
E. formosa will use all larval stages and the pupal stage as food, whereas with *T. vaporariorum* the second larval instar is preferred. *E. formosa* prefers to host feed on tobacco whitefly larvae rather than the larvae of the glasshouse whitefly.

The behaviour of *E. formosa* when parasitizing the tobacco whitefly is in outline the same as when parasitizing the glasshouse whitefly.
A parasitized pupa of *B. tabaci* is less clearly recognized, however, since the pupa remains a transparent brown leaving the developing parasite within clearly visible. Parasitic wasps that develop on the tobacco whitefly are slightly smaller than those that develop on glasshouse whitefly (corresponding to the difference in size between the two hosts), and also lay fewer eggs per day. A comparison of the biology of *Encarsia formosa* on glasshouse whitefly and on tobacco whitefly is given in table 4.6.

Table 4.6. Comparison of biology of *Encarsia formosa* on glasshouse whitefly and on tobacco whitefly at 21°C and relative humidity of 65% on poinsettia (Hoddle et al., 1996)

	E. formosa on glasshouse whitefly	*E. formosa* on tobacco whitefly
Stage for egg laying	L3 en L4	L3 en L4 (vóór popvorming!)
Eggs per ♀ per day	5.0	5.9
Total no. eggs per ♀	59.2	51.3
Stage for host feeding	L2	alle stadia
Number of larvae eaten per day	2.8	2.3
Total reduction (no. larvae killed per day per ♀)	95	69
Lifespan (days)	11.9	8.7
Egg-adult development time	24.5	29.8

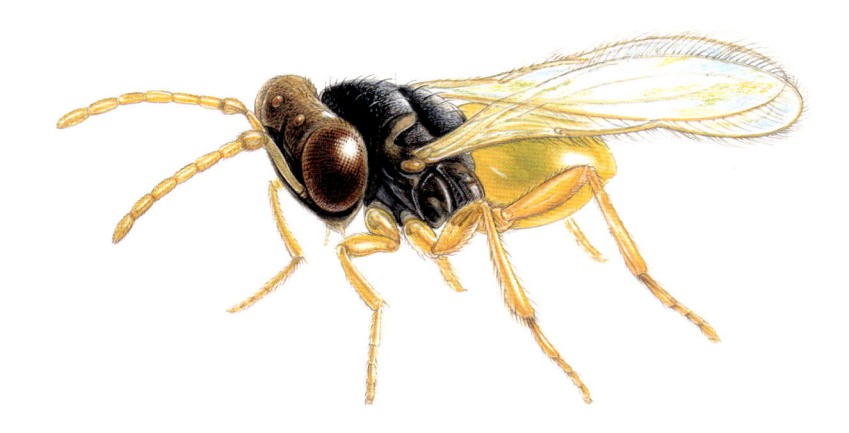

- **Parasitic wasp**
- **Better at higher temperatures and better parasitization of *Bemisia tabaci* than *Encarsia formosa***
- **Less sensitive to chemical pesticides than *Encarsia formosa***

Eretmocerus eremicus

Eretmocerus eremicus is a parasitic wasp that parasitizes *Bemisia tabaci* and *Trialeurodes vaporariorum* (among other hosts), and belongs to the family Aphelinidae within the superfamily Chalcidoidea. This species originally comes from the south-west of the United States. All *Eretmocerus* species are parasites of whitefly.

Before it was known that this species was *E. eremicus*, it was referred to as *E. nr. californicus* (nr being an abbreviation of "near", meaning that it strongly resembled *E. californicus*). This species, also often referred to as *Eretmocerus* sp. and *Eretmocerus haldemani*, is an ectoparasite. The egg is laid beneath the whitefly larva, but as soon as the egg hatches, the wasp larva creeps into its host. This parasitic wasp has several hosts, such as *Aleyrodes* sp., *Bemisia tabaci*, *Trialeurodes vaporariorum* and *Trialeurodes abutilonea*. It has been used for the biological control of whitefly since 1994.

Population growth

As with all insects, the development time for *E. eremicus* is primarily determined by temperature. Table 4.7 shows the figures for development at different temperatures. These figures can vary, however, because the age of whitefly larvae at the time of parasitization also influences the total time taken for development: eggs do not hatch until their larval host has reached the fourth instar. The development time for *Eretmocerus eremicus* is slightly longer than for *Encarsia formosa*, but because *Eretmocerus eremicus* parasitizes whitefly at an earlier instar, the adults emerge earlier.

Following emergence the females can begin parasitizing new hosts almost immediately. Mating is necessary in this species, otherwise only male progeny are produced. Whilst a population of *Encarsia formosa* consists almost entirely of females, a population of *Eretmocerus eremicus* usually consists of 60% females and 40% males.

Although *Eretmocerus eremicus* seems to parasitize *B. tabaci* better than does *Encarsia formosa*, it can also produce a great number of progeny on *T. vaporariorum*. A female can lay between 50 and 200 eggs. *Eretmocerus eremicus* has a shorter lifetime than *Encarsia formosa*, but during its first days as an adult the female lays more eggs. One advantage of *Eretmocerus eremicus* is that its activity is maintained at higher temperatures (30 - 40°C). Table 4.7 provides data relating to the population growth of *Eretmocerus eremicus*.

As with *Encarsia formosa*, *Eretmocerus eremicus* can also host-feed. In fact, this is necessary for the production of eggs. Because the ovipositor of *Eretmocerus eremicus* is not very strong, whitefly can only be pierced via the excretory opening. The fluid from this wound is ingested whilst the larva, which is no longer suitable for parasitization, dies. This host-feeding is practised to such an extent that a single young female *E. eremicus* can kill up to 30 *T. vaporariorum* larvae per day by this means. Field observations in The Netherlands have shown that *E. eremicus* is more resistant to chemical pesticides than *Encarsia formosa*.

Before a female *E. eremicus* commits to laying an egg beneath the host,

Table 4.7. The population growth of *Eretmocerus eremicus* (data from Koppert (tobacco) and Hoddle et al., 1996 (poinsettia))								
	\multicolumn{6}{c	}{temperature (°C)}	crop	host*
	14	20	22	24	25	30		
egg-adult development time (days)								
	44		22		19		tobacco	T.v.
				20			poinsettia	T.v.
				16			poinsettia	B.t.
eggs laid/♀/day								
		10.4			14.2	17.3	tobacco	T.v.
				8			poinsettia	T.v.
				3			poinsettia	B.t.
lifespan (days)								
		13.5			4.8	2.5	tobacco	T.v.
				20			poinsettia	T.v.
				10			poinsettia	B.t.

*: B.t. = *Bemisia tabaci*
T.v. = *Trialeurodes vaporariorum*

she first probes it with her antennae to determine whether the larva satisfies certain requirements. If satisfactory, she will deposit her egg. If there are too few acceptable host, further eggs may be deposited beneath the same host (super-parasitism). In this case, however, only a single parasitic wasp develops in each parasitized whitefly.
The searching behaviour of this species of wasp also appears to be influenced by the whitefly's host plant; on plants with hairy leaves the wasp has more difficulty seeking out its host.
Several differences between the behaviour of *Encarsia formosa* and that of *Eretmocerus eremicus* are set out in table 4.8.

Table 4.8. Comparison of *E. formosa* and *E. eremicus* (Vet & van Lenteren, 1981, Enkegaard, 1993, Koppert B.V.)

		E. formosa	E. eremicus
sex ratio (% females)		98	60
development time	17°C	32	48
(days)	25°C	18	19
lifespan (days)	17°C	44	30
	25°C	12	5
	> 30°C	0-1	few
parasitization		egg inside larva	egg beneath larva
preferred larval stage		L3/L4	L2/L3
eggs /day on T.v *	17°C	8.3 (166 in total)	7.5 (150 in total)
	25°C	13	14
	28°C		17
eggs /day on B.t **	22°C	6	
	25°C		5
	28°C		10
		given choice, first T.v.	
host feeding		+	++
sensitivity to pesticides		++	+

*: T.v.= *Trialeurodes vaporariorum*
**: B.t.= *Bemisia tabaci*

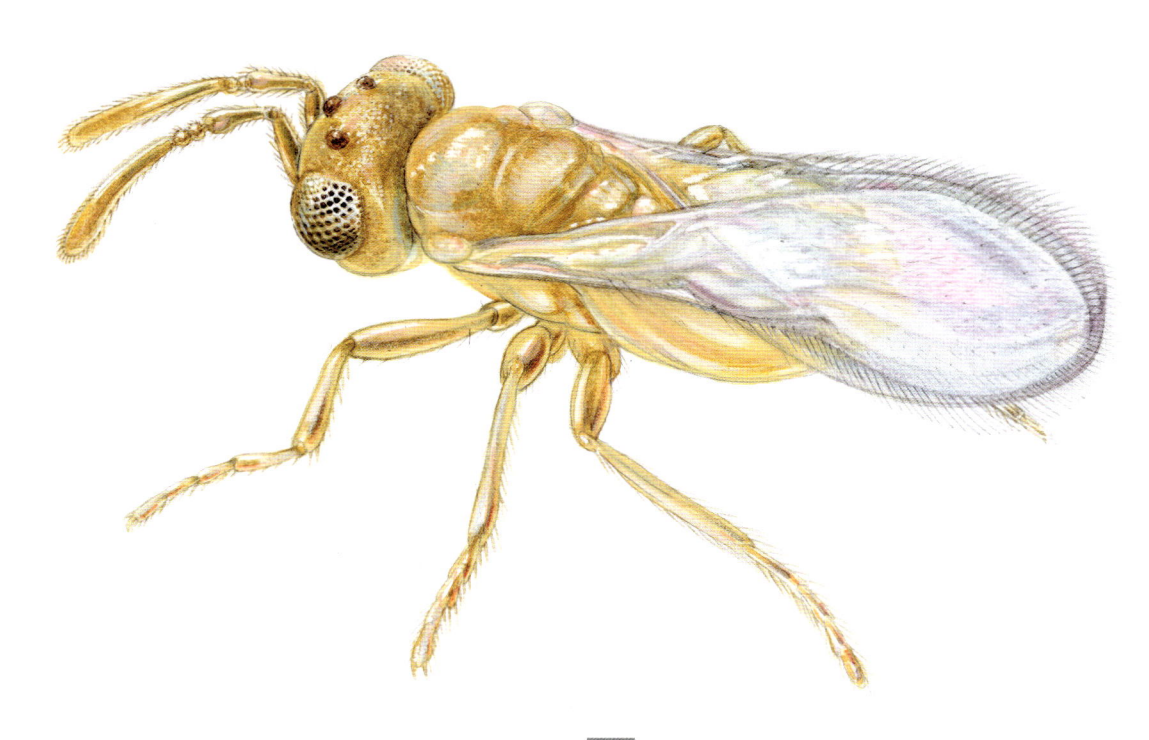

Life-cycle and appearance
Eretmocerus spp.

The development of *Eretmocerus* spp. consists of an egg stage, 3 larval instars (including a prepupal stage following the third instar), a pupa and finally the adult wasp. In the choice and parasitization of whitefly larvae, *Eretmocerus* spp. differ markedly from *Encarsia formosa*. For egg-laying, the second and third larval instars of whitefly are preferred, whereas the third and fourth instars are preferred by *Encarsia formosa*. The female deposits her egg beneath the larva, whereas *Encarsia formosa* injects her egg into the larva (see plate 4.31).

At first, the pear-shaped egg is transparent, but after 2 to 3 days it turns light brown. Although the parasitic wasp seeks out the younger stages of its whitefly host to parasitize, the egg only hatches once the whitefly has developed to the fourth larval instar. This gives the wasp the best chance of survival. After hatching, the parasitic wasp larva penetrates the host to complete its development. The remnants of the egg remain visible on the underside of the larva as a brown circle (plate 4.32).

The pear-shaped larva begins by making a small hole in the host with its jaws. The penetration of the host by the parasite larva takes about 2 days. When penetration is complete, the larva moults to the second instar and begins to feed on its host. During the third instar, the larva largely fills the host pupa. The prepupal stage is a period in which the changes from larva to pupa take place; the pupa is the stage of major morphological changes leading to the adult form. During this development, the pupal cuticle of the whitefly turns an even yellow (see plates 4.33 and 4.34).

The adult parasitic wasp can be clearly seen through the pupal cuticle before it emerges from its own pupal case. With the help of a microscope, the sex of the adult can be determined on the basis of the colour and form of the antennae (see plate 4.35). The adult parasitic emerges from a round exit hole which it gnaws in its pupal casing (see plate 4.36).

The body of the female is a lemon-yellow with green compound eyes and three red ocelli (simple eyes). Females are approximately 0.75 mm in length. The male is a dark yellow to brown colour, also with green compound eyes and red ocelli, and is slightly smaller than the female. Males and females are best distinguished by the form of their antennae; female antennae consist of five segments, male antennae have only three, the last of which is enormously enlarged. Individual species of *Eretmocerus* are difficult to tell apart.

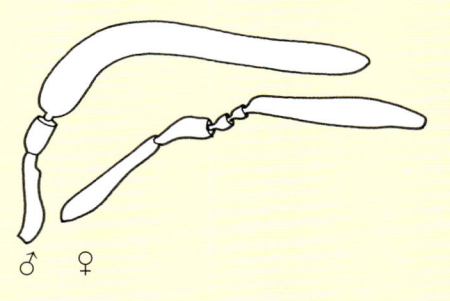

4.35. Male and female antennae of *Eretmocerus*

4.31. Parasitizing female of *Eretmocerus eremicus*

4.32. Remains of an egg on the underside of a larval *T. vaporariorum*

4.33. Parasitized pupa of *Trialeurodes vaporariorum*

4.34. Parasitized pupa of *Bemisia tabaci*

4.36. Empty, parasitized pupa of *Bemisia tabaci*

- Parasitic wasp
- Parasitizes *Bemisia tabaci* but not *Trialeurodes vaporariorum*
- Occurs naturally in Mediterranean regions, but is also released commercially

Eretmocerus mundus

Eretmocerus mundus occurs naturally in Mediterranean regions and will appear spontaneously when *Bemisia tabaci* is present. This chalcid wasp exclusively parasitizes the tobacco whitefly *Bemisia tabaci*, and not the glasshouse whitefly or any other whitefly species.
This parasitic wasp has been used for the biological control of *Bemisia tabaci* since 2002.

Population growth
An adult female *E. mundus* lays between 80 and 250 eggs during her lifetime. Like *E. eremicus*, the female selects the second or early third instar whitefly larvae to parasitize. Development from egg to adult normally takes 2 to 3 weeks depending on temperature and the whitefly larva's stage of development when parasitized. Under sub-optimal conditions, development can take as long as one month or more. Adult females live for one to two weeks at higher temperatures. Apart from temperature, the growth of a population also depends on the crop (see table 4.9). A population of *E. mundus* consists of approximately 40% males and 60% females. As with *E. eremicus*, host feeding also takes place.

4.36.a Pupa

4.36.b Adult

Table 4.9. The population growth of *Eretmocerus mundus* at 25°C. (Koppert B.V.)

	sweet pepper	tomato
development time (egg-adult)	15.9	16.2
egg-laying		
total eggs laid per female	171	148
eggs per female per day	18.5	19.2
lifespan (days)	9.7	7.3
host feeding		
host larvae per female	15.6	10.4
host larvae per female per day	1.6	1.5
% survival (egg-adult)	73.2	63.7

- **Predatory bug**
- **Nymphs and adults feed on all stages of whitefly, but prefer eggs and larvae**
- **Polyphagous, thus effective against other pests**

Macrolophus caliginosus

Bugs of the genus *Macrolophus* are polyphagous predators that belong to the family Miridae (subfamily Dicyphinae, suborder Heteroptera). The genus *Macrolophus* occurs in the Mediterranean region, including southern France, Italy, Spain and the Canary Islands. The bugs are mainly found on Solanaceae, particularly tomato and tobacco, but also on other crops.

M. caliginosus is polyphagous, but shows a clear preference for whitefly. The bug occurs spontaneously in tomato crops grown in tunnels and ensures effective control of whitefly. In Europe, the species has been used for the biological control of glasshouse whitefly and tobacco whitefly since 1994.

Population growth

Population growth depends particularly on temperature, but the nature and supply of food also plays a role. Development is slow; it takes at least 10 days for eggs to hatch and at least 19 more days before the adult emerges from the last moult. The threshold temperature for development is between 10 and 15°C. At 10°C development stops, and temperatures above 40°C are lethal for nymphs.

Temperature also influences egg hatching. At low temperatures, eggs do not hatch for at least 2 months, which is unsurprising as *M. caliginosus* over-winters in the egg stage. At higher temperatures, there is a considerable mortality among eggs (only 42% hatch at 30°C).

Mating takes place within 3 days of the last moult. Three to 6 days after mating the first eggs are laid. The egg-laying period varies from 85 days at 15°C to 23 days at 30°C.

Egg laying capacity depends on temperature and food supply and varies enormously from one individual to another. At 20°C, females lay an average of 270 eggs, with a maximum of 400 eggs, equating to 4 - 5 eggs per day. At 25°C an average of 120 eggs are laid. Egg-laying also varies from one day to another (from 0 to 14 eggs) and is positively correlated with food supply. On a diet of aphids, females lay fewer eggs than when they have a supply of whitefly. When provided exclusively with spider mites as a food source even fewer eggs are laid. When no prey is available they can feed only on plant sap, and egg-laying continues for some time, albeit at a much lower rate.

Under laboratory conditions of constant climate and optimal food supply, the lifespan of females varies from 90 days at 10 - 20°C to 40 days at 25 - 30°C. Under the same conditions, males live longer than females. The sex ratio is approximately 1:1.

M. caliginosus over-winters on various host plants, mainly in the egg stage, and withstands low temperatures well. Nymphs can tolerate a temperature of 6°C for a month.

In the autumn, a population of *M. caliginosus* can completely collapse as a result of infection by species of Entomophthorales, entomopathogenic fungi which can naturally occur in glasshouses (see plate 4.41, p. 78). A description of these fungi can be found in chapter 5.

Data relating to population growth of *M. caliginosus* are given in table 4.10.

Table 4.10. The population growth of *Macrolophus caliginosus* at a relative humidity of 60-75% (Fauvel et al., 1987)

	temperature (°C)						
	10	15	20	25	30	crop	prey *
ontwikkelingsduur (dagen)							
egg		36.9	18.3	11.4	10.6	tomato	**
nymph	257	57.8	29.4	18.9	18.7	tomato	T.v.
total			47.7	30.3	29.3	tomato	
mortality (%)							
egg-nymph		12.8	11.5	23.5		geranium	
			76.9	44.4	57.6	tomato	
nymph-adult	41.4	13.8	15.4	11.8	46.4	tomato	T.v.
egg-laying capacity (eggs per ♀)							
	23	151	268	122	87	geranium E.k eggs	
lifespan (days)							
♀		94	111	85	40	40	geranium E.k eggs
♀		210	123	116	79	53	geranium E.k eggs

*: E.k. = *Ephestia kuehniella* (Lepidoptera sp.)
 T.v. = *Trialeurodes vaporariorum*
**: at 15°C on geranium

Life-cycle and appearance
Macrolophus caliginosus

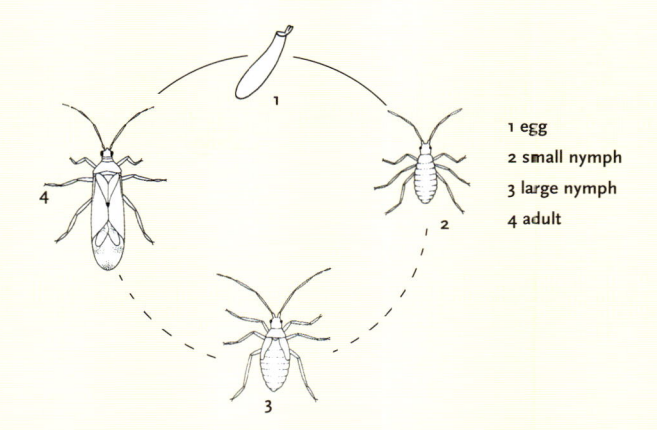

1 egg
2 small nymph
3 large nymph
4 adult

The life-cycle of *M. caliginosus* consists of an egg stage, 5 nymphal instars and the adult insect.

Adult females are 3.0 - 3.6 mm long and the males between 2.9 - 3.1 mm. They are of slim build with long legs and antennae. The bugs are green except for occasional small black markings; the first antennal segment is black, and a black stripe runs behind the eyes. The first part of the forewings is a transparent, soft green with a small black spot in the middle. The hind-part is colourless with a vague brown marking. *M. caliginosus* has large brown, hemispherical compound eyes that are clearly visible on the sides of the head. The bug has piercing-sucking mouthparts. Females have a large, rounded abdomen within which the ovipositor is visible. Males are somewhat smaller than females and have a thin abdomen.

Females lay their eggs in older leaf veins, leaf stalks and in the main stem. No eggs are laid in the topmost 10 - 15 cm of young shoots. The eggs themselves are not visible to the naked eye. Under a microscope, one can see a miniscule pennant projecting from the leaf surface, and often several eggs are laid together.

First and second instar nymphs are a yellow-green colour. Older nymphs are a similar green to the adults, but lack the black antennal segments and black stripe behind the eyes. The wing buds are visible in older nymphs.

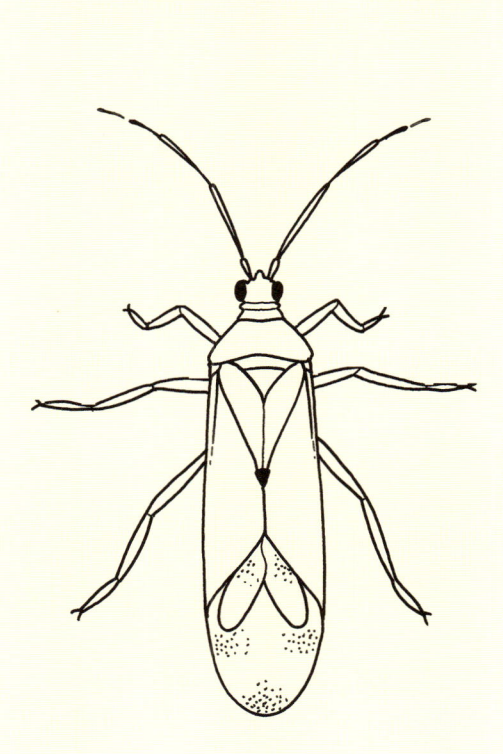

4.37. Excised egg

4.38. Nymph

4.39. Adult

4.40. Head of an adult

Feeding behaviour

Although *M. caliginosus* prefers to feed on whitefly, it will also eat aphids, spider mites, moth eggs and caterpillars, as well as the larvae of miners and thrips. When the bug is present in a crop in large numbers, it can contribute substantially to the control of these pests. The bug has even been introduced into tomato crops as a supplementary biological control against spider mite.

All stages of whitefly are eaten, but the preference is for eggs and larvae. An adult bug can consume 30 - 40 whitefly eggs per day, 15 - 20 pupae or 2 - 5 adults. When a bug feeds on a whitefly egg, larva or pupa, it sucks out the body contents leaving only the empty cuticle. This sometimes collapses, but is usually left completely intact. If the prey has not been entirely consumed, the brown remains are left behind.

Although plant sap is necessary for a population of *M. caliginosus* to develop well, it is insufficient for full development of a population. When females are only able to feed on plant sap they lay far fewer eggs, and the nymphs (if similarly deprived of living prey), will die. Consumption of plant sap, however, can cause damage to crops such as cucumber, certain varieties of tomato and gerbera. Conditions that promote the onset of such problems include a shortage of prey, a high population of bugs (more than 100 - 150 per plant), dull weather, a weak crop, and certain plant varieties such as the more delicate types of tomato. In certain fruiting vegetables there may be fruit deformations during setting, and in gerbera misshapen blooms may appear. The use of this bug in gerbera is therefore not recommended, care should be taken in cucumber and in cherry tomato crops.

Searching behaviour and dispersal

Nymphs and adults can move very rapidly. When danger threatens they run across the plant in search of a hiding place. The adults are mainly found in growing shoots and along the stalks. They are also good fliers and are thus able to disperse with ease. The nymphs are mainly found on the underside of leaves.

The bugs have good vision and show highly directed searching behaviour. When a suitable prey item is located, it is pierced with the mouthparts and the contents sucked out. The bugs appear to have no trouble with the glandular hairs of tomato. They make no distinction between parasitized and non-parasitized whitefly larvae, preying on both equally. However, as soon as the whitefly pupae turn either black (if they contain *Encarsia formosa*) or yellow (if they contain *Eretmocerus eremicus*) they are less frequently attacked.

4.41. *Macrolophus caliginosus* infected with Entomophthorales fungus

4.42. *Macrolophus caliginosus* in action

- Entomopathogenic fungus
- Mostly infects small larvae, but larger larvae, pupae and adults can also be affected
- Can be used together with other natural enemies of whitefly

Verticillium lecanii

Verticillium lecanii is a commonly occurring fungus that affects various organisms including arthropods. It was first described in 1861 and has been observed on various species of insect, especially aphids, scale insects and thrips, and also on spiders, mites and nematodes. It also occurs as a saprophyte (an organism that lives on dead organic matter) and is readily isolated from soil, or as a hyperparasite on rusts and mildew and even on other entomopathogenic fungi. *Verticillium lecanii* belongs to the class Deuteromycetes (= Fungi Imperfecti, the asexually reproducing fungi) and to the order Moniliales. It is a naturally occurring, widespread fungus in temperate, subtropical and tropical regions of the world. In the tropics, insect populations are regularly and naturally infected, but in temperate regions infection is almost exclusively a glasshouse phenomenon. *V. lecanii* is a specific parasite and poses no danger to plants, birds, fish or mammals.

V. lecanii was first observed on whitefly in 1915, when it was described under the name *Cephalosporium lefroyi*. Subsequently the fungus was observed on different occasions on whitefly and repeatedly described under different names, *e.g. Cephalosporium lecanii*.

The species *V. lecanii* consists of a multiple complex of different, mainly entomopathogenic strains that were previously believed to be different species. These strains differ very little in external appearance, but do have a different host range. Some mycologists are convinced that this *V. lecanii* complex should be further split into several species, and as a result one often encounters different names for the fungus in the literature.

Because of its superior activity, the "whitefly strain" of the fungus has been used in the biological control of whitefly since 1980, and occasionally also in the control of thrips.

The spread of infection

V. lecanii is dispersed in nature by means of conidiospores. Unlike other entomopathogenic fungi, *V. lecanii* is able to form spores on living insects, although this is rare; the spores are usually formed once the fungus has killed the insect, or on insects that have died from other causes. *V. lecanii* cannot be spread quickly or efficiently within a pest population because the spores are not air born. The spores have a coat of slime and are dispersed either by mechanical means or by water. In particular, this occurs through the splashing of water droplets, by direct contact with a contaminated organism or by means of vector organisms. The latter can occur, for example, if conidiospores stick to ants, whereupon they may be transported considerable distances. Mostly, however, the infection is only spread to non-infected insects in the immediate environment. Although it would seem theoretically likely for an epidemic to erupt in the insect population, in practice this only occurs at very high infection densities and in highly favourable climatic conditions (*i.e.* high humidity). In practice, such situations hardly ever occur. For the fungus to become effectively established, the application must be repeated several times until almost all the insects have been killed. If there is a pest resurgence, further applications will be needed.

4.43. Infected *Trialeurodes vaporariorum*

Effect and appearance
Verticillium lecanii

Entomopathogenic fungi infect their hosts by penetrating the insect epidermis. This form of infection is uniquely characteristic of fungi, as other insect pathogens such as bacteria and viruses infect their host via the alimentary canal. The *V. lecanii* spore germinates on the insect and the fungal threads, or hyphae, usually grow on the honeydew that the whitefly secretes, or on the carbohydrates that are added to the commercial product. After this, the insect itself can become infected. However, the fungus can also direct its hyphal growth into the insect immediately without this external phase. Further internal development of the fungus then kills the insect. Although the precise cause of death is difficult to ascertain, several factors are involved:
- mechanical obstruction of respiratory tubes and blood circulation,
- exhaustion of reserves and a disruption of organ function,
- the production of specific proteins by the fungus within the insect.

When whiteflies are infected by *V. lecanii*, they die before the fungal growth is obvious. Dead larvae and pupae sometimes appear light to dark brown in colour, sunken and lacking their usual sheen. Eventually the fungal hyphae emerge through the insect epidermis and form spores outside the body. The infection can then spread to other insects.

The first symptoms of the fungal infection can be seen on whitefly after seven to ten days. Two weeks after spraying the effects are clearly visible. In the case of an infestation of thrips the fungal action is slower to take effect because of the greater mobility of these insects. After three weeks, however, the effect on the population is clearly visible.

The fungus has a white cottony appearance. Under the microscope, its white threads (hyphae) can be seen with almost perpendicular branches. The hyphae are 1-3 µm thick. The side branches are separate spore-forming cells, the phialides, which produce globules at their tips containing one or more conidiospores (at most about 25), surrounded by a slimy substance. The spores themselves are cylindrical or elliptical with rounded tips.

There are several strains of *V. lecanii*, such as the "whitefly strain" and the "aphid strain". These differ in the range of their effects, but there are also morphological differences such as the size of the spores, which can vary from 2.3 - 10 µm in length and from 0.9 - 2.6 µm in width.

The "whitefly strain" of the fungus mainly infects the larvae of whitefly, but at higher relative humidities the fungus also kills pupae and adults. Eggs are not affected. In the case of thrips, the fungus mainly kills larvae, but adults and pupae may also be affected. The eggs, which are found in the leaf tissue, are hardly ever infected. Favourable conditions for the growth and development of *V. lecanii* are temperatures between 15 and 28°C and a relative humidity of 80% or above. An application of the fungus will be most effective under these conditions. Although in glasshouses whitefly can also be naturally infected by *V. lecanii*, spontaneous epidemics very rarely occur, and natural biological control by the fungus is therefore hardly effective.

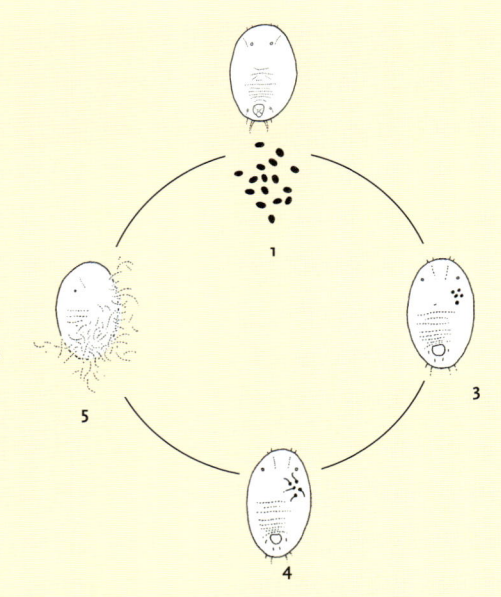

1 spores of *Verticillium lecanii*
2 L2 whitefly
3 L2 infected by spores of *Verticillium lecanii*
4 growing hyphal threads of the fungus
5 external sporulation

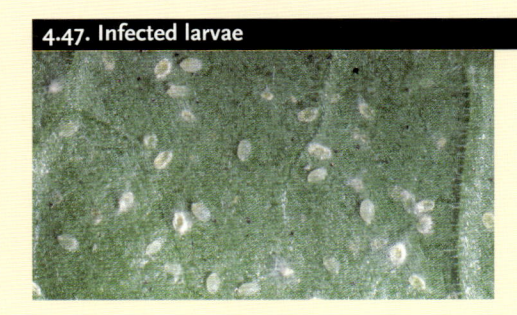

4.44. Spores

4.45. Infected pupa and adult

4.46. Infected larva

4.47. Infected larvae

Knowing and recognizing **Whiteflies**

Biological control of whitefly

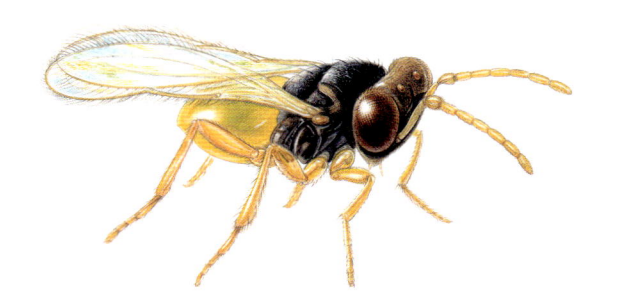

OVERVIEW OF THE BIOLOGICAL CONTROL OF WHITEFLY				
product name	natural enemy	controlling stage	stage of whitefly controlled	crop
EN-STRIP	*Encarsia formosa*	adult	larva	all crops
ERCAL	*Eretmocerus eremicus*	adult	larva	all crops
ENERMIX	*Encarsia formosa* + *Eretmocerus eremicus*	adult	larva	all crops
MIRICAL	*Macrolophus caliginosus*	nymph and adult	all stages	possibility of crop damage should be considered
MYCOTAL	*Verticillium lecanii*	spore	larva, sometimes pupa and adult	all crops
BEMIPAR	*Eretmocerus mundus*	adult	larva of *B. tabaci*	all crops
BEMIMIX	*Eretmocerus eremicus* + *Eretmocerus mundus*	adult	larva	all crops

For years *Encarsia formosa* was the only natural enemy introduced to combat whitefly. Immediate and regular introductions must be made as soon as the first whiteflies are found; some recommend that the wasp should be introduced preventatively before the first whitefly is detected. To ensure successful control, each crop calls for a different schedule of application regarding the number of introductions and the intervals between them. As the wasp is highly sensitive to chemical pesticides, it is important to exercise extreme caution when dealing with diseases and pests. *E. formosa* has more difficulty controlling tobacco whitefly than glasshouse whitefly and has a preference for the latter.
E. formosa is supplied in the form of pupae attached to cardboard cards with integral hanging hooks. The cards are hung among the plants, and the pupae soon hatch to release adult parasitic wasps. This system has the advantage that it is simple to administer and the material is easily transported. For low temperatures, cards are supplied with half the normal number of wasps, and the number of distribution points is multiplied. The international trade name of the product is EN-STRIP. At higher temperatures and in cases where corrective treatment with chemicals is necessary, the parasitic wasp *Eretmocerus eremicus* has the advantage over *Encarsia formosa*. When both species of parasitic wasp are introduced, *E. formosa* out-competes *Eretmocerus eremicus* in the spring, but from the beginning of summer onwards the situation is

reversed and *E. eremicus* begins to take over. *E. eremicus* is supplied in the form of parasitized whitefly pupae on cards under the international trade name ERCAL.

A combination of *Eretmocerus eremicus* and *Encarsia formosa* is supplied in the form of parasitized whitefly pupae on cards under the name ENERMIX. *Eretmocerus mundus* is used exclusively against *Bemisia tabaci*. In common with the other whitefly parasites, the product contains pupae stuck on to cards. It is sold under the international trade name BEMIPAR. In cases where both whitefly species occur, the product should be combined with *E. eremicus*. A mix of both parasites is sold under the trade name BEMIMIX.

The predatory bug *Macrolophus caliginosus* is much used in tomatoes, but also in cucumber, sweet pepper and other crops. Both the glasshouse and tobacco whitefly are kept well under control by the bug. At high densities the bug can even contribute to the control of other pests. As the bugs cannot be used in all countries, and must not remain present on some products, *e.g.* tomatoes destined for the U.S., the use of these natural enemies depends on the country in which they will be used and the destination of the product.

Its slow development at low temperatures means that the bug should be introduced as early as possible in the season together with *E. formosa*. However, a source of prey must be available, otherwise the bug will have little chance of developing. Because there is often a shortage of food in the spring, eggs of the meal moth *Ephestia kuehniella* are supplied as an alternative food source. These are scattered over the leaves at regular intervals. These eggs are sold in plastic bottles under the international trade name ENTOFOOD. The eggs are small, white, oval and are sterilised so that no caterpillars can develop from them.

Care is necessary when using *M. caliginosus*. The bugs are capable of building up a large population on various crops, and their habit of sucking plant sap means that there is a danger of crop damage. An example of this is gerbera; the bugs feed on flower buds causing misshapen flowers to form. Problems have also arisen several times in truss and cherry tomato varieties. If the population density reaches high levels during the course of the season and the bugs run short of food, they feed more heavily on the plant with the result that some of the fruit fails to set.

M. caliginosus adults and nymphs are supplied in vermiculite in shaker bottles under the international trade name MIRICAL.

The product MYCOTAL contains the spores of a strain of *Verticillium lecanii* that is highly effective against whitefly (both *Trialeurodes vaporariorum* and *Bemisia tabaci*) and to some extent against thrips. This strain was isolated in 1979 from an infected whitefly from a commercial cucumber company in the north of England. MYCOTAL is supplied as a wettable powder containing the conidiospores of the fungus. By spraying MYCOTAL, the spores are introduced directly on to the pests where they can then germinate. More mobile insects, such as thrips, can also pick up spores in the crop.

MYCOTAL is fairly specific and has hardly any effect on other insects. The fungus can therefore be applied without danger in crops where *Encarsia formosa*, *Phytoseiulus persimilis* or other natural enemies are being used. Only under conditions that are ideal for the fungus (*i.e.* 100% relative humidity and around 20 - 25 °C) are beneficial insects and mites likely to be affected. MYCOTAL is often used as a corrective measure. However, it can also be used as the main weapon in a whitefly control programme. The success of MYCOTAL depends on the climate in the crop, and not on the crop per se. In addition, care needs to be exercised in the handling of other products such as fungicides, as these may have an adverse effect on the action of the fungus. The need for high relative humidity can be reduced by making use of an adjuvant based on vegetable oil that protects the spores from drying out. The name of this product is ADDIT. As with all pesticides, MYCOTAL should only be used in strict accordance with the instructions for use as set out on the label.

5. Thrips and their natural enemies

As a result of changes in methods of cultivation and the arrival in western Europe of the western flower thrips (*Frankliniella occidentalis*), thrips are now a major problem in glasshouse horticulture. When crops were grown in soil, many overwintering thrips were killed by sterilizing the ground. With the increased popularity of artificial substrates, soil sterilization is less common, and thrips can thus easily overwinter and attack young plants early in the season. The presence of thrips early on in the crop calls for quick action.

Until the early 1980's, the most significant species of thrips in glasshouse horticulture was the onion thrips, *Thrips tabaci*. This species had been a problem in glasshouses for many years, especially in the cucumber. However, since the mid 1980's *Frankliniella occidentalis* has become the most troublesome species. Originally imported from America, it has caused enormous problems world-wide. This is mainly due to the biology of the two species; the onion thrips lives mainly on the leaf, where it is easy to control, whereas the western flower thrips prefers a more hidden habitat and is much less sensitive to pesticides. Unlike the onion thrips, the western flower thrips does not enter diapause.

Another species that has appeared more frequently in recent years, causing damage in both vegetable and ornamental crops, is *Echinothrips americanus*. The biological control of this species is also difficult, as the available natural enemies are not very effective against it.

Apart from the thrips species mentioned so far, the rose thrips (*Thrips fuscipennis*) can cause occasional problems in glasshouse crops, while the glasshouse thrips (*Heliothrips haemorrhoidalis*) and the banded glasshouse thrips (*Hercinothrips femoralis*) cause problems particularly in ornamental crops. The latter two species are imported mainly on tropical ornamental crops. Although they cause similar damage to *E. americanus* and can easily be confused with it, they are much less common and are not indigenous.

All these species will be discussed and compared in this chapter, in addition to *T. palmi* (the melon thrips), a thrips species which, although not yet permanently established in Europe, has already caused major problems, and the palm thrips (*Parthenothrips dracaena*), a species that occurs mainly on indoor plants. With the increased use of biological control in ornamental horticulture and a corresponding reduction in the use of chemical pesticides, these species occasionally cause problems.

Several other thrips species that should be mentioned are unable to establish themselves in the glasshouse, but are sometimes found on sticky traps. As they can often be easily confused with species that do cause damage, they serve to confuse the pest situation in the glasshouse. The main culprit is *Thrips major*, a species that can only be distinguished from *T. fuscipennis* by experts. The cereal thrips *Limothrips cerealium* is also often found on sticky traps in large numbers when hay is cut anywhere in the vicinity of the glasshouse. Finally, *Thrips flavus* and *T. nigropilosus* are also sometimes found on traps, and can easily be confused with *T. palmi* and other similar species.

Biological control of thrips (namely *T. tabaci*) has been practised since 1980, when the predatory mites *Amblyseius cucumeris* and *Amblyseius barkeri* were introduced. The use of *A. cucumeris* in sweet pepper proved to be particularly successful, with experiments in cucumber cultivation also producing positive results. However, when *F. occidentalis* entered The Netherlands in 1983, it soon became apparent that *A. cucumeris* was not capable of keeping this species under control, and an alternative or supplementary control agent was needed. The reason for this problem lay in the fact that the strains of predatory mite employed at the time entered diapause, whereas *F. occidentalis* does not. There was thus nothing to control the thrips during the winter period. Nowadays there are diapause-free strains of *A. cucumeris*, but even these are not always capable of providing adequate control. In such situations, predatory bugs of the genus *Orius* are an effective back-up. Furthermore, in pollen-producing crops and at low relative humidities, the predatory mite *A. degenerans* is even better at controlling thrips than *A. cucumeris*, and is also much more visible in the crop. In ornamental horticulture, the predatory mites *Hypoaspis aculeifer* and *Hypoaspis miles* are also used as supplementary control measures. These predatory mites live in the soil, and amongst other food sources they eat thrips pupae. The fungus *Verticillium lecanii* can also contribute to the biological control of thrips, as can certain other naturally occurring fungi (Entomophthorales) in the autumn. All of these organisms are discussed in this chapter. The spider mite predator *Amblyseius californicus*, the predatory bug *Macrolophus caliginosus*, the lacewing *Chrysoperla carnea* and various parasitic nematodes can all contribute to the control of thrips, and are discussed elsewhere in this book.

In pollen-producing crops the control of thrips is no longer a major problem, thanks to the results achieved by the combination of the various species of predatory bugs and predatory mites. In crops without pollen-producing flowers, such as cucumber and many ornamental crops, control is not yet perfect. Continuing research is therefore still directed toward the discovery of new natural enemies.

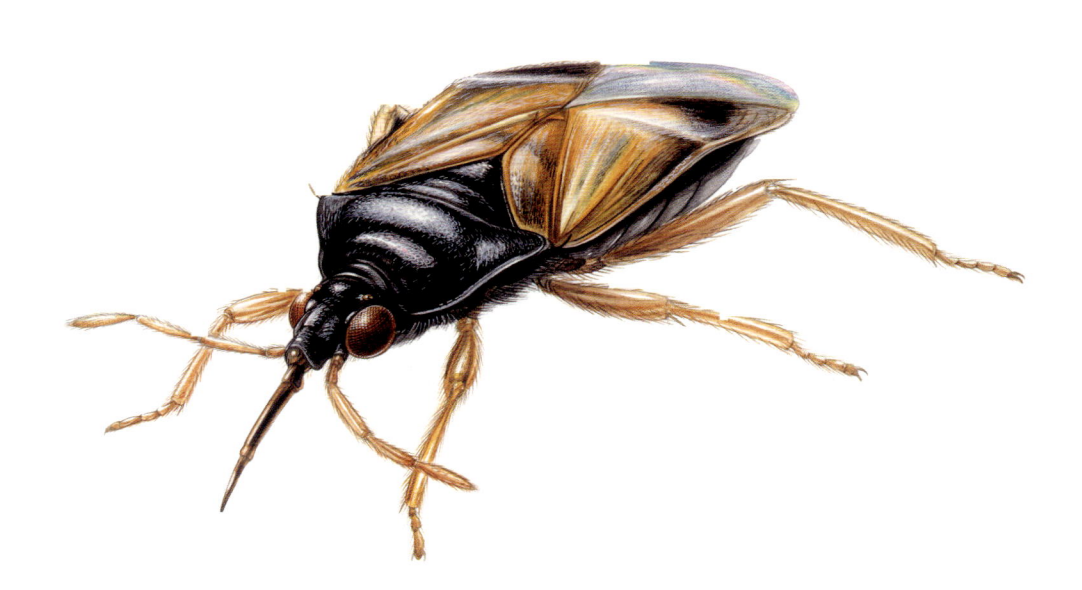

Knowing and recognizing Thrips

Thrips

5.1. Frankliniella occidentalis

5.2. Echinothrips americanus

The following thrips species are found in glasshouses:
Frankliniella occidentalis
Thrips tabaci
Echinothrips americanus
Thrips fuscipennis
Heliothrips haemorhoidales
Hercinothrips femoralis

Thrips form the order Thysanoptera, a name literally meaning "fringed wings", and referring to the eyelash-like fringe of hairs along both edges of the thin wings. Due to the bladder-like growths on the feet, they were formerly also known as Physopoda. They are more popularly known as thunderflies or stormflies, because of their habit of swarming in close weather. "Thrips" is both the singular and the plural. Thus you can have several thrips or a single thrips; there is no such thing as a "thrip". There are some 5,000 known species, most of which are small, thin insects with a prominent head. They are found all over the world, and although the tropics are particularly rich in species, there are also many in the temperate regions and even a few that inhabit the arctic zone. Most are either harmless or cause little damage, but in Europe there are a few dozen harmful thrips species, fewer than 10 of which occur in glasshouses.

Thrips are the smallest winged insects (0.5 - 14 mm), with the largest species mainly found in the tropics. In temperate regions they are no larger than 1 - 2.5 mm.

All the species of thrips found in glasshouses belong to the family Thripidae, a family that also contains the species that cause most damage.

Lifecycle and appearance
Thrips

1 egg
2 larva 1
3 larva 2
4 prepupa
5 pupa
6 adult

A thrips develops through six stages: the egg, two larval instars, a prepupa and a pupal instar, and finally the adult insect. The eggs are kidney-shaped and have a white or yellow shell. Before a female deposits an egg, she first makes an opening in the plant tissue. The eggs are laid in leaves, flower petals and in the soft parts of stalks. On the leaves of sweet pepper, these egg-laying sites are recognizable after a while as wart-like growths (see plate 5.3), whereas on cucumber and many other crops the egg sites are not recognizable.

Immediately after hatching (plate 5.4) the larvae begin to feed on plant tissue on the underside of the leaf. They feed on all aerial parts of the plant and are fairly mobile. The larvae have a smaller head than the adults, have fewer antennal segments and lack wings. A first instar larva is smaller and lighter in colour than a second instar larva, the latter being roughly the same size as the adult.

At the end of the second instar, the larva falls to the ground to pupate. Pupation occurs in damp places or in natural crevices in the ground up to 15 mm beneath the surface. The prepupal and pupal instars are recognizable by their developing wing buds. Pupae are often a slightly lighter colour than larvae. Compared with the prepupa, the pupa has longer, more developed wing buds and longer antennae that are curved back over the head. The prepupal and pupal instars do not feed and although capable of movement, do so only if disturbed. In the adult both pairs of wings are fully developed. Only at this stage can the particular species of thrips be identified on the basis of form, colour and pattern of hairs.

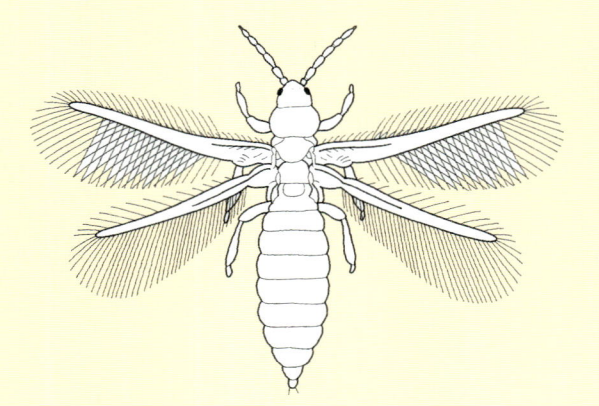

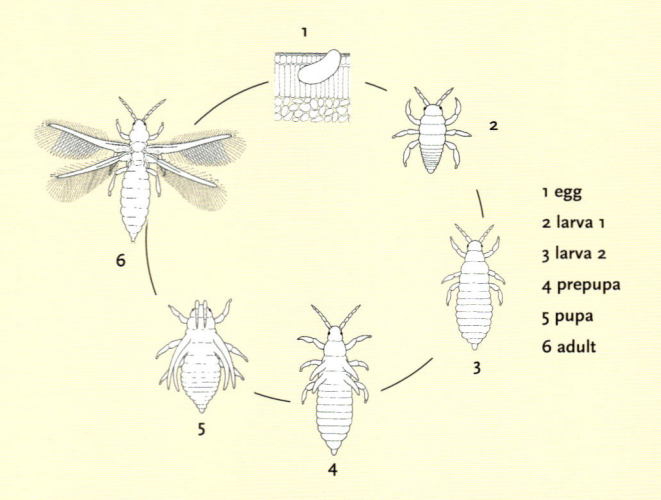

5.3. Egg-laying sites of thrips in sweet pepper

5.4. Thrips larva emerging from egg

5.5. Larva of *Frankliniella occidentalis*

5.6. *Frankliniella occidentalis*

Reproduction

Reproduction in thrips can be either sexual or asexual. In general, unfertilized females produce male offspring, while fertilized females produce both male and female progeny. In those species where males are rare, females are produced parthenogenetically, *i.e.* from unfertilized eggs, such as in *Thrips tabaci*.

Overwintering

In winter, larvae and adults seek warm places such as plant debris, cracks in the wall, in the fabric of the glasshouse or in the ground, ready to re-emerge when a new crop arrives and the temperature rises. Of those thrips that cause damage to glasshouse crops, *T. tabaci* and *T. fuscipennis* are able to overwinter outdoors (on onions and leeks). This is not known to occur in *F. occidentalis*. *Echinothrips americanus* is unable to overwinter outside in temperate regions.

Pest development and dispersal

An infestation of thrips can begin through insects being introduced into the glasshouse along with plant material. Later in the season, adults may fly in from outside. In addition, thrips can overwinter inside the glasshouse, in crevices and fissures and other similar places.
A thrips infestation often begins on a few plants and spreads gradually through the glasshouse. The first thrips can usually be detected beside the main path and in particularly warm places such as around the gables or near to heating pipes.
The spread of thrips can be both an active (flight) and/or a passive process (floating on air currents, or via the movement of plants, materials and humans).

Damage

Thrips cause damage to the plant by piercing the cells of the surface tissues and sucking out their contents, causing the surrounding tissue to die. The resulting silver-grey patches on leaves and the black dots of their excreta indicate their presence in the crop. The silver-grey discolouration is caused by the pierced epidermal cells becoming detached from the underlying leaf parenchyma, producing an airspace beneath the surface. At a later stage the empty cells become desiccated and the adjacent cells turn brown.
The vigour of the plant is also reduced by loss of chlorophyll. With a serious infestation the leaves themselves can shrivel, and there can be varying levels of fruit damage depending on the species of thrips and their population density. The damage to ornamental crops is of various kinds; flowers can be seriously damaged, while leaves too are often damaged and misshapen.
Thrips are also responsible for the transmission of viruses, the best known of which is tomato spotted wilt virus (TSWV), mainly transmitted by *F. occidentalis*. The issue of virus transmission is dealt with in more detail below in the description of this thrips species. Plate 5.7 shows different forms of infestation.

5.7. Damage caused by thrips

Overview (cucumber)

Detail of leaf damage (cucumber)

Leaf damage due to tomato spotted wilt virus (sweet pepper)

Damage to fruit (cucumber)

Damage to fruit (sweet pepper)

Flower damage (gerbera)

- Very common species
- Has been a problem since the reduction in insecticide use against whitefly and spider mite
- Affects leaves particularly

Thrips tabaci
Onion thrips

Apart from the arctic zone, the onion thrips (*Thrips tabaci*) is distributed throughout the world. It is found outdoors on cotton, onion, leeks and other crops, and in glasshouses on cucurbits, tomato, sweet pepper, aubergine, roses, chrysanthemum, gerbera, carnations, bulbs and many others.

T. tabaci has become more of a problem since the reduction in the use of broad-spectrum pesticides which had previously kept it under control.

Life-cycle and appearance
The eggs of *T. tabaci* are very small, approximately 0.2 mm long, kidney-shaped, white-yellow in colour, and are found within the leaf tissue. A young larva is roughly 0.4 - 0.6 mm long and a light colour, sometimes even white, with a large head and bright red eyes. In the second larval instar the insect is roughly 0.7 - 0.9 mm long and light yellow to yellowish green in colour. The colour of the adult depends on the food source. Females are greyish yellow to brown and 0.8 - 1.2 mm long, although when newly emerged from pupae they are somewhat lighter. The rare, wingless males are smaller and lighter than females.

Population growth
Population growth is highly dependent on temperature. Development stops at around 11.5°C, and increases between 16 and 28°C. *T. tabaci* can consume spider mite eggs, which has a positive effect on population development, lifespan and egg-laying. More importantly, however, is the fact that thrips can take cover in spider mite webs, where they are virtually inaccessible to natural enemies. The presence of spider mite in a crop can therefore promote the growth of a thrips population. A population of *T. tabaci* contains very few males, and thus reproduction is generally asexual with unfertilized females producing female progeny. At 25°C a female lays 2 - 5 eggs per day, and may lay between 70 to 100 eggs in total. Data relating to the population growth of *T. tabaci* on cucumber are given in table 5.1.

Damage
T. tabaci is found mostly along the main leaf veins, with most damage being evident where these veins join. Thus, in cucumber the fruit is only damaged if the crop is heavily infested.

In some countries the onion thrips has been identified as an important vector of tomato spotted wilt virus. Although the virus was discovered in temperate regions some decades ago, it was never a significant problem until the arrival of the western flower thrips. One explanation is that there may be different strains of *T. tabaci* with different vector capacities, and the tropical strain may be a better vector of TSWV than the strain found in temperate regions. It may also be that *T. tabaci* is only an effective transmitter of the virus at very high population densities, which are very rarely achieved in temperate regions.

Distribution and dispersal
T. tabaci colonizes all parts of the plant, preferring the underside of young leaves in the top of the plant, but can also be found on shoots and flowers.

Table 5.1. The population growth of *Thrips tabaci* on cucumber at 25°C (van Rijn et al., 1995)

development time (days)	
egg	3.9
larva 1	2.1
larva 2	3.2
prepupa	1.1
pupa	2.4
total egg-adult	12.7
po-period *	1.9
egg to egg	14.6
egg-laying/ ♀ /day	2-5
egg laying period (days)	20

*: po-period = pre-oviposition period, *i.e.* period from becoming adult to first egg-laying

- Major problem in many crops
- Highly polyphagous
- Transmits viruses and affects fruit and flowers

Frankliniella occidentalis
Western flower thrips

Ever since the mid 1980's, the western flower thrips (*Frankliniella occidentalis*) has been a major problem. The species originates from the west coast of north America, but has spread almost world-wide due to the intensive international trade in ornamental crops. It is difficult to ascertain now exactly how *F. occidentalis* arrived in Europe. There is a suspicion that it had been present in Europe for several years before it was recognized as a pest. *F. occidentalis* was first encountered in Europe in Germany, and was subsequently found in The Netherlands in 1984 in a rose crop. Since 1986 it has been spread widely in infested chrysanthemums. Because its life-cycle includes no diapause, this species can begin to build up very early in the season. The natural enemies available at the time would be in diapause, and were therefore unable to tackle the thrips early enough to ensure good control.
The western flower thrips can be found on a wide variety of plants, including many vegetable and ornamental crops in glasshouses, and on various weeds. It is an especially significant pest in cucumber, sweet pepper, aubergine, most ornamental crops, cut flowers and pot plants. In the summer *F. occidentalis* can also be found on many outdoor plants.

Life-cycle and appearance
F. occidentalis has roughly the same appearance as *T. tabaci*. The larvae are generally slightly more yellow, even orange-yellow. However, to distinguish the two species with certainty adult specimens need to be examined under a microscope. The most reliable indicator is the number of antennal segments, which can be clearly seen: there are 8 in *F. occidentalis* and 7 in *T. tabaci*. Other characteristics which distinguish the western flower thrips from the onion thrips are body colour, body length, and the presence of certain diagnostic hairs.
A population of *F. occidentalis* consists of both males and females. Females are larger (1.3 - 1.4 mm) and darker than the males and can vary in colour from yellow to dark brown. In addition, the female abdomen narrows to a point with a clearly visible ovipositor, whereas the male abdomen has two visible orange patches.
As in the onion thrips, western flower thrips generally pupate on the ground, although pupae can also be found on leaves, in flowers or in other, usually sheltered places. Plate 5.8 shows an adult female.

Population growth
The population growth of *F. occidentalis* depends mainly on temperature, but is also dependent on other environmental factors such as relative humidity and the species of host plant. There is still a great deal left to discover about the influence of these factors.
Data relating to the development time of *F. occidentalis* on cucumber at 25°C are given in table 5.2, p. 90. These data correspond closely with those for *T. tabaci*, and show that development is most rapid at 30°C, whilst above 35°C and beneath 10°C development ceases. At 18°C development takes roughly twice as long as at 25°C.
It appears that *F. occidentalis* develops much faster on flowers than on leaves, and is thus found preferentially in flowers.

5.8. *Frankliniella occidentalis*

Reproduction in *F. occidentalis* can be either sexual or asexual. Unfertilized females produce male progeny while fertilized females produce males and females in the ratio of 1 ♂ : 2 ♀. At the beginning of the season more males are found in the glasshouse than females, whereas later in the year the percentage of females overtakes that of males.
On cucumber, females (whether fertilized or not) produce three eggs per day at 25°C. If the thrips have access to pollen the number of eggs laid can be considerably higher. In cucumber at 25°C and under optimal conditions, a population can double in about four days.
If the development time of *F. occidentalis* on cucumber under long day-length conditions is compared with that of *T. tabaci* on the same crop, there are hardly any differences. However, *F. occidentalis* has greater resistance to insecticides and is better able to protect itself against attack by natural enemies due to its larger size. Further, in pollen producing crops, *F. occidentalis* makes better use of this food source as it often inhabits flowers. This in turn has a positive effect on population growth (see table 5.4). Under short day-length conditions, *F. occidentalis* has an advantage over diapausing *T. tabaci*.
A temperature of around 25°C is optimal for the population growth of *F. occidentalis* (see table 5.3). At higher temperatures mortality rises rapidly and lifespan declines sharply.
The presence of spider mite also has a positive effect on the population

Table 5.2. The development time of *Frankliniella occidentalis* on cucumber at 25°C (van Rijn et al., 1995)

development time (days)	
egg	2.7
larva 1	2.3
larva 2	3.7
prepupa	1.1
pupa	2.7
egg-adult	12.5
po-period*	1.8
egg-egg	14.3

*: po-period = pre-oviposition period, *i.e.* period from becoming adult to first egg-laying

Table 5.4. The development time from egg to adult (in days) of *Frankliniella occidentalis* on different food sources (Fransen et al., 1993)

crop	temperature (°C)	
	20	25
gerbera leaf	22.2	16.5
gerbera pollen	18.9	13.3
gerbera flower	19.5	12.8

Table 5.3. The population growth of *Frankliniella occidentalis* on chrysanthemum at different temperatures (Robb, 1989)

	temperature (°C)				
	15	20	25	30	35
development time (days)					
egg	10.1	6.6	3.2	2.5	2.4
larva 1	5.6	2.9	1.7	1.3	1.4
larva 2	11.5	9.5	4.8	2.6	3.3
prepupa	3.6	2.2	1.1	0.9	1.0
pupa	8.6	5.1	2.7	2.0	1.9
egg-adult	39.4	26.3	13.5	9.3	10.0
po-period*	6.4	2.1	1.7	1.6	1.4
egg-egg	45.8	28.4	15.2	10.9	11.4
mortality during development to adult (%)					
	13.7	13.2	8.9	10	27.1
lifespan ♀ (days)					
	46.3	75.2	31.4	12.7	9.5
eggs hatched/♀					
	50.5	125.9	135.6	42.0	5.1

*: po-period = pre-oviposition period. *i.e.* period from becoming adult to first egg-laying

growth of this species. As in *T. tabaci*, this is due to the consumption of mite eggs and to the protection against natural enemies afforded by spider mite webs. *F. occidentalis* is also capable of preying on *T. tabaci*.

Overwintering
F. occidentalis remains active as long as the temperature in the glasshouse is sufficiently high. In glasshouses that only have frost protection or are unheated in the winter, larvae and adults search for sheltered places to spend the coldest periods. During this time, development virtually stops. When the temperature rises, *F. occidentalis* becomes active once more, and can thus cause problems early in the new season. It is not yet fully known whether they can overwinter outside, and if so, how they do it.

Damage
Western flower thrips feed on developing plant tissues such as growing tips and flower buds more than does the onion thrips. When these tissues develop further the leaves and flowers appear grossly deformed. Severely infested flower buds may not open at all. Fruit can also be damaged, even at low densities, giving rise to deformities such as the "pig-tail" fruit sometimes found in cucumber crops. Thus, much lower populations of *F. occidentalis* are tolerated in cucumber than *T. tabaci*. In ornamental crops even very low numbers of thrips can cause damage by transmitting viruses, reducing aesthetic value by damaging blooms (*e.g.* roses), or by pollen transfer (*e.g.* saintpaulia).

The western flower thrips is the most important vector of both tomato spotted wilt virus (TSWV) and impatiens necrotic spot virus (INSV). TSWV (which mainly causes damage in warm climates) and INSV are both tospoviruses, all of which are transmitted by thrips. Both these viruses affect a wide range of plants, and often a single host plant may be infected by both viruses.

The effects of TSWV appear regularly in both vegetable and ornamental crops, and are familiar to many growers. Initially INSV was encountered only occasionally, but over recent years there has been an increase in the number of cases. The virus is mostly observed in floral crops, but infections can also occur in sweet pepper.

Although *T. tabaci* can transmit TSWV, as *F. occidentalis* is a more effective vector of the virus and because their populations generally exceed those of other thrips species, in many regions it is only since the arrival of *F. occidentalis* that problems with TSWV have appeared. Both larvae and adults can transmit the virus, although only first instar larvae can acquire it. Thus an adult can only transmit the virus if it ingested it during its first larval instar. The virus is not passed on by females to their progeny. Because an infected plant cannot be cured it is of the utmost importance to prevent infection as far as possible. Plate 5.7, p. 87 shows damage symptoms of TSWV.

Distribution and dispersal

F. occidentalis is mainly present in the uppermost parts of the plant. In plants with relatively soft leaves, such as gerbera and chrysanthemums, the thrips may be encountered both on lower and upper leaf surfaces. In plants with tougher leaves such as roses, yucca and orchids, they exhibit a marked preference for the growing tips. As soon as flower buds form, the thrips migrate there. They congregate preferentially in open flowers with pollen, as this is a highly attractive food source. In a rose crop, where only a few flowers are in bloom at any one time, these will be the flowers where the thrips are found; no damage will be evident at this stage in young flowers or buds. *F. occidentalis* are patchily distributed throughout the glasshouse, becoming active and leaving their hiding places in the early hours of the morning.

- Resembles the onion thrips
- Massive invasions of glasshouses from outside can occur in August and September
- Does not often cause problems

Thrips fuscipennis
Rose thrips

Thrips fuscipennis (the rose thrips) occurs naturally in Europe, and occasionally causes damage in sweet pepper, aubergine and roses. The adults are polyphagous flower visitors that can enter glasshouses, whereas the larvae are found mainly on woody perennials outside. Adults that fly into the glasshouse can be seen on the sticky cards and on flowers. There is little evidence of development in glasshouse crops and larvae of this species are therefore rarely seen under glass. Sweet pepper is particularly vulnerable to invasion from outside, and this can cause some damage. Although the species does not really constitute a problem for many growers, where it is a problem it can attack crops year after year. *T. fuscipennis* is difficult to control biologically, and can thus complicate the control of other thrips.

Most attacks by *T. fuscipennis* are probably wrongly identified. Another very similar species that enters the glasshouse from outside and can regularly be observed on sticky cards is *T. major*. However, this species is apparently unable to establish in glasshouses. In addition, the cereal thrips *Limothrips cerealium* can also be caught on sticky cards in large numbers if hay is being made anywhere in the vicinity although, like *T. major*, it causes no damage to the crop. The most important consequence of these species is the confusion they cause.

Life-cycle and appearance

The life-cycle of *T. fuscipennis* corresponds with those of the other thrips species. *T. fuscipennis* closely resembles *T. tabaci*, as the male is yellow to yellowish-brown, and the female is brown with a dark brown or even black abdomen. Females are generally darker than those of the onion thrips, although the antennae and hairs are very similar. The larvae of the two species differ very little. A rose thrips population consists of both males and females, in contrast to a population of onion thrips which consists mostly of females.

Damage

The pattern of damage caused by *T. fuscipennis* is roughly the same as that caused by the onion thrips.

- Occurs on many crops, but does not always cause damage
- Pupation takes place on the underside of the leaf
- Biological control is still difficult

Echinothrips americanus

Echinothrips americanus is an increasing problem in glasshouse cultivation. It originates from the eastern United States, stretching through several states including Florida. The species also occurs in Bermuda and Canada.

Initially only ornamental crops were affected, but nowadays this thrips species is also responsible for damage in sweet pepper, cucumber and aubergine. Because this thrips cannot be effectively controlled biologically, its presence can complicate the biological control of other pests in the crop.

The species has rapidly spread via the trade in foliage plants belonging to the family Araceae (dieffenbachia, homalomena, philodendron and syngonium). Because real damage is only caused to a few crops, *E. americanus* can spread through glasshouses virtually undetected, assisted by the fact that it has such a broad host range. Damage is not restricted to dicotyledons; monocotyledons such as grasses are also attacked. The species lives exclusively on the leaf, and although plants with tough leaves seem to be attacked preferentially, other crops can also be affected. The species is often present on weeds in glasshouses.

E. americanus can be controlled using chemical pesticides, but so far biological control has proved difficult. The reason is that the predatory mites introduced to control other thrips are not equipped to deal with the much larger larvae of this species. In addition, *E. americanus* is usually found low down in the crop whereas the predatory mites and bugs used for controlling other thrips species are mainly found higher up.

Life-cycle and appearance

Unlike the other thrips species discussed, *E. americanus* completes its entire development from egg to adult on the plant, and more specifically on the leaves. Eggs are deposited in the leaf tissue, while subsequent stages are found on the upper and undersides of the leaf. The larvae, prepupae and pupae of *E. americanus* are a white or light yellow colour and prefer the undersides of leaves. Thus pupation occurs on the leaf rather than on the ground. The (pre)pupae remain immobile on the leaf tissue and only move if startled. Adults are found on both upper and undersides of the leaves. This species of thrips is relatively large. Both males and females are dark brown to black in colour with orange pigmentation between the segments, and the dark wings are white at the base. This colouration distinguishes the species from most other thrips species commonly found in glasshouses. Plate 5.9 shows an adult *E. americanus*.

Two species with which *E. americanus* is easily confused are *Heliothrips haemorrhoidalis* and *Hercinothrips femoralis*, both of which are occasionally found in ornamental crops. However, adults of *Hercinothrips femoralis* have banded wings and *Heliothrips haemorrhoidalis* lacks the row of black hairs on its wings.

Population growth

The population growth of *E. americanus* is dependent on the temperature of the crop (see table 5.5); relative humidity appears to play a lesser role. Below 0°C the thrips cannot survive for more than a few hours, and even then only as the older instars. Overwintering outside is therefore impossible in temperate climates.

Below 15°C most juvenile individuals fail to complete their development, and above 35°C not a single thrips survives. At unsuitable temperatures hardly any eggs are laid, and those that are fail to hatch. However, a few individuals can survive a temperature of around 4°C for 2 weeks. Reproduction may be either sexual or asexual. Depending on the crop, a female lays approximately 80 eggs over a period of roughly 40 days, depending on the crop.

A population always consists of more females than males, but the actual ratio can vary enormously, depending on conditions.

Damage

E. americanus is highly polyphagous, but in those crops in which the species occurs, it does not always cause economic damage.

Adult and larval feeding causes the usual symptoms of thrips damage to leaves - the patchy silvering of the leaf in places where the contents of cells have been sucked out.

In sweet pepper the damage can be seen as the discolouration of older leaves at the bottom of the plant. At higher levels of infestation the leaves can even shrivel entirely and fall off, while the outside of ripe fruit takes on a silvery sheen.

Most ornamental crops, however, show no serious damage.

Distribution and dispersal

All stages of *E. americanus* are found on the plant. They have a strong preference for the underside of leaves, but can also be found on the upper surface and sometimes even in flowers. *E. americanus* stays mainly along the veins of the leaf, where it can remain immobile for hours at a time. With increasing numbers the insect spreads over the whole leaf. If disturbed, the thrips immediately becomes active.

An infestation of *E. americanus* in sweet pepper begins locally low down in the crop on the undersides of leaves, while adults are not particularly good fliers. Therefore finding infestations requires careful inspection. Biological control is also complicated by the fact that thrips predators such as *Orius* spp. prefer to inhabit the upper regions of the crop. During the normal routine of crop maintenance it is unlikely that an infestation will be noticed before it reaches serious proportions.

Table 5.5. The development time of *Echinothrips americanus* on impatiens and chrysanthemum (Oetting and Beshea, 1993) and in cucumber and sweet pepper (Opit, 1996)

crop	RH* (%)	temperature (°C)	development time (days)					
			ei	larva 1	larva 2	prepupa	pupa	total
impatiens		20	15.3	4.2	7.6	1.9	4.8	33.8
		25	7.7	2.6	3.0	0.9	1.5	15.7
		30	5.8	2.5	1.2	0.6	1.7	11.8
		25/20**	8.8	3.6	5.8	1.6	3.1	22.9
chrysanthemum		20						36.8
		25						22.6
cucumber	45%	23/19**	15.6	3.6	2.1	-------5.2-------		26.5
sweet pepper	45%	23/19**	15.0	6.0	5.5	-------5.2-------		31.7

*: RH = relative humidity
**: day/night temperature

5.9. *Echinothrips americanus*

In the following table, the characteristics of the thrips species that are most important in glasshouses will be covered. Although most of the species included have already been dealt with, *Thrips palmi* is included for comparative purposes, and *Parthenothrips dracaena* as it is a species that may cause more problems in the future. *T. palmi* has not yet become established in Europe, but due to the large-scale economic damage it has caused in other parts of the world a strong phyto-sanitary policy has been implemented to prevent it becoming established. Because the species spreads rapidly over the world via plant material, it is often found in import inspections. It is easily confused with indigenous species such as *T. flavus* and *T. nigropilosus*, which can sometimes be seen on sticky traps.

Overview of important thrips species

	Frankliniella occidentalis	*Thrips tabaci*	*Thrips fuscipennis*, *Thrips major*
body colour	light yellow-dark brown	yellow to dark brown; sometimes two-toned, in which case thorax lighter than abdomen	♂: yellow to yellow-brown ♀: yellow to dark brown with a dark brown or even black abdomen
body length ♀	1.2-1.4 mm	0.8-1.2 mm	1.0-1.2 mm
antennae	8 segments	7 segments	7 segments
head	• 2 long hairs between ocelli (red) • 1 long, and numerous short hairs behind compound eyes	• 2 short hairs between ocelli (red) • numerous short hairs behind compound eyes	pattern of hairs as in *T. tabaci*
thorax			
forewing			
location in crop	mainly in the upper part of the plant, preferably in flowers and specifically in open flowers with pollen, pupation in soil and on the plant (in hidden places)	all parts of the plant, preferring the undersides of young leaves, often sitting close to veins, pupation in soil and on the plant (in hidden places)	*T. fuscipennis* rarely on leaves, mainly only adults in flowers, *T. major* is only found outdoors and on traps

	Thrips palmi	*Echinothrips americanus*	*Parthenothrips dracaenae*
body colour	white / yellow	dark brown to black, orange colour between segments, strong sculpturing	light to dark brown, dark markings on wings. Striped appearance with strong sculpturing
body length ♀	1.0-1.2 mm	1.3-1.6 mm	1.4-1.6 mm
antennae	7 segments	8 segments	7 segments
head	pattern of hairs as in *T. tabaci*		
thorax	as in *T. fuscipennis* and *T. major*		
forewing			
location in crop	mainly on leaves, seldom in flowers or ripe fruits, pupation occurs in the soil	all stages on the leaf, and sometimes on fruits low in the crop, adults on upper and underside of the leaf, other stages mainly on undersides, pupation occurs on underside of the leaf	all stages on the leaf, including pupae

Knowing and recognizing Thrips

Natural enemies of thrips

The most important natural enemies of thrips in glasshouses are:
Amblyseius cucumeris
Amblyseius degenerans
Hypoaspis spp.
Orius spp.
Verticillium lecanii
Entomophthorales spp.

5.11. *Amblyseius cucumeris*

5.12. *Amblyseius degenerans*

5.13. *Hypoaspis miles*

5.14. *Orius laevigatus*

5.15. *Verticillium lecanii*

5.16. *Entomophthorales sp.*

The most important enemies of thrips are predatory mites (specifically *Amblyseius cucumeris* and *Amblyseius degenerans*) and *Orius* predatory bugs. *Amblyseius* spp. mainly eat small thrips larvae, while *Orius* spp. can also consume adults. Both predatory bugs and predatory mites are often introduced to control thrips.
Predatory *Hypoaspis* mites are mainly used in the cultivation of ornamental crops, and will consume thrips pupae.

Entomopathogenic fungi can also parasitize thrips, and thus the naturally occurring Entomophthorales moulds that can appear in the autumn can remove a large proportion of the thrips population. In addition, *Verticillium lecanii*, mostly used to combat whitefly, can also contribute significantly to the control of thrips.

- Predatory mite
- Commercially introduced
- Nymphs and adults eat first instar thrips larvae preferentially

Amblyseius cucumeris

The predatory mite *Amblyseius cucumeris* (also known as *Neoseiulus cucumeris*) is widely used for the biological control of thrips in glasshouse crops. This predatory mite belongs to the family Phytoseiidae within the order Acarina and the class Arachnida (spiders and related arthropods).
In view of the fact that the chemical control of thrips interferes with the biological control of whitefly, spidermite and aphids, and, more importantly, given that it does not produce the desired results, alternatives to agro-chemicals have long been sought. *A. cucumeris* proved to be a suitable candidate. The use of biological control in sweet pepper was further encouraged by the United States' demand that all imported sweet pepper should be free from any chemical residues. Thus from 1985, *A. cucumeris* was introduced in sweet pepper crops to control thrips.
During the same period, the possibility of controlling thrips in cucumber with the help of *A. barkeri* (*Neoseiulus barkeri*) was investigated, a species closely resembling *A. cucumeris*, but this did not give suitable results. However, when *F. occidentalis* became a problem, both species of predatory mite were used as a matter of urgency. Currently, many growers of vegetable and ornamental crops have employed *A. cucumeris* on a large scale. *A. barkeri* occurs spontaneously in glasshouses and in the course of the season can sometimes replace the introduced *A. cucumeris*.

Table 5.6. The development time (Gillespie & Ramey, 1988) and daily egg-laying rate (van Houten et al., 1995) of *Amblyseius cucumeris* on first instar larvae of *Frankliniella occidentalis* on bean

temperature (°C)	20	25	30
development time (days)			
egg	2.9	3.1	1.9
larva	1.4	1.2	0.4
protonymph	3.2	2.4	2.0
deutonymph	3.6	2.1	2.0
total egg-adult	11.1	8.8	6.3
egg-laying/♀/day			
		2.2	
		2.1*	

* with sweet pepper pollen instead of thrips larvae as food supply

Population growth

The population growth of *A. cucumeris* depends on temperature and on the nature and availability of prey and other food sources within the crop. The relative humidity also has considerable influence on the growth of a population of this predatory mite; low humidity has a negative effect on egg-laying, hatching, larval survival, adult longevity and development time.
The minimum temperature for development is around 8°C. At 35°C, no more than 50% of eggs hatch, and 90% of those that do hatch die within 2 days. Protonymphs do not survive at this temperature.
Table 5.6 shows data relating to the population growth of *A. cucumeris*. Mating is essential for reproduction in predatory mites. For optimal egg-laying, this mating must be repeated several times. Approximately 2 days after adult emergence, egg-laying commences and lasts for roughly 20 days, after which the mites die.
A relative humidity of 65% is critical for *A. cucumeris* eggs: only 50% survive at this level of humidity. At 60% or less, all the eggs shrivel. At a relative humidity in excess of 70% and at normal temperatures, more than 90% of eggs hatch. Temperature has hardly any effect on egg mortality at these humidity levels. However, a relativity humidity of less than 65% in the glasshouse is not necessarily fatal if the predatory mites sit on healthy leaves, since the microclimate at the leaf surface can be more favourable than the general climate of the glasshouse. Thus, in the winter period the microclimate among the cucumber leaves largely protects eggs against desiccation. If the glasshouse temperature is high and the relative humidity low, the crop can be severely affected by spider mite or thrips. If it is a heavy infestation the leaves turn necrotic, reducing evaporation. This damage results in a higher temperature and lower relative humidity at the leaf boundary layer. Such changes in the microclimate at the leaf surface are detrimental to the population of predatory mites. If the crop is not too badly affected by pests, the predatory mite will also be protected against desiccation during the summer months by the microclimate around the leaf. In crops where the microclimate at the leaf surface is much less clement (as in sweet pepper and rose) the situation is rather different. It is important for mites that predate thrips that they feed on pollen and spider mites, not only to be able to survive in the absence of thrips, but also to provide the nutrition essential for reproduction.

Life-cycle and appearance
Amblyseius cucumeris

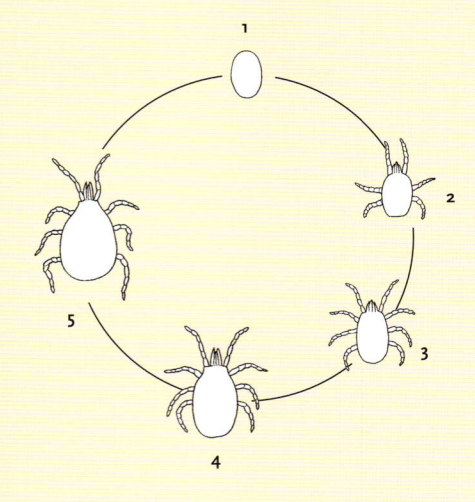

1 egg
2 larva
3 protonymph
4 deutonymph
5 adult

The life cycle of *A. cucumeris* consists of the following stages: egg, a larval instar, two nymphal instars and finally the adult.

Exactly where eggs are laid depends on the crop. In cucumber and sweet pepper, they are laid on the underside of the leaves on top of the hairs in the axils of main and ancillary veins. The eggs are oval and slightly smaller than the eggs of the spider mite predator *Phytoseiulus persimilis*, with a diameter of approximately 0.14 mm, and are white in colour (see plate 5.17).

The larvae, which have six legs, remain close to the place where they hatched from the egg. They are the same colour as the eggs and only very slightly larger. They do not feed and, like the larvae of *P. persimilis*, they are inactive. The nymphs are larger than the larvae and are pale brown. Males are smaller than females, and like the adults, both protonymphs and deutonymphs (the first and second nymphal instars) are extremely mobile and actively search for and ingest food. Nymphs and adults have four pairs of legs, the first pair being used as feelers.

Adult mites resemble adult *P. persimilis* to some extent, but are generally lighter in colour and have a flatter, more elongated body. They are roughly 0.4 mm in length. However, the spider mite predator has longer legs and a rounder body. Thrips-predatory mites are difficult to spot in the crop. Plates 5.18 and 5.19 show adult *A. cucumeris*.

5.17. Eggs

5.18. Adult

5.19. Preying adult

Knowing and recognizing **Thrips**

Overwintering
The strain of *Amblyseius cucumeris* employed in biological control does not enter diapause and can therefore be applied under short day-length conditions.

Feeding behaviour
Predatory mites pierce their prey and suck out the contents (see plate 5.20). *A. cucumeris* preys on several other organisms such as spider mite, tarsonemids, the larvae of spider mite-predatory mites and even their own larvae. They also feed on pollen, which is useful in a sweet pepper crop as the predatory mite can establish in the crop before the pest appears. This is not possible in crops such as cucumber that produce no pollen.

The success of the thrips-predatory mites is mainly dependent on the size of the available prey larvae. First instar thrips larvae are taken much more often than second instar larvae, and smaller thrips species more easily than larger ones. Environmental factors such as the crop, climate and use of pesticides also have a significant impact on success. At 25°C under optimal conditions, an adult predatory mite will consume about 6 first instar thrips larvae per day.

A thrips larva often tries to defend itself against attack by a predatory mite by violently jerking the abdomen, and for this reason the larger, more powerful larvae are much less frequently taken than their smaller counterparts. Thus, in the case of *Frankliniella occidentalis* it is almost exclusively first instars that are preyed on. In addition, when disturbed, they can secrete fluid from a gland at the rear of the abdomen with which they foul an attacker, which will quickly lose interest in the thrips in order to clean itself. Nymphs of the predatory mite have more trouble in capturing prey than do adults, and thrips larvae killed by an adult female mite are sometimes partly consumed by the nymphs.

The quantity of thrips larvae eaten also depends on the crop. Predatory mites find it much easier to attack thrips larvae on a sweet pepper leaf, for example, than on a cucumber leaf.

The availability of suitable prey depends on the age structure of a prey population and can therefore be lower than the total thrips population might suggest. For various reasons (*e.g.* a treatment with chemical pesticides) a large proportion of the population may be in the same developmental stage and perhaps unsuitable for consumption. In such cases, alternative sources of food are essential to sustain the predatory mite population.

5.20. Predation of a thrips larva by *Amblyseius cucumeris*

- **Predatory mite**
- **Commercially introduced**
- **Larvae, nymphs and adults feed preferentially on first instar thrips larvae**

Amblyseius degenerans

Amblyseius degenerans is not indigenous to north-west Europe. Its natural habitat is in Mediterranean countries, such as Israel, Italy, Turkey, Portugal and Greece, as well as large parts of Africa, Madeira, the former USSR and Hong-Kong. The species is also known as *Iphiseius degenerans*, and can be found on many plants, including trees. These predatory mites run around freely on the foliage and make no attempt to hide. They live on thrips larvae, spider mite, and various species of pollen. Like *Amblyseius cucumeris* and *Phytoseiulus persimilis*, this species of predatory mite also belongs to the family Phytoseiidae. It has been used for the biological control of thrips larvae since 1994.

Population growth

A. degenerans is used particularly in sweet pepper crops where it appears to develop and reproduce better than *A. cucumeris*. This is mainly due to the fact that this species of predatory mite is more frequently found in flowers, where it can take advantage of both thrips larvae and pollen. Further, this predatory mite performs better in conditions of low humidity; at a relative humidity of 50%, the hatching rate for eggs is still 50%, whereas *A. cucumeris* would require a relative humidity of 65%. The development time is shorter than for *A. cucumeris*, and at 25°C development from egg to adult takes 6 days.
Reproduction is dependent on temperature and food supply. *A. degenerans* females must mate several times for maximum egg production, and the sex ratio is on average 1:1.
A. degenerans does not enter a diapause and can therefore be introduced under short day-length conditions.
Population growth progresses just as well if thrips are present as when the food source is pollen alone.
Table 5.7 gives data relating to the population growth of *A. degenerans*.

Feeding behaviour

A. degenerans preys particularly on first instar thrips larvae, but also eats spider mites and pollen. It can consume 4 - 5 first instar thrips larvae per day. However, this predatory mite dislikes leaves that are heavily covered with webs and is thus not an effective predator of two-spotted spider mite. All mobile instars feed, although feeding is not essential in the larval stage. Whenever they are present in high density, more prey are killed than is necessary for optimal development.

Searching behaviour and dispersal

A. degenerans moves rapidly, actively searching over leaves and in flowers. It spreads with ease from plant to plant, and is quite conspicuous in the crop. They can sometimes be found in large numbers in the flowers of sweet pepper plants.

Life-cycle and appearance
Amblyseius degenerans

The developmental stages are the same as for other predatory mites, namely egg, larva, protonymph, deutonymph and adult. The eggs are deposited on the undersides of leaves, and in sweet pepper, they are placed mainly on the leaf hairs next to the veins. The eggs are transparent white at first, gradually turning brown.
The larvae have three pairs of legs and a brown X-shaped dorsal marking. Over the course of time their colour changes to a dark brown, and under certain conditions (although exactly which are unknown), they turn brown-red.
The adult predatory mite is almost black, spherical in shape and approximately 0.4 mm long. The males are somewhat smaller and slimmer than the adult females, but more difficult to distinguish from immature females.

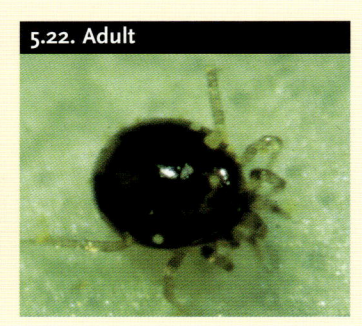

5.21. Nymphs

5.22. Adult

Table 5.7. The population growth of *Amblyseius degenerans* at 25°C and 75-90% relative humidity on *Tetranychus pacificus* and bean (Takafuji & Chant, 1976)

development time (days)	
egg	2.2
larva	1.0
protonymph	1.2
deutonymph	1.3
total	5.7
po-period *	2.3
lifespan ♀ (days)	53.3
egg / ♀	67.8
egg / ♀ / day	2.2

*: po-period = pre-oviposition period, *i.e.* period from becoming adult to first egg-laying

- **Ground-living predatory mites**
- **May occur naturally, but are also introduced**
- **Nymphs and adults feed on thrips pupae**

Hypoaspis spp.

Hypoaspis miles and *Hypoaspis aculeifer* are predatory mites that occur naturally in Europe, north America and Japan. Like other species of predatory mite, these too belong to the suborder Mesostigmata, although they belong to the family Laelapidae, not the Phytoseiidae. Both species are polyphagous, living on ground-dwelling organisms such as springtails, the larvae of fungus gnats, beetles and flies, as well as nematodes and other mites, including *Rhizoglyphus* spp. (bulb mites) and *Tyrophagus* spp. (storage mites). Both *Hypoaspis* spp. also feed on thrips pupae and are therefore introduced for this purpose, especially in ornamental crops. Plate 5.23 shows *Hypoaspis* preying on a thrips pupa. These species are discussed in detail in chapter 11.

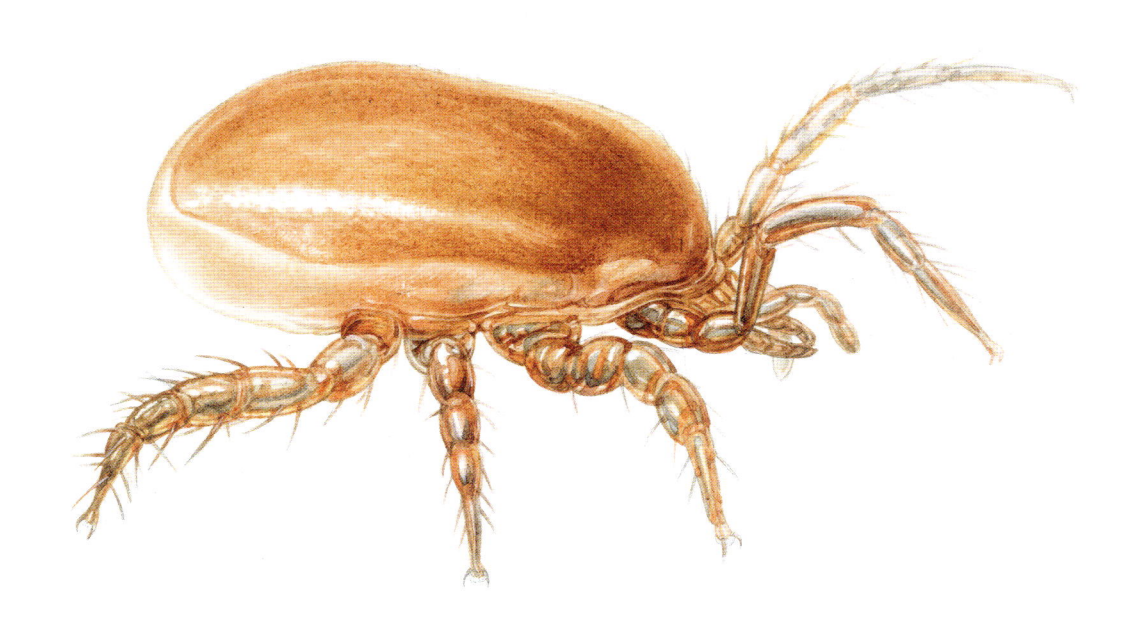

5.23. Predation of a thrips pupa by *Hypoaspis*

Overview of important species of predatory mite

	Amblyseius cucumeris (A.c.)	*Amblyseius barkeri (A.b.)*	*Amblyseius degenerans (A.d.)*	*Amblyseius californicus (A.cal.)*	*Phytoseiulus persimilis (P.p.)*	*Hypoaspis miles (H.m.)*	*Hypoaspis aculeifer (H.a.)*
adult appearance	beige-pink, sometimes with X-shaped orange marking; 0.4 mm; depressed posture (short legs); many short hairs	red-brown sometimes with X-shaped orange marking; 0.4 mm; depressed posture (short legs); many short hairs	almost black; 0.4 mm; spherical and smooth appearance	transparent white with x-shaped orange marking on back; 0.4 mm; resembles *Amblyseius cucumeris* and *Amblyseius barkeri*	light red, 0.6 mm; tall, on long legs; few very long hairs	dull brown, 0.5-1 mm; more rounded than *H.a.*, pointed carapace, soft hairs on legs; jaws much larger than those of *H.a.*	shiny brown, 0.5-1 mm; slimmer than *H.miles*, round carapace, thorn-like hairs on legs
food	- thrips larvae - pollen - spider mites - storage mites - broad and cyclamen mites	- thrips larvae - pollen - spider mites - storage mite - broad and cyclamen mites (sometimes)	- pollen - thrips larvae - spider mites - broad and cyclamen mites (sometimes)	- spider mites - thrips larvae - pollen - storage mites - broad and cyclamen mites	spider mites	ground-dwelling organisms: incl. thrips pupae, sciarid fly larvae and straw mites	ground-dwelling organisms incl. thrips pupae, sciarid fly larvae and straw mites
RH-sensitivity	at < 65% RH < 50% egg hatching	more sensitive to low RH than *A.c.*	at < 50% RH < 50% egg-hatching	at < 70% RH < 50% egg-hatching	particularly at higher temperatures, sensitive to low RH. At < 65%RH < 50% egg-hatching	needs damp ground	needs damp ground
ease of observation in crop	difficult	difficult	very easy	difficult	rather conspicuous	difficult, but sits nearer to the surface than *H. aculeifer*	difficult, more concealed than *H. miles*
sensitivity to chemical pesticides	sensitive	sensitive	very sensitive	less sensitive than other predatory mites	sensitive	only sensitive to pesticides if these penetrate through to ground level	only sensitive to pesticides if these penetrate to ground level
reproductive organs female							
male jaws			remark: good at seeking out flowers; thus more often in contact with pollen and western flower thrips than *A.c.* or *A.b.*	remark: a useful supplement for control of spider mite particularly at higher temperatures and low prey density			

- Predatory bugs
- Introduced, but sometimes also occur naturally
- Nymphs and adults feed on all stages of thrips

Orius spp.

Bugs form the suborder Heteroptera within the order Hemiptera. Other hemipterans include aphids and cicadas. In the glasshouse the predatory bugs of the family Anthocoridae (flower bugs), mainly *Orius* spp. and sometimes *Anthocoris* spp. are frequently encountered. Members of this family are polyphagous predatory insects feeding on thrips, aphids, mites, and other small arthropods. They can attack all stages, including adults (mainly of soft-bodied species), larvae, nymphs and eggs. Sometimes a predatory bug may puncture human skin, causing a sharp stinging sensation. Most species can occasionally feed on plant sap, but without causing real damage to the plant. Both *Orius* and *Anthocoris* spp. are widely distributed and often form an important element in an integrated IPM programme. *Anthocoris* spp. are generally highly polyphagous bugs, whereas *Orius* spp., although they are also polyphagous, often show a strong preference for a particular species of prey.

The development of the different species varies considerably. Some species have one generation per year while others may have two, three or even more. There is a high degree of correlation between prey density and the size of predatory bug populations.

Hungry predatory bugs increase their chances of discovering prey by moving more rapidly, actively searching in parts of the plant where prey might previously have been expected, and intensifying their search in places where they have previously found prey. Adult predatory bugs are mainly found in flowers, with the immature stages usually on leaves. Predatory bugs generally overwinter as adult females, which can be found in sheltered places such as hollow stalks, under dead leaves or under bark. Whether an individual enters diapause or not in the autumn depends on various factors, such as the species, area of origin, food supply, temperature and day-length.

Predatory bugs of the genus *Orius*, of which some 70 species are known, occur all over the world and are found in both wild vegetation and in horticultural crops. They are mainly found in flowers.

Since 1991, *Orius* spp. have been employed to combat thrips, and are used specifically against *F. occidentalis*. At present, three species are used to control thrips: *O. laevigatus*, *O. majusculus* and *O. insidiosus*, an American species used in the United States. In southern Europe, the naturally occurring *O. albidipennis* can aid control.

Population growth

Temperature and food supply are the main factors influencing the reproduction and development of a population of *Orius*, with higher temperatures and a good quality food supply accelerating population growth. The crop and the relative humidity are less important. The development times of various *Orius* spp. have been investigated under different conditions, and results show that inter-specific differences are minimal.

The presence of pollen enhances the rate of development and also improves survival rates in some *Orius* spp.. Although all species develop better when there is an additional living food source available, *Orius* cannot survive solely on plant sap.

Mating in *Orius* often occurs very soon after reaching maturity. Two or three days after mating, egg-laying begins.

The lifespan of the adult is three to four weeks.

Overwintering

Most temperate zone species of *Orius* are liable to enter diapause. If nymphs mature under short day-length conditions, after reaching the adult stage they lay no eggs before first passing through diapause. In north-west Europe, fertilized females enter their winter quiescence from September / October, hiding under tree bark, in small crevices in trees and other plants, and also in any layers of straw. In the spring, from mid April, they re-emerge from this diapause and become active once more. The day length at which females enter diapause depends on species and geographical location.

Feeding behaviour

All stages capture and kill small, softer-bodied insects and mites, which are pierced with the specially adapted proboscis and sucked empty. All stages of thrips are preyed on. At higher prey densities, more are killed than are needed as food. Predatory bugs do not hesitate to take their own species as food, and other suitable insects are also sometimes consumed.

Predatory mites and predatory bugs can coexist, and although they often compete for food, at other times they complement each other because each performs best under different conditions and in different parts of the crop.

Searching behaviour and dispersal

Orius bugs are fast-moving opportunists. Prey is chiefly discovered by olfactory or tactile senses, and not by sight. The area probed by the predator is the area that lies within reach of its antennae and depends on the length of the antennae and the angle at which they are held.

Predatory bugs react to any movement of their prey, and adult specimens in particular react very rapidly to any disturbance. At any threat of danger they immediately fly off, crawl away or simply fall to the ground. They are fairly good fliers and readily move around, and are thus quickly able to find new concentrations of prey.

In pollen-producing crops the adults are often found in flowers, whereas the nymphs are mostly located on leaves. Because of its glandular hairs, tomato is an unsuitable crop for *Orius*.

Life-cycle and appearance
Orius spp.

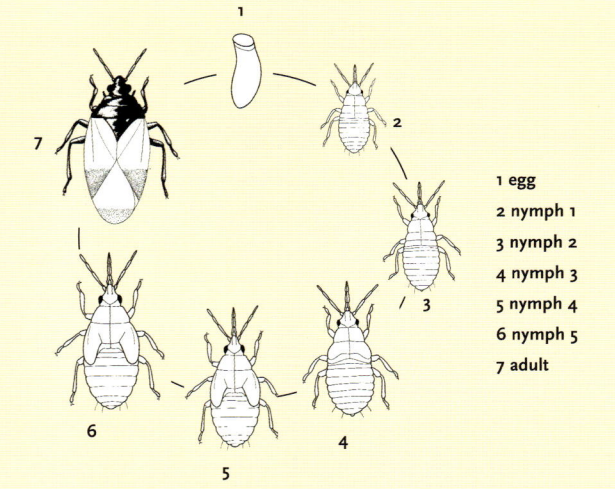

1 egg
2 nymph 1
3 nymph 2
4 nymph 3
5 nymph 4
6 nymph 5
7 adult

The life-cycle of an *Orius* predatory bug consists of seven developmental stages: the egg, five nymphal instars and the adult bug.

A newly laid egg is about 0.4 mm long and 0.13 mm wide and colourless, later developing a milky white hue. The eggs are embedded in the leaf tissue, although the site chosen on the leaf by different species varies. They are often found in the leaf stalk or in the main vein on the underside of the leaf, but sometimes also in parts of the flower or the inflorescence. Because the eggs lie almost parallel with only their tips visible where they stick out from the leaf tissue, they are very hard to see. If they are incompletely imbedded, the red eyes and orange body of the developing nymph within the egg will become visible. Eggs are sometimes laid in groups of a few eggs together, but are mainly deposited separately.

When the nymphs hatch from their eggs (plate 5.25) they are shiny and almost colourless, but within a few hours they become yellow.

The colour of the nymphs depends on the instar and on the particular species, and varies from totally yellow to totally brown. The red eyes are conspicuous and easily recognized in all instars. In the second instar the wings buds begin to develop, but only in the fifth nymphal instar do they become clearly visible as wing buds.

Immediately after the final moult the adult predatory bugs are yellow, but after a few hours they acquire their characteristic colours. The wings take about an hour before they are fully unfolded. Adult specimens are generally brown to black with lighter grey-white or brown areas on the wings. The appearance of the female and male predatory bug is more or less the same, although there are slight differences in the abdomen. The female abdomen is slightly larger and more robust than the male, and is symmetrical, which is not the case in the male. The size varies from 1.5 - 3 mm, depending on the species.

There may be differences in the colour of the wings between different *Orius* species, although the best way of separating species is by comparison of the male genitalia. However, this is specialised and time-consuming work.

5.24. Eggs of *Orius majusculus*

5.25. Hatching nymph of *Orius*

5.26. Nymph of *Orius laevigatus*

5.27. *Orius majusculus*

5.28. *Orius majusculus*

Orius laevigatus & Orius majusculus

Orius laevigatus is at present the most widely used Orius species for the biological control of thrips, although O. majusculus is also used. Orius laevigatus is widely distributed in the Mediterranean region and in north Africa, whereas O. majusculus occurs very commonly in central and southern Europe, and Asia Minor. O. majusculus may also spontaneously enter glasshouses, particularly in July and August.

Appearance
Nymphs of the different species of Orius can only be distinguished from one another in the later instars. Nymphs of O. majusculus are dark brown in the fourth and fifth instars, whereas the nymphs of O. laevigatus remain yellow until the last nymphal instar, acquiring orange-brown spots in the later stages.
Adults of O. majusculus are slightly larger than those of O. laevigatus and lack the black areas on the wings. They are 2.6 - 3 mm long, whereas adults of O. laevigatus are 1.4 - 2.4 mm in length.

Population growth
Population growth in O. laevigatus is mainly dependent on temperature and food supply. Unlike O. majusculus, the development time of the strain introduced into glasshouses is hardly affected by day-length. The point at which development ceases is around 11°C. Data relating to the population growth of O. laevigatus are given in table 5.8 (1, 2 and 3). Egg-laying is strongly dependent on the nutritional value of the prey, and can reach a maximum of approximately 165 eggs per female.
Mating begins on the day of adult emergence from the last moult and takes place day and night. Most females lay eggs at temperatures between 20 and 30°C. When the temperature falls to 15°C, around 50% of females die without having produced eggs, while those that do produce eggs lay a reduced number.

Overwintering
O. laevigatus used in glasshouses do not enter diapause and can therefore be introduced throughout the year as long as sufficient food is available. O. majusculus do go into diapause and can therefore only be successfully used from spring onward. If this species is released earlier in the year, a generation of nymphs is produced but the adults that emerge then go into diapause.

Feeding behaviour
Under normal conditions O. laevigatus females (both nymphs and adults) consume around 12 thrips (larvae or adults) per day. Plate 5.29 shows a feeding O. majusculus.
O. laevigatus is more of a flower visitor and therefore makes better use of pollen, whereas O. majusculus is found more generally distributed over the whole plant.

Table 5.8/1. The population growth of *Orius laevigatus* at different temperatures with butterfly eggs as prey (Aulazet et al., 1984)

	temperature (°C)			
	15	20	25	30
development time (days)				
egg	11.7	6.4	4.6	3.3
nymph 1	9.4	4	2.9	2
nymph 2	6.2	2.9	2.2	1.2
nymph 3	6	3	1.8	1.2
nymph 4	7.4	3.4	2.1	1.8
nymph 5	14	7.1	4.1	2.9
total	54.7	26.8	17.7	12.4
eggs/♀	62	137	157	152
hatched eggs (%)	78	73	87	84
lifespan ♀ (days)	78	50	39	21

Table 5.8/2. The population growth of *Orius laevigatus* at 20 and 25°C (Vacante *et al.*, 1995 and 1997)

	temperature (°C)		prey*
	20	25	
development time nymph 1 to adult (days)		17.7	sppo
		14.5	E.k.
		14.6	sppo + E.k.
survival (%) of nymph 1 to adult		65	sppo
		85	E.k.
		90	sppo + E.k.
		0	sppl
lifespan ♀ (days)	47		E.k.
lifespan ♂ (days)	24		E.k.
sex ratio (♀:♂)	1:1		E.k.

*: E.k. = *Ephestia kuehniella* eggs
sppo = sweet pepper pollen
sppl = sweet pepper plant

5.29. Preying *Orius majusculus*

Table 5.8/3. The population growth of *Orius laevigatus* with different insects as prey (Tawfik, 1973; Tommasini & Nicoli, 1993)

	temperature (°C)		prey
	25	26	
eggs/♀	23.3		T.u.
	113		A.g.
		55.6	F.o.
		119	E.k.
lifespan ♀ (days)			
		18	F.o.
		39	E.k.

*: E.k. = *Ephestia kuehniella* eggs
T.u.= *Tetranychus urticae*
A.g.= *Aphis gossypii*
F.o.= *Frankliniella occidentalis*

Overview of important *Orius* species

	O. majusculus	*O. laevigatus*
subgenus	Heterorius	Orius
geographical distribution	very common in central Europe, also present in southern Europe and Asia Minor	widespread in the Mediterranean and northern Europe
appearance		
	adult length: 2.6-3 mm wing tip colour: light, faint	**adult** length: 1.4-2.4 mm wing tip colour: terminal portion grey-brown and clearly darker than the rest
	 **nymph** dark brown	**nymph** yellow, with orange spots in the last instar 
natural/introduced:	introduced, but can also appear naturally	introduced, but can also appear naturally

5.30 5.31 5.32 5.33

- ■ Entomopathogenic fungus
- ■ Commercially introduced but can also occur naturally
- ■ Particularly affects larvae, and to a lesser extent adults

Verticillium lecanii

Verticillium lecanii is a common fungus which attacks arthropods among other organisms. The fungus is discussed in detail in chapter 4. The "whitefly" strain of the fungus also has an effect on thrips, causing high mortality when conditions are favourable (high humidity and a temperature between 18 and 25°C). Since none of the other natural enemies of thrips are affected by the fungus it can be used to great advantage in combination with predatory mites and predatory bugs. Plate 5.34 shows a thrips infected by *Verticillium lecanii*.

5.34. Thrips infected by *Verticillium lecanii*

- ■ Entomopathogenic fungi
- ■ Occur naturally
- ■ Appear in the late summer and autumn

Entomophthorales

In the late summer and autumn the numbers of thrips can be sharply reduced by insect-parasitic fungi. These are moulds of the order Entomophthorales, belonging to the class of Zygomycetes. The Entomophthorales are distributed world-wide and the majority are parasites of insects and mites. Many different arthropods are infected, including many agricultural and horticultural pests. For this reason, for several decades efforts have been made to cultivate these fungi as a means of biological control, but so far without success. Even if a culture method is found, it would still remain to be seen if they were active by the time they were introduced into the crop.

In nature, these fungi can destroy an entire population of their host within a few days. They appear in September, sometimes in August. An attack presents a characteristic picture that should not be confused with the moulds of several less specific fungi that can appear at high humidities. *Neozygites parvispora* is an important cause of mortality in thrips outdoors, but this species is rarely seen in glasshouses and when it does occur it very rapidly proceeds to its resting spore stage. Another species, *Entomophthora thripidum*, is found on thrips in glasshouses, infecting its host in the late summer and autumn and remaining active until the end of the cropping period. At room temperature the life-cycle of *E. thripidum* lasts roughly 4 days, most of this period being spent in the insect. A spore formed on the insect (a primary spore) develops a secondary spore within 24 hours that is discharged. This secondary spore is infectious. It is not clear where the first infectious material comes from, but presumably it is from outside, either carried on air currents or introduced by infected insects.

Larval and adult thrips are all susceptible to infection, but the number of spores produced by the fungus depends on the instar affected. The spore-bearing fungal hyphae all emerge at roughly the same time. It is known that *E. thripidum* infects only the abdominal organs of the thrips, and killed adult thrips are characterized, among other symptoms, by a split in the abdomen (see plate 5.35).

5.35. Thrips infected by *Entomophthora thripidum*

Knowing and recognizing Thrips

Biological control of thrips

OVERVIEW OF THE BIOLOGICAL CONTROL OF THRIPS				
product name	natural enemy	controlling stage	stage of thrips controlled	crop
THRIPEX(-PLUS)	Amblyseius cucumeris	nymph and adult	particularly small thrips larvae	all crops
THRIPOR	Orius spp.	nymph and adult	all stages	all crops
THRIPANS	Amblyseius degenerans	nymph and adult	particularly small thrips larvae	crops with pollen
MYCOTAL	Verticillium lecanii	spore	larvae and to lesser extent adults	all crops
ENTOMITE	Hypoaspis aculeifer/miles	nymph and adult	thrips pupae	soil cultures

Predatory mites are the most widely employed natural enemies of thrips. In crops that produce pollen, such as sweet pepper, the predatory mites *Amblyseius cucumeris* and *A. degenerans* are introduced at an early stage so that a large population of these predators can develop on pollen and be ready to suppress the thrips as soon as they appear. The predatory mites remain present in the crop throughout the period of cultivation, even when thrips have been reduced to very low levels.
As in crops like cucumber that produce no pollen the predatory mite is unable to develop a population as explosively as the thrips, it is necessary to deal with such an infestation at an early stage. If the thrips population is to be held at a low level, a preventive strategy of regularly introducing large numbers of predatory mites from the very beginning of the culture is needed.
The predatory mites can be introduced into many vegetable and flower crops to control thrips.
A. cucumeris is supplied in bags, in shakers or buckets. Using shakers they can be distributed evenly and simply over the crop. The bags can be hung on plants, with each containing a culture of predatory mites that

can spread to establish themselves in the crop over several weeks. The international trade name of the product is THRIPEX for *A. cucumeris* in shakers or buckets and THRIPEX-PLUS for *A. cucumeris* in bags.
A. degenerans is only supplied in shakers under the international trade name THRIPANS.

Orius spp. are in most cases used in combination with predatory mites. *O. laevigatus*, a real flower visitor, can establish itself before thrips arrive in most crops that produce pollen. *O. majusculus*, a species which is more generally distributed over the entire plant, is released in crops that do not produce pollen, such as cucumber, as soon as thrips are observed. The choice of species also depends on the time of introduction, since *O. majusculus* can be introduced later than *O. laevigatus* on account of the difference in sensitivity to day-length. Predatory bugs are supplied as adult specimens and nymphs in shakers under the name THRIPOR (THRIPOR *laevigatus* and THRIPOR *majusculus*). With these shakers they can be evenly distributed over the crop.

The fungus *Verticillium lecanii* also attacks thrips. The fungus has no effect on natural enemies and can therefore be usefully employed as a back-up when predatory bugs and/or predatory mites no longer give adequate control. The dependence of the fungus on high humidity can be reduced by using a specific plant-based oil that protects the spores against desiccation. The international trade name of the fungal preparation is MYCOTAL, and the name of the oil-based adjuvant is ADDIT.

In ornamental crops, the predatory mites *Hypoaspis aculeifer* and *H. miles* are widely employed. These mites consume thrips pupae found on the ground. *Hypoaspis* spp., under the name ENTOMITE (ENTOMITE *miles* and ENTOMITE *aculeifer*), are sold in shakers containing all stages of the predatory mites along with carrier material.

So far, there has been little experience of successful biological control of *Echinothrips americanus*. The larvae of this thrips species are generally larger than the larvae of the other species discussed, in addition to which they are found more on the underside of leaves. Although *Orius* spp. and *Macrolophus caliginosus* can probably contribute to the biological control of this species, neither is wholly satisfactory. When introduced preventatively, *M. caliginosus* can prevent *Echinothrips americanus* populations from becoming established.

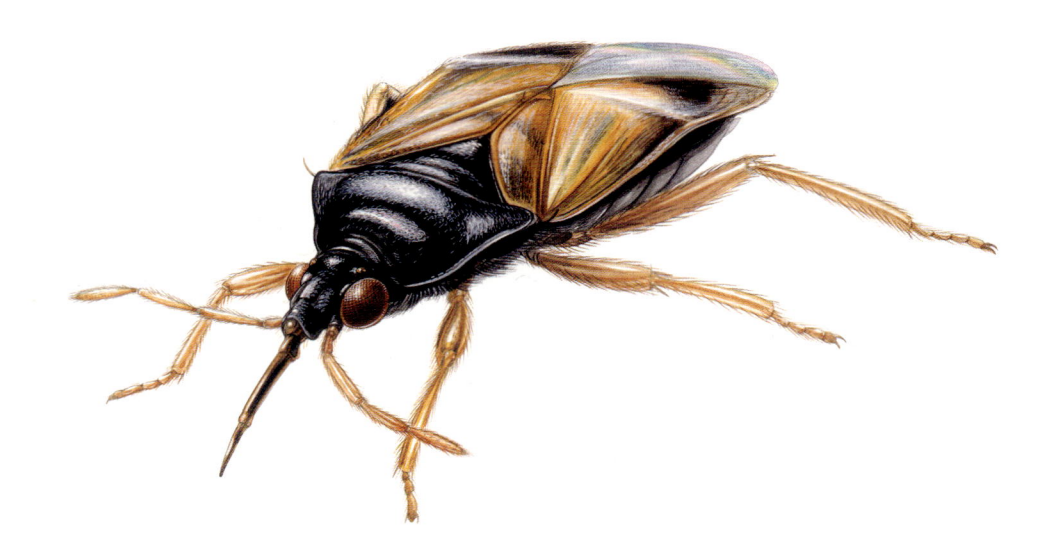

6. Leaf miners and their natural enemies

There are many species of leaf miner affecting various crops in temperate parts of the world. Under natural conditions, the larvae of these species are parasitized by parasitic wasps and so cause few problems. However, the use of chemical insecticides kills these natural enemies, allowing the leaf miner population to erupt into serious numbers. In addition, the pesticides used to combat leaf miners disrupt the biological control of other crop pests.

Leaf miners of the genus *Liriomyza* in particular have become a problem. Three species are now pests in Europe, of which the first, the tomato leaf miner *Liriomyza bryoniae*, is an indigenous species. Since 1976, *Liriomyza bryoniae* has been a pest of fruiting vegetables and ornamental crops. The other two species were imported from subtropical areas and have prospered under European glasshouse conditions. The Florida leaf miner (*L. trifolii*) has caused problems in glasshouses since the early 1980's, with the pea leaf miner (*L. huidobrensis*) first found in northern Europe in 1989. These species are resistant to many insecticides and can cause serious problems in various vegetable and ornamental crops.

The biology of *L. bryoniae*, *L. trifolii* and *L. huidobrensis* is described below. The chrysanthemum leaf miner, *Chromatomyia syngenesiae*, which mainly causes problems on ornamental plants is described later on, followed by an account of two indigenous parasitic wasps, *Dacnusa sibirica* and *Diglyphus isaea*, both of which are used in the biological control of these species. A third parasite, *Opius pallipes*, is also discussed. This wasp can also occur in glasshouses where it parasitizes the tomato leaf miner. Finally the predatory bug *Macrolophus caliginosus* can also make a significant contribution to the control of leaf miners. This insect is described in chapter 4.

Knowing and recognizing Leaf miners

Leaf miners

6.1. Liriomyza bryoniae

6.2. Liriomyza trifolii

6.3. Liriomyza huidobrensis

6.4. Chromatomyia syngenesiae

The most important leaf miners in crops under glass are:
Liriomyza bryoniae
Liriomyza trifolii
Liriomyza huidobrensis
Chromatomyia syngenesiae

Leaf miners belong to the order Diptera (the true flies) and form the family Agromyzidae. This is a family of small flies whose larvae mainly tunnel into the leaves of plants, creating 'mines'.

The leaf miner species that cause damage in horticultural crops are mostly polyphagous; that is, they feed on many crops. However this is not universal among the Agromyzidae. Of the approximately 2,500 species in this family, only eleven are truly polyphagous, five of which belong to the genus *Liriomyza*. The species that cause most damage all belong to this genus and are very common in temperate regions. The species that are the most important pests in European glasshouses are *L. bryoniae*, *L. huidobrensis* and *L. trifolii*. Larvae of the genus *Liriomyza* are distinct from the larvae of other leaf miners in that they pupate in the ground and not in the 'mines' in the leaf. The 300 or more different species of *Liriomyza* are very similar.

In ornamental crops the chrysanthemum leaf miner *Chromatomyia syngenesiae* can also be found. This is an indigenous species which attacks almost all crops of the order Compositae (such as gerbera and chrysanthemum).

The four species of leaf miner discussed in this chapter are compared in the table on page 120.

Life-cycle and appearance
Leaf miner

1 egg
2 larva
3 pupa
4 adult

The life-cycle of a leaf miner consists of the egg stage, three larval instars, a pupal instar and the adult fly.

Adult leaf miners are small yellow and black coloured flies, at most only several millimetres long. Because the colour of the adults of different *Liriomyza* spp. can vary it is often difficult to distinguish one from another. In cases of doubt, microscopic examination of the male genitalia and other laboratory techniques will give a definitive identification. The larvae can be used to determine species, but the characteristics are not completely reliable. When an adult female feeds or lays eggs, she bores a hole using her toothed ovipositor, usually in the upper side of the leaf. Often no egg is laid in this hole, but the plant exuding sap is consumed. In this case the hole is known as a feeding spot, since it is visible to the naked eye as a spot on the leaf. If the female lays an egg in this hole it is referred to as an egg spot. Feeding spots are round, whereas egg spots are oval and hard to detect. Males have no ovipositor and are therefore dependent on females to create feeding spots. Both males and females also feed on nectar from flowers and on honeydew.

When the larva hatches from the egg, it begins to eat into the leaf at once, tunnelling down into the mesophyll tissue where damage is caused by extensive mines, but leaving the outer layers of the leaf and stalk intact. Thus the larvae do not come into contact with the outside air. Apart from their increasing size, the three larval instars are difficult to distinguish. If the leaf is too small to provide sufficient food for larval development the larva can migrate through the leaf stalk to another leaf, but a larva cannot enter another leaf from outside.

In *Liriomyza* species, before pupating the grown larva cuts a sickle-shaped exit hole in the leaf with its mouthparts. After roughly one hour the larva crawls out of the leaf and falls to the ground. This occurs in the early morning. The larva crawls into the ground (to a depth of approximately 5 cm) or in the case of an artificial substrate between the folds of the plastic culture, and pupates. A small percentage of the larvae remain hanging on the leaf and pupate there, sometimes on the upper surface but more usually on the underside. The late third instar larva, when it has emerged from its tunnel just prior to pupating, is known as a prepupa. This period lasts only a few hours.

In contrast to the larvae of *Liriomyza* spp., the larvae of the chrysanthemum leaf miner pupate in the mine.

The colour of the pupae of the four species can vary from light yellow to black.

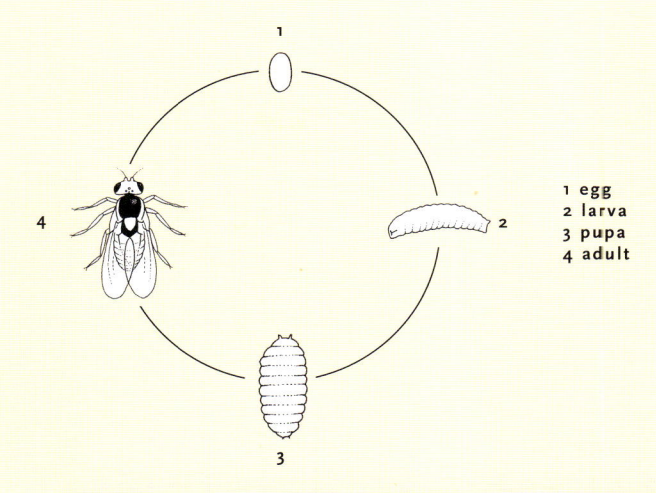

6.5. Egg spots

6.6. Larva

6.7. Pupae

6.8. Adult

Population growth
Temperature, light, the species and quality of host plant, the presence of parasites, population density and (to a lesser extent) humidity all influence the development of a population of leaf miners.
A leaf miner population usually consists of 50% females. Adults become active at sunrise and they are most active during the morning. Activity ceases at dusk.
Mating takes place within 1 – 2 days of the adults emerging. A single mating is sufficient to fertilize all the eggs laid, although maximum fecundity requires multiple matings. Unfertilized females lay no viable eggs.

Overwintering
Few adults develop during the winter, either because pupae enter diapause or because of retarded development. *L. bryoniae* and *L. huidobrensis* pass through diapause and can also overwinter outdoors, unlike *L. trifolii*. *C. syngenesiae* can also overwinter outdoors.

Damage
Leaf miners cause damage to plants both directly and indirectly. The most direct damage is caused by the larvae mining the leaf tissue which can lead to desiccation, premature leaf-fall and cosmetic damage. In (sub)tropical areas this can lead to burning in fruit such as tomato and melon. Loss of leaves also reduces yield. In full-grown plants of fruiting vegetable crops, however, a considerable quantity of foliage can be lost before the harvest is affected. The size of a leaf tunnel depends on the stage of development of the leaf, the species of host plant and the species of leaf miner. The older larvae make wider tunnels. Feeding spots made by adult females can also reduce yield, although with the exception of ornamental crops, this is usually of less significance. Seedlings and young plants can be totally destroyed as a result of the direct damage caused by leaf miners. The effects of leaf miners can be seen in plate 6.9.
The relationship between population size, leaf damage and yield reduction varies according to the season, culture method and the susceptibility of the host plant. This susceptibility can also vary considerably from one cultivar to another. So far, researchers have been unable to predict yield losses on the basis of the visible damage.
Indirect damage arises when disease causing fungi or bacteria enter the plant tissue via the feeding spots.

6.9. Damage caused by leaf miner

Overview (tomato)

Egg- and feeding-spots (tomato)

Mines (tomato)

Mines (cucumber)

- A commonly occurring pest
- Can remain present in the glasshouse the whole year
- Occurs on many crops

Liriomyza bryoniae
Tomato leaf miner

Liriomyza bryoniae (the tomato leaf miner) has many host plants and has caused crop damage in many parts of the world, including North Africa, Europe and northern Asia. In southern Europe the insect is found outdoors but in the rest of Europe it only occurs in glasshouses.
The tomato leaf miner has been found in glasshouses for decades, but has become a common pest since the 1960's. It is also a pest in sweet pepper, lettuce, melon, chrysanthemum and gerbera. The insect can be readily controlled using chemical pesticides, but this produces problems in glasshouses where other pests are controlled biologically.

Life-cycle and appearance
Eggs of *L. bryoniae* are on average 0.12 x 0.27 mm long, milky white and oval.
A larva which has just emerged from the egg is transparent and about 0.5 mm long. An early second instar larva is a rather dirty white colour, about 1.0 mm long, with the gut and its contents clearly visible. In the third larval instar the larva remains this dirty white colour but has a yellow head. The gut contents are visible and green-black in colour. Fresh from its moult, a young third instar larva is about 2 mm long, but soon increases in length and girth.
The colour of the pupa varies from golden yellow to dark brown, and even to black (plate 6.10). A pupa measures an average 0.9 x 2 mm. The adult insects are small, shiny yellow and black coloured flies (plate 6.11). They have a yellow head with red eyes and a conspicuous clear yellow dorsal thoracic shield. A male measures about 1.5 mm, whereas females measure 2 to 2.3 mm. The female has a clearly visible black spot on the abdomen. The ventral surface and legs of the adult are generally light yellow.

Population growth
The population growth of *L. bryoniae* depends, among other factors, on temperature and the host plant. The lower threshold for development is between 8°C and 10°C. Above 35°C development ceases. Data relating to population growth are given in table 6.1.
Mating usually takes place 1 - 2 days after the adult emerges from the pupa, and is followed by egg-laying. The number of eggs laid is particularly dependent on temperature and host plant. The sex ratio is roughly 1:1.
Adult males have a shorter lifespan than females, living for only a few days.
Although many overwintering pupae will die, it has been shown that a proportion are able to survive over winter in temperate climates.

Damage
The pattern of damage caused by *L. bryoniae* larvae depends on the host plant. In general, they tunnel over the whole leaf, the tunnels becoming wider as the larvae grow larger. The larvae mostly remain active in the upper mesophyll tissue of the leaf.

6.10. *Liriomyza bryoniae* pupae

6.11. *Liriomyza bryoniae*

Table 6.1. The population growth of *Liriomyza bryoniae* in tomato at 60-70% relative humidity (Minkenberg, 1990; Nedstam, 1985)

temperature (°C)	12	15	18	19,5 (22-16)*	20	25	30
development time (days)							
egg	13.1	6.1	6.9	4.0	4.2	3.0	3.0
larva 1		4.6		2.0	3.3	1.4	
larva 2		3.7		3.1	2.5	2.0	
larva 3		4.0		3.0	2.7	1.6	
egg to pupation	32.2		16.0				6.2
pupa		22.2		14.4	13.9	9.2	
egg to adult	77.8	40.6	32.4	26.5	26.6	17.2	14.2
po-period ** (days)		2.5		1.8	1.6	1.1	
mortality (%)							
larva 1		61		6	3	3	
larva 2		20		6	3	13	
larva 3		4		4	3	5	
pupa		37		23	29	18	
total		80		38	35	34	
eggs/♀							
		92		65	144	163	
eggs/♀/day							
		6.7		7.7	15.2	23.3	
lifespan ♀ (days)							
		13.6		6.9	9.0	6.6	

* day/night temperature
** po-period = pre-oviposition period, *i.e.* period from becoming adult to first egg-laying

Table 6.2. The population growth of *Liriomyza trifolii* in tomato at 60-70% relative humidity (Minkenberg, 1990)

temperatuur (°C)	15	19,5* (22-16)*	20	25
development time (days)				
egg	6.6	3.8	3.1	2.7
larva 1	3.3	3.4	2.8	1.4
larva 2	3.7	2.2	2.1	1.4
larva 3	3.7	2.4	2.3	1.8
pupa	26.8	16.8	15.0	9.3
egg to adult	44.1	28.6	25.3	16.6
po-period ** (days)	2.4	1.8	1.2	1.3
mortality (%)				
egg	23	12	20	21
larva 1	45	5	22	18
larva 2	29	2	9	19
larva 3	0	1	1	9
pupa	8	20	7	15
total	73	36	48	60
egg/♀				
	5	34	79	59
egg/♀/day				
	0.8	4.5	5.9	9.1
lifespan ♀ (days)				
	6.5	7.4	14.4	5.6

* day/night temperature
** po-period = pre-oviposition period, i.e. period from becoming adult to first egg-laying

- Has caused problems since 1982
- Appears mainly in the summer
- Occurs on many crops

Liriomyza trifolii
American serpentine leaf miner

The american serpentine leaf miner *Liriomyza trifolii*, a leaf miner originating in north America, has been known in northern Europe since about 1976, and is thought to have been introduced on infected chrysanthemum cuttings imported from Florida. Until 1980 the insect was only seen in chrysanthemum and gerbera, but has since become a pest in vegetable crops. *L. trifolii* is highly polyphagous and thus lives on many host plants such as chrysanthemum, gerbera, gypsophila, celery, sweet pepper, pea, bean and potato. Today the insect is widespread across the whole world. It is insensitive to many chemical pesticides and is capable of rapidly developing resistance to new products.

Life-cycle and appearance
The appearance of the american serpentine leaf miner is very similar to that of the tomato leaf miner. At first the eggs are oval and transparent, but later become cream-coloured. They are about 0.1 - 0.2 mm long. The larvae, unlike those of *L. bryoniae*, are completely ochre-yellow in colour, particularly in the last instar. The pupae and adults of the two species, however, are virtually indistinguishable, although adult american serpentine leaf miner have a less shiny appearance than those of *L. bryoniae*.

Population growth
In common with *L. bryoniae*, the development time of *L. trifolii* depends mainly on temperature and host plant. The total development time is much the same in both insects (see tables 6.1 and 6.2, p.116).
The lower threshold for development in *L. trifolii* lies between 8 and 10°C depending on the crop.
The number of eggs laid by a female of *L. trifolii* depends on the temperature and the crop (see table 6.2).
The sex ratio of *L. trifolii* varies from about 2:1 (♀:♂) to 1:1. Male adults have a shorter lifespan than females, living for only a few days.
It is evident from table 6.2 that *L. trifolii* develops rather more slowly than *L. bryoniae* on tomato at normal temperatures. This species also has a higher rate of mortality and a lower reproductive capacity. As a result, under temperate conditions, the population growth of *L. bryoniae* on tomato will be faster than that of *L. trifolii*. *L. bryoniae* is therefore a pest in tomato, whereas *L. trifolii* is less frequently a problem in this crop. In warmer regions the situation is rather different, as *L. trifolii* has a higher optimal temperature.
On chrysanthemum, *L. trifolii* lays far more eggs than on tomato (275 - 300 on chrysanthemum as opposed to a maximum of 80 on tomato). The lifespan is also several days longer on chrysanthemum than on tomato.

Damage
The damage caused by *L. trifolii* is very similar to that caused by *L. bryoniae*. It is often impossible to determine which insect is present on the basis of the mines and feeding spots. In gerbera, however, a larva of the american serpentine leaf miner eats its way outwards around the egg spot, so that its mines join to form small plates. In various other crops one finds intermediate forms of tunnelling between these 'plate mines' and normal mines, making it an unreliable criterion on which to identify the species.

6.12. Larva

6.13. Damage in aubergine

- Present in northern Europe since 1989
- Can appear throughout the year
- Highly polyphagous

Liriomyza huidobrensis
Pea leaf miner

The pea leaf miner, *Liriomyza huidobrensis*, occurs in a variety of vegetable and ornamental crops. It originates from South and Middle America, from which since 1985 it has spread over parts of the United States and Hawaii. In 1989 the insect was brought into northern Europe by unknown means, and its subsequent spread was via plant material. This species can be particularly damaging to leaf vegetables and causes problems in lettuce grown on open ground.

Many commonly used insecticides are ineffective against this leaf miner, although biological control can be successful.

6.14. *Liriomyza huidobrensis*

Life-cycle and appearance

The life-cycle of *L. huidobrensis* is virtually the same as that of both other species of leaf miner discussed here. This species is closely related to the tomato leaf miner. The adult of *L. huidobrensis* is slightly larger (1 - 2 mm), darker in colour, but also has the bright yellow thoracic shield. Larvae are slightly whiter than those of the tomato leaf miner (milky to yellowish white) and are at most 3.5 mm in length. A binocular microscope is usually necessary to distinguish the two species as larvae. Pupae are about 2.2 mm long. Pupae in a single crop and of the same age can nevertheless vary enormously in colour, from pale yellow to black.

Plate 6.14 shows an adult *L. huidobrensis*.

Population growth

Under favourable conditions a population of *L. huidobrensis* can quickly build up. In outline, the relevant data concerning population growth are the same as for the tomato leaf miner and the Florida leaf miner, although possibly development in this species is faster at higher temperatures than in the other *Liriomyza* species. At around 19°C the egg stage lasts 2 days, the larval period 6 days, pupa 8 days, with a pre-oviposition period of 1 - 2 days. At this and slightly warmer temperatures the life-cycle lasts 2 to 2.5 weeks

It appears that this species is slightly more resistant to cold than the Florida leaf miner and that pupae can overwinter in a cold, empty glasshouse for at least 9 weeks, probably longer. Although in Europe many such overwintering pupae die (mortality is about 90%), a proportion do survive. Several data relating to the population growth *L. huidobrensis* are given in table 6.3.

Damage

The kind of damage caused by *L. huidobrensis* is distinct from that caused by other leaf miners. The female lays her eggs in the leaf tissue on the upper surface of the leaf. Mining often begins on the upper surface of the leaf, but thereafter the larva tunnels down toward the underside. The mines of *L. huidobrensis* often run along mid-veins and side-veins but can also run irregularly over the leaf. When more mines appear on a leaf, a large 'plate-mine' may be formed. These are mostly located at the base of the leaf. Because the greater part of the mine is on the underside of the leaf, the damage is less conspicuous and its effect are discovered later than with other leaf miners. This is particularly the case in summer. In the winter months, in many crops the mines are found more on the upper side of the leaf.

L. huidobrensis makes an enormous number of feeding spots on the leaves. The number of spots depends on the crop and varies from an average of 650 per female in tomato to more than 1,300 per female in chrysanthemum. At high levels of infestation, *L. huidobrensis* causes greater damage to a crop than *L. trifolii*.

Distribution and dispersal

In the glasshouse, a population of the pea leaf miner is well dispersed, even when there are few individuals present. This differs from the distribution of *L. trifolii* which usually occurs in localized concentrations.

Table 6.3. The population growth of *Liriomyza huidobrensis* on chrysanthemum at 26.7°C and 50-60% relative humidity (Parrella & Bethke, 1984)

development time (days)		mortality (%) during development	
egg	3.0		64.0
larva	4.7		
pupa ♀	9.3	lifespan ♀ (days)	
pupa ♂	8.5		12.2

- On the increase in recent years
- Particularly found in ornamental crops
- Pupates inside the mine

Chromatomyia syngenesiae
Chrysanthemum leaf miner

The chrysanthemum leaf miner (*Chromatomyia syngenesiae*), a species that is indigenous to Europe, is known to attack plants belonging to the family Asteraceae, of the order Compositae. The chrysanthemum leaf miner is especially important as a pest of chrysanthemum, gerbera and various other ornamental crops, and also in lettuce where it can cause serious damage. *Chromatomyia syngenesiae* was previously known as *Phytomyza syngenesiae*, and earlier still as *P. atricornis*.

The species is widespread in Europe, particularly in the more northern countries. Its original area was probably West Europe from where it spread to the other areas. It is now also found in North America, Japan, New Zealand and Australia.

The chrysanthemum leaf miner is highly susceptible to pesticides, and its increase in recent years is probably due to the reduction in the use of these chemicals.

Life-cycle and appearance

This species can be distinguished from the other economically important leaf miners by the fact that it pupates inside the leaf, in the mine (see plate 6.15).

The eggs are oval, white and smooth, measuring 0.35 x 0.15 mm. A full-grown larva is 3.5 mm long and greenish white in colour. The pupa is oval-shaped, slightly flattened, greyish white to light brown in colour, 2.5 mm long at most, and is clearly visible in the mine if the leaf is held against the light. The adult eventually emerges through a round hole at the tip of the pupa. Adults are about 2 mm long with a greyish white body and light yellow patches on both sides and the ventral surface. They are paler than the other species of leaf miner and are not shiny. They are easily distinguished from the other species of leaf miner because they lack the typical yellow dorsal shield (see plate 6.16), and as a result are also more easily overlooked. They travel only very short distances, usually only a few centimetres at a time.

Population growth

The development time, as in the other leaf miner species, is mainly dependent on the temperature and host plant. At a lower temperature the development time is slightly shorter than in other species.

When food is present, adult females can live for 2 - 3 weeks, but in the absence of food they live only 3 - 4 days. Each female lays about 75 eggs. Mating is essential for the production of viable eggs, and although unfertilized females may lay eggs, these do not hatch.

An infestation can occur at any time in cultures that are maintained all the year round. Chrysanthemum cuttings can be affected at an early stage. Infestations can build up outside throughout the summer and in this period a glasshouse may be infected by adults flying in from outside. Damage seems to decline in the autumn. If there are no parasites present, the population can expand very rapidly. It is known that scores of larvae can develop on a single leaf. Out of doors, the first generation of flies appears in May and the second at the end of July. The last generation each year overwinters in the pupal stage.

Damage

The extent of damage varies from one season to another, but the reason for this is unknown. The larva tunnels into the leaf creating mines, mostly near the mid-vein. There are often several mines on a single leaf, which may intersect many times to create the effect of a large plate mine. This is particularly the case in gerbera.

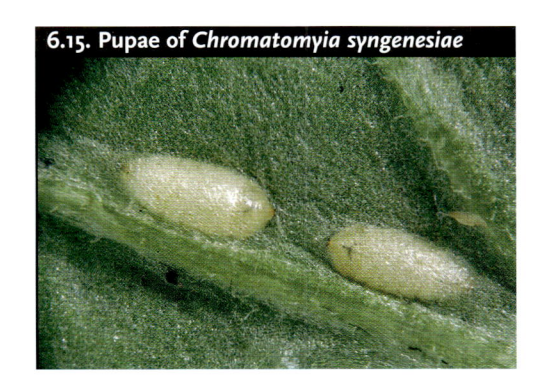

6.15. Pupae of *Chromatomyia syngenesiae*

6.16. *Chromatomyia syngenesiae*

Differences between various leaf miner species

	Liriomyza bryoniae tomato leaf miner	*Liriomyza trifolii* american serpentine leaf miner	*Liriomyza huidobrensis* pea leaf miner	*Chromatomyia syngenesiae* chrysanthemum leaf miner
adults				
size	1.5-2.3 mm	1.5-2.3 mm	1.7-2.5 mm	1.5-2.5 mm
colour	glossy, black and yellow, yellow thoracic shield	mat, grey-black and yellow, no thoracic shield	glossy, black and yellow, appears darker than other species, yellow thoracic shield	mat, grey-black and light yellow, appears paler than other species, no yellow shield
larvae				
colour	transparent to dirty white, in last instar a yellow head (green gut contents visible)	ochre yellow, not transparent (gut contents not visible)	milk-white to cream, not transparent (gut contents not visible)	creamy white
pupae				
colour	golden yellow to dark brown or even black	mostly yellow-brown	pale yellow to black mostly yellowish to reddish brown	yellow to dark brown
location	in the ground or hanging on a leaf	in the ground or hanging on a leaf	in the ground or hanging on a leaf	in the mine
mines				
form and location	over the whole leaf on one side (usually the upper side)	sometimes round feeding spots, not along veins, on one side (usually the upper side)	along the vein, on both sides, most visible on the underside, sometimes plate mines	usually plate mines, over the whole leaf
optimal temperature for development	20-25°C	± 27°C (25-30°C)	± 23-27°C	20-25°C
important host plants	vegetable and ornamental crops	vegetable and ornamental crops	particularly (leaf) vegetable crops	ornamental crops, particularly chrysanthemum and gerbera
period of occurrence	whole year	in the summer	infrequent, but can occur throughout the year	whole year
susceptibility to insecticides	susceptible	highly resistant	highly resistant	susceptible

Knowing and recognizing **Leaf miners**

Natural enemies of leaf miners

6.21. Dacnusa sibirica

6.22. Diglyphus isaea

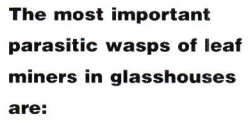

6.23. Opius pallipes

The most important parasitic wasps of leaf miners in glasshouses are:

Dacnusa sibirica
Diglyphus isaea
Opius pallipes

The most important wasps that parasitize leaf miners in glasshouses are *Dacnusa sibirica*, *Diglyphus isaea* and *Opius pallipes*.
Other natural enemies of leaf miners are known, but these three parasitic wasps have been found to be most suitable for biological control in glasshouses. *D. isaea* and *D. sibirica* are both released in the glasshouse, whereas *O. pallipes* occurs naturally.
Like all parasitic wasps, these three species belong to the order Hymenoptera and parasitize the larval stage of their host. They are native to this part of Europe. *D. isaea* is an ectoparasite, which means that it lays its egg beside the host, whereas the other two species are endoparasites, laying their egg inside the host.
D. isaea and *D. sibirica* parasitize all four species of leaf miners found in glasshouses in Europe, while the parasitization of *L. trifolii* by *O. pallipes* is inhibited by the leaf miner larva encapsulating the parasitic wasp egg and rendering it harmless.
The predatory bug *Macrolophus caliginosus* can also contribute to the control of leaf miners by preying on their larvae. If there are numerous leaf miners, many predatory bugs and a shortage of whitefly, the leaf miner population can be considerably reduced. This predatory bug is discussed in chapter 4.

- Ectoparasite; females lay their eggs beside the leaf miner larvae
- May be introduced, but also occurs naturally
- Also kills leaf miner larvae by host feeding

Diglyphus isaea

Diglyphus isaea is an ectoparasite (a parasite that lays eggs beside, not inside the host) belonging to the family Eulophidae. *D. isaea* is native to Europe, North Africa and Japan, but has now been introduced in other regions throughout the world. *D. isaea* can parasitize many species of leaf miner, and has been used commercially since 1984. From May onwards, *D. isaea* can appear spontaneously in the glasshouse.

Population growth
Above 20°C a population of *D. isaea* grows faster than that of either its hosts or of endoparasitic wasps. For this reason, at higher temperatures this ectoparasite gives better control of leaf miners. The sex ratio of this wasp can vary widely. One or two days after emerging from the pupa, once she has fed sufficiently on host larvae, the female parasite begins to lay eggs. The number of eggs laid depends particularly on the temperature.
D. isaea is indigenous to Europe and enters glasshouses naturally from May onwards. This parasitic wasp is capable of overwintering outdoors. Table 6.4 presents data relating to the population growth of *D. isaea*.

Parasitization
The presence of *D. isaea* in the crop is signalled by short mines in the leaves that contain dead larvae. The parasitic wasp punctures mainly first and second instar larvae in order to feed on their body fluids. This is known as host feeding, and is necessary for the production of eggs.
A certain density of leaf miner larvae is therefore necessary for *D. isaea* to be able to reproduce.
Once parasitized by the wasp, the leaf miner larvae cease feeding. Shortly before the host is finally inactivated it expels the gut contents; a pierced larva can thus often be recognized by an extra quantity of excreted frass. Although the development of the leaf miner ceases at this point, it does not die immediately.
Endoparasites of leaf miners die if the host is parasitized by *D. isaea*. As a result, the latter is capable of totally supplanting *O. pallipes* and *D. sibirica*. Data relating to the control of *L. bryoniae* by *D. isaea* are presented in table 6.5.

Table 6.4. The population growth on tomato of *Diglyphus isaea* (D.i.), and for comparison of its hosts *Liriomyza trifolii* (L.t.) and *Liriomyza bryoniae* (L.b.) at different temperatures and 60-80% relative humidity (Minkenberg, 1990)

	temperature (°C)			host
	15	20	25	
development time (days)				
egg-adult D.i.	26.0	16.6	10.5	L.t.
egg-adult D.i.	25.5		9.8	L.b.
egg-adult L.t.	44.1	25.3	16.6	
egg-adult L.b.	40.6	26.5	17.2	
po-period * D.i.	1.4	2.2	1.1	
% mortality (egg-adult)				
D.i.	54	18	23	L.t.
lifespan				
D.i.	23	32	10	L.b.
L.b.	13.6	9.0	6.6	
eggs/♀				
D.i.	293	286	209	L.b.
L.b.	92	144	163	
eggs/♀/day				
D.i.	12.7	9.0	18.9	L.b.

* po-period = pre-oviposition period, i.e. period from becoming adult to first egg-laying

Table 6.5. The control of *Liriomyza bryoniae* by *Diglyphus isaea* at different temperatures (Minkenberg, 1990)

temperature (°C)	15	20	25
total number of eggs per female	293	286	209
host feeding (number of dead larvae)	192	70	73
lifespan (days)	23	32	10
host feeding per day	8	2	6
+ eggs laid per day	13	9	19
= total reduction per day	21	11	25

Life-cycle and appearance
Diglyphus isaea

1 adult
2 egg
3 larva
4 pupa

Because *Diglyphus isaea* is an ectoparasite, all instars (egg, larva, pupa and adult) develop outside the host.

A female wasp first paralyzes the leaf miner larva and then lays a single egg (occasionally more) next to it. In general, the wasp selects late second and third instar larvae to parasitize. An egg has an elongated shape, measuring 0.3 x 0.1 mm, and is transparent white in colour. It lies alongside the paralyzed leaf miner larva and is often difficult to spot.

When the parasite larva hatches, it remains beside its host, punctures it and sucks it empty from the outside, as a result of which the leaf miner larva dies. A few days after being parasitized, the leaf miner larva turns flaccid and brown. Older larvae of *D. isaea* leave their dead host and crawl back into the mine to pupate. There are three larval instars: the first is colourless and transparent, the second is yellow and semi-transparent with a brown fat body, and the third larval instar is blue-green (plates 6.24 and 6.25).

The full-grown larva usually pupates in the mine some distance from the dead host. The pupa lies between the upper and lower epidermis of the leaf which are held apart by six pillars made from its faeces or frass. This is probably done to prevent the leaf collapsing should it dry out.

At first, the pupa is green with red eyes (plate 6.26) subsequently turning black, and measuring about 1.5 mm. The appearance of the adult is already clearly visible. Pupae are often visible through the leaf.

The parasitic wasp escapes from the mine by making a round hole through the upper epidermis. These holes are the evidence that a parasitic wasp rather than a leaf miner has come from the mine. This is also seen in the pillars that remain visible as black spots in the leaf. The adult is metallic green to black in colour and unlike *O. pallipes* and *D. sibirica* has short antennae (plate 6.27). The female is generally rather larger than the male and can be recognized by a broad black stripe over the hind leg, whereas males have two small black bands.

Differences between the various instars of *D. sibirica*, *O. pallipes* and *D. isaea* are set out in the table on page 127.

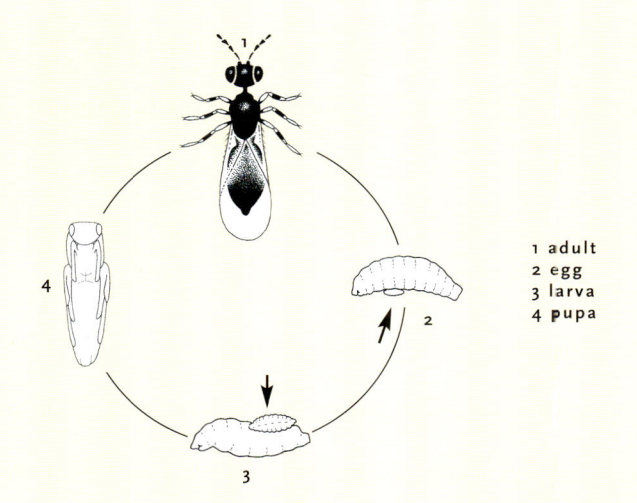

6.24. Leaf miner larva with young larva

6.25. Leaf miner larva with older larva

6.26. Young pupa

6.27. Adult

- Endoparasites: females lay their eggs in the larvae of leaf miners
- Both species occur naturally, but *Dacnusa sibirica* is also introduced
- Mainly effective at lower temperatures

Dacnusa sibirica & Opius pallipes

Both *O. pallipes* and *D. sibirica* belong to the superfamily Ichneumonoidea and the family Braconidae (the braconid wasps). *D. sibirica* belongs to the subfamily Alysiinae, one of the largest braconid subfamilies. *O. pallipes* belongs to the Opiinae, a subfamily closely related to the Alysiinae. Both species occur naturally in temperate regions and can appear spontaneously in glasshouses in all seasons. Since 1981, *D. sibirica* has been introduced to combat leaf miners.

Population growth

The development time and reproductive capacity of *O. pallipes* are very similar to *D. sibirica*. Both parasites are good controllers of leaf miner at lower temperatures. The population growth of both these parasitic wasps is faster than that of the leaf miners. Both parasites are able to overwinter in the glasshouse within the leaf miner pupae. Natural control is thus possible in the spring.

The development time of both parasites is shorter than that of the leaf miners. The time taken for eggs and first instar larvae to develop depends to a large extent on the age of the host. In young hosts, these two stages last longer; the parasite only moults for the first time after the host has pupated. The adults mate immediately after emerging, following which the females can lay eggs. As well as temperature, the reproductive capacity depends on the nature of the crop. For example, *D. sibirica* lays fewer eggs on chrysanthemum than on lettuce. The sex ratio can vary, but under normal conditions is roughly 1:1. The reproduction of *O. pallipes* is very similar to that of *D. sibirica*. Various data concerning population development are given in table 6.6.

Parasitering

Although both *O. pallipes* and *D. sibirica* are highly efficient at detecting a plant affected by leaf miners, *O. pallipes* is slightly more efficient. Both species, once such a plant has been found, locate a suitable host very rapidly. They use the antennae and ovipositor to find a mine and a leaf miner larva. *O. pallipes* and *D. sibirica* both display host discrimination; that is, they can distinguish between parasitized and non-parasitized larvae. *D. sibirica* is even capable of distinguishing a leaf that has already been visited from one that has not. At high rates of parasitization, super-parasitization may occur when more than one egg is laid in a host larva. In such cases only one parasitic wasp will emerge from the leaf miner

Table 6.6. The population growth of *Dacnusa sibirica* (D.s.), *Opius pallipes* (O.p.) and *Liriomyza bryoniae* (L.b.) on tomato at a relative humidity of 60-80% (22°C: Hendrikse et al., 1980, other temperatures: Minkenberg, 1990; Nedstam, 1985)

temperature (°C)	12	15	18	20	22	25	27	30
development time egg-adult (days)								
L.b.	77.8	41.4	32.4		19.6		15.1	14.2
D.s.	54	32.1	26.6		15.7		13.4	12.8
O.p.					18.3			
lifespan adult (days)								
L.b.		13.6		9.0	8.7	6.6		
D.s.		20.2			11.4	7.4		
O.p.					10.6			
Number of eggs/♀								
L.b.		92		144	67.3	163		
D.s.		225		94	55.6	48		
O.p.					37.0			
Number of eggs/♀/day								
L.b.		6.7		15.2		23.3		
D.s.		11.4		14.2		8.6		

Life-cycle and appearance
Dacnusa sibirica & *Opius pallipes*

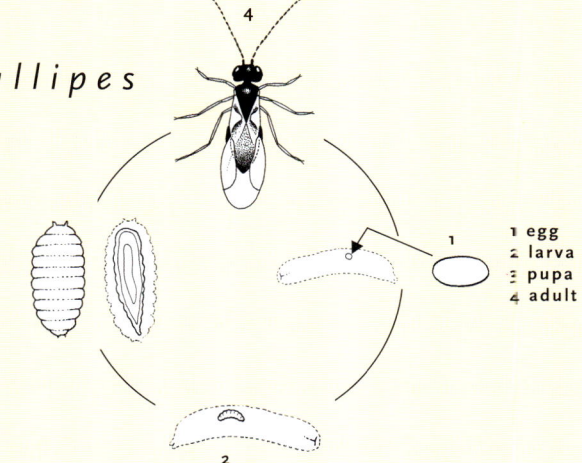

1. egg
2. larva
3. pupa
4. adult

The life-cycles of *Dacnusa sibirica* and *Opius pallipes* are identical and consist of an egg, a larva, a pupa and an adult.

In order to see the eggs and larvae of these endoparasites, the parasitized leaf miner larvae must be dissected; it is impossible to tell whether a leaf miner larva or pupa has been parasitized from external appearance.

Although all larval instars can be parasitized, both parasites show a preference for first and second instars. Their eggs are laid in the larvae and the adults emerge from the host pupae.

A newly laid egg of *O. pallipes* is transparent grey and elongated, whereas the egg of *D. sibirica* is oval and clear white (plate 6.28). After two days the egg of *O. pallipes* is much larger and has assumed a more oval shape, so that it is now more difficult to distinguish from an egg of *D. sibirica*.

The larvae of the two parasitic wasps can be clearly distinguished by the jaws and shape of the head. *D. sibirica* has a small, pointed head with small jaws (plate 6.29), whereas *O. pallipes* has a broad head with large jaws. Both larvae are transparent while the jaws are reddish brown. The larvae of the parasitic wasps only reach full development after pupation of the leaf miner larva. Identification at the pupal stage is impossible.

The adult parasitic wasps emerge from the leaf miner pupa by biting a hole in the pupal cuticle with their jaws. The orientation of both parasites within the host pupa is the same as that of the leaf miner itself, the head of the parasite thus developing where the head of the leaf miner originally was. Adults of *O. pallipes* and *D. sibirica* strongly resemble each other; they are both dark to black and 2 - 3 mm long (plates 6.30 and 6.31).

The wasps have long antennae with at least 16 segments. The two species can be distinguished by a difference in the venation of the forewing, and by the same criterion they can both be distinguished from *Aphidius* species, another genus of braconid wasps which parasitize aphids. The table on page 127 presents data for the various instars of *D. sibirica*, *O. pallipes* and *Diglyphus isaea*.

6.28. Egg of *Dacnusa sibirica*

6.29. Larva of *Dacnusa sibirica*

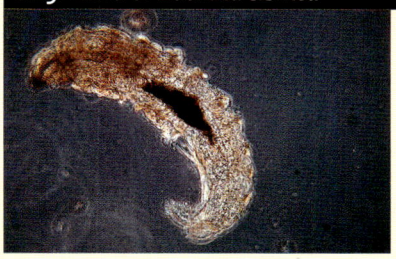

6.30. *Dacnusa sibirica*

6.31. *Opius pallipes*

pupa; although all the eggs hatch inside the host larva, the young larvae attack each other using their jaws. Wounded or dead larvae are then recognized by the host immune system and are encapsulated. Eggs are apparently not attacked.

O. pallipes is not very efficient at parasitizing middle-sized or large larvae. Their egg-laying is probably disrupted by the greater mobility of the larger larva, and further hindered by their thicker cuticle. *D. sibirica* has less trouble, and the larger larvae are certainly more easily located by the parasitic wasps.

The Florida leaf miner, *L. trifolii*, encapsulates the eggs of *O. pallipes*, rendering it an ineffective control measure for this species of leaf miner. Although *D. sibirica* also has more difficulty with the larvae of *L. trifolii* than with the larvae of *L. bryoniae*, it is perfectly capable of developing in the larvae of both species.

Differences between various stages of
Dacnusa sibirica, Opius pallipes & Diglyphus isaea

	Dacnusa sibirica	**Opius pallipes**	**Diglyphus isaea**
egg	- laid in leaf miner larva - oval - wholly grey-white	- laid in leaf miner larva - elongated form - white internally with transparent outer layer	- lies against the leaf miner larva in the leaf - elongated form - transparent white
larva	- in the larva of the leaf miner - older larvae have yellow contents 1: transparent tail 2: pointed head with dark coloured jaws jaw detail	- in the larva of the leaf miner - older larva almost transparent - blunt head with conspicuous red-brown jaws jaw detail	- young larva lies alongside the leaf miner larva - older larva crawls into the mine of leaf miner larva to pupate - transparent dark grey and orange-yellow in mid-region, the older the larva, the more the grey colour turns light green
pupa	no visible difference between *Dacnusa sibirica* and *Opius pallipes*	no visible difference between *Dacnusa sibirica* and *Opius pallipes*	first colourless, later light green with red eyes and finally black with red eyes
adult	♂ long antennae ♀ right wing detail overlapping hind wings in natural position with characteristic venation	long antennae right wing detail overlapping hind wings in natural position with characteristic venation	short antennae right wing detail

Knowing and recognizing Leaf miners

Biological control of leaf miners

OVERVIEW OF THE BIOLOGICAL CONTROL OF LEAF MINER

product name	natural enemy	controlling stage	host feeding	crops and remarks
MINEX	Dacnusa sibirica (90%) + Diglyphus isaea (10%)	adult	yes	all crops at low temperatures and infestation levels
DIMINEX	Dacnusa sibirica (50%) + Diglyphus isaea (50%)	adult	yes	all crops with increasing leaf miner infestation
MIGLYPHUS	Diglyphus isaea	adult	yes	all crops with increasing leaf miner infestation
MINUSA	Dacnusa sibirica	adult	no	all crops at lower temperatures and low leaf miner infestation

The extent to which parasitic wasps are active in the glasshouse is difficult to determine on site. Leaf samples are therefore collected and examined in the laboratory to identify the species of leaf miners and parasites present, and the level of parasitization. *D. sibirica* is especially recommended for situations where both the infestation of leaf miners and the temperature are relatively low, as in winter and spring. *D. isaea* is used at higher densities of leaf miner and at higher temperatures. Often the introduction of *D. isaea* is started in the spring to give the parasite time to build up a population in the glasshouse.

The parasitic wasps are supplied as adults in bottles, which allows for an even distribution in the glasshouse. The wasps can be used to control leaf miners in very many crops.

The parasitic wasps can be supplied as separate species under the international trade names MINUSA (*Dacnusa sibirica*) and MIGLYPHUS (*Diglyphus isaea*). Two products are available containing a mixture of both species, under the international trade names MINEX (90% *D. sibirica* and 10% *D. isaea*) and DIMINEX (50% *D. sibirica* and 50% *D. isaea*).

7. Aphids and their natural enemies

Aphids feed on plant sap. As a consequence of this, and their enormous reproductive capacity, many species of aphids inflict serious damage on a variety of crops. Furthermore, they are notorious virus vectors.

For a long time, in crops where natural enemies had been released against other pests, aphids could be controlled effectively by means of selective chemical pesticides that had very few adverse effects on biological control. But increasing resistance to these pesticides meant that a different approach was called for, the use of natural enemies being the obvious alternative. The biological control of aphids was thus first introduced into Dutch glasshouses in 1988.

A number of different species of aphids occur in glasshouses, each calling for a slightly different control strategy. In this chapter, we shall first look at the life habits and characteristics of aphids in general before discussing in more detail the important species in glasshouse horticulture. In an overview of these species, which also includes several less significant species, their specific characteristics will be described. This account is followed by a discussion of various predators, parasitic wasps and fungi that can contribute to biological control, together with information on hyperparasites of parasitic wasps whose presence can have a deleterious effect on aphid control.

Orius spp. that also contribute to the control of aphids are dealt with in chapter 5.

Knowing and recognizing **Aphids**

Aphids

7.1. *Myzus persicae*

7.2. *Aphis gossypii*

7.3. *Macrosiphum euphorbiae*

7.4. *Aulacorthum solani*

The most important species of aphids encountered in glasshouses are:

Myzus persicae subsp. *persicae*
Myzus persicae subsp. *nicotianae*
Aphis gossypii
Macrosiphum euphorbiae
Aulacorthum solani

Aphids form a single, very large group of insects: the super-family Aphidoidea, belonging to the order Hemiptera and the sub-order Homoptera. The aphids discussed here all belong to the family Aphididae, a family containing many species that cause damage in cultivated crops. The most significant aphids that occur in glasshouses are:
- *Myzus persicae* subsp. *persicae* (the peach potato aphid) and *Myzus persicae* subsp. *nicotianae* (the tobacco aphid) on various vegetable and ornamental crops,
- *Aphis gossypii* (the cotton aphid) mainly on Cucurbitaceae, but also on chrysanthemums, kalanchoë and sweet pepper,
- *Macrosiphum euphorbiae* (the potato aphid), mainly on Solanaceae and various ornamental crops,
- *Aulacorthum solani* (the glasshouse potato aphid), mainly on Solanaceae and various ornamental crops, e.g. begonia.

These are all polyphagous species, *i.e.* species with a wide range of host plants. They will be dealt with in some detail below.

In ornamental horticulture in particular there are several other aphid species that may cause damage. A few of these species are covered in the overview on page 140, where two species of grain aphid (*Sitobion avenae* and *Rhopalosiphum padi*) used in the open culture of natural enemies are also dealt with.

Population growth

The growth of aphid populations depends on several factors, including the species and quality of host plant, climatic conditions, population density and the presence of natural enemies. Under optimal conditions, 6 or 7 days is sufficient for young aphids to reach maturity.
The rate of reproduction among aphids can be rapid (see figure 7.5), and there are several reasons for this:
- they reproduce largely parthenogenetically,
- they are viviparous, *i.e.* young aphids develop within the female parent,
- under glasshouse conditions no males develop.

Because of their short generation time, the rapid rate of reproduction and the fact that under glass their reproduction is often continuously asexual, an infestation of aphids can very quickly reach serious levels. An individual female aphid produces between 40 and 100 progeny at a rate of 3 to 10 per day over several weeks. Because no fertilization is required, the development of offspring can begin immediately an aphid is born. Thus, by the time an aphid is fully grown, several young nymphs are already developed and ready to be born.

7.5. Asexual reproduction in aphids

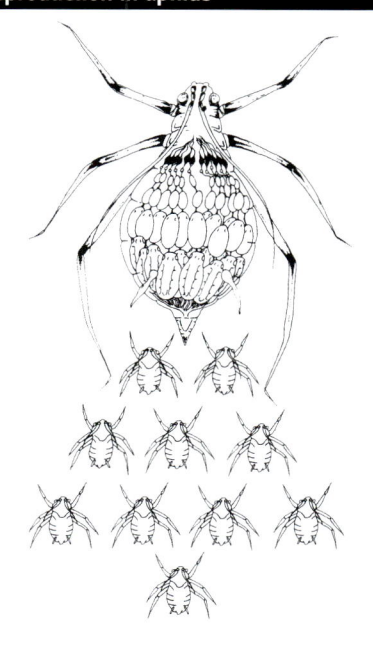

Damage

Depending on the species, aphids can inflict various kinds of damage on a crop (see plate 7.6):
- nymphs and adults extract food materials from the plant and disturb the balance of growth hormones. As a result, the plant's growth is retarded giving rise to deformed leaves or even, if the infestation occurs early enough in the season, the death of young plants. Retarded growth and defoliation reduce yield,

7.6. Damage caused by aphid infestation

Deformed growing tip of cucumber resulting from an infestation of *Aphis gossypii*

Shed aphid skins on tomato

Sweet pepper affected by toxic substances secreted by the glasshouse potato aphid

Leaf affected by cucumber mosaic virus

Fruit affected by cucumber mosaic virus

Life-cycle and appearance
Aphids

Aphids have a complex life-cycle, with both winged and wingless forms of adults of the same species developing depending on the conditions. Wingless adult aphids are known as apterous, and winged adults as alate. Alate aphids have two pairs of wings, one of which is much larger than the other.

Figure 7.7 shows a drawing of an apterous adult, showing the cauda and the siphunculi (or cornicles) which, together with the frontal tubercles on the head are used for identification purposes.

The following factors are responsible for the development of the alate form:
- crowding,
- reduction of the quality of the plant (which depends, among other factors, on the water and fertilizer regime),
- temperature and day-length,
- genetic factors.

Crowding however, is the most important factor leading to the development of alates, and at high densities they migrate to other plants. This polymorphism allows aphids to react optimally to changes in environmental conditions. The differences between alate and apterous aphids are set out in table 7.1.

Alate forms are able to fly considerable distances in order to found new colonies. With the help of rising atmospheric thermals and high altitude airstreams alate aphids can be carried enormous distances (1,300 km has been known!). After descending on to a crop, a brief exploratory probe into the phloem of the plant is conducted, and if unsatisfactory the search is continued by means of short flights until a suitable host plant is found.

For a large part of the season an aphid population consists of viviparous females. Because reproduction is asexual, the progeny of a single female are genetically identical to the mother. In other words, they are clones. Because there is no genetic recombination, different characteristics such as colour forms or resistance to pesticides remain unchanged and unsorted. The young aphids are born as developed nymphs and at once begin to feed on plant sap. They grow rapidly and moult four times before becoming adult, with the conspicuous white cuticles shed at each of these moults betraying their presence in the crop. There are two types of aphid: those species that change host plant in the winter and those species that do not. Species that alternate host plants reproduce asexually during the summer on their summer host plant, while in the winter they migrate to their winter host plant where they reproduce sexually and lay eggs that overwinter. The summer host plants are either herbaceous or woody, while the winter host plants are hardy, woody perennials. Aphids that do not alternate host plants also mate in the winter and lay eggs which overwinter. Where mating occurs and eggs are laid the life-cycle is termed holocyclic (complete). Under glass, however, the alternation of host plant and egg-laying may not take place. In this case, reproduction continues through the winter by parthenogenesis, with viviparous unfertilized females continuing to produce new generations of females. This is known as an anholocyclic (incomplete) life-cycle.

Table 7.1. Differences between alate and apterous aphids

characteristics	alate	apterous
wings	present	absent
antennae	long	short
thorax	large (houses wing muscles)	normal
pigment	many (black)	few
larval period	long	short
reproductive period	long	short
number of off-spring	few	many
tolerance of hunger	strong	weak
lifespan	long	short

7.7. Apterous form, side view

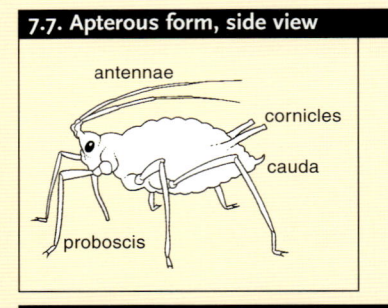

7.8. *Aphis nasturtii* with moulted cuticles

7.9. *Myzus persicae* with offspring

7.10. Colony of *Aphis gossypii*

Life-cycle and appearance
Aphids

Life-cycle of an aphid with host alternation (after Jones, 1942)

winter host

summer host

a	colony founder (fundatrix)
b	viviparous female (apterous)
c	viviparous female (spring migrant)
d	viviparous female (summer migrant)
e	sexual aphid producing progeny
f	viviparous female (alate)
g	egg-laying female
h	male
i	egg
X	mating

Life-cycle of an aphid without alternation of host plants (after Jones, 1942)

- plant sap has a low protein content but is rich in sugars. Aphids therefore need to extract large quantities of sap in order to get sufficient protein. As a consequence the excess sugar is secreted in the form of honeydew, making the crop and its fruit sticky. Black fungal moulds (*Cladosporium* spp.) grow on this honeydew, contaminating fruit and ornamental crops and rendering them unsuitable for market. At the same time, photosynthesis in the leaves is reduced, affecting production,
- the aphid's saliva can induce strong "allergic" reactions such as malformations of the growing tips,
- pathogenic organisms and particularly viruses are introduced. Viruses are mainly transmitted by the winged individuals. Potato virus Y (PVY) is thus transmitted by aphids in tomato, as is the cucumber mosaic virus (CMV) in cucumber.

An aphid colony has a clear effect on a plant. A growing plant will translocate more food materials to an affected part in order to maintain growth, which of course further advantages the aphid colony.

Distribution and dispersal
Usually the first aphids found in a crop are sparsely scattered, but due to their rapid reproductive rate these soon form dense colonies. If discovered in time, however, these can be controlled by local treatment. Once the population becomes too large, winged individuals are formed that are able to disperse throughout the crop.
Aphids react to the colour of plants. The leaf colour gives the aphid information on the age and quality of the leaf, and since young leaves are generally the most suitable food source, yellowish-green is an especially attractive colour. Scent only plays a part at distances of less than one meter.
An aphid will first test a leaf that looks suitable by probing with its proboscis and making a hole. Only after feeding on the exuded phloem sap is the aphid able to judge its nutritional value.

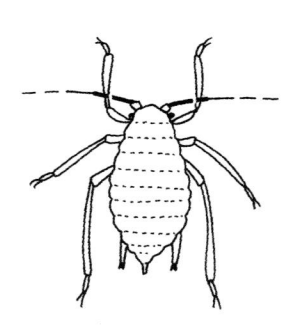

apterous form

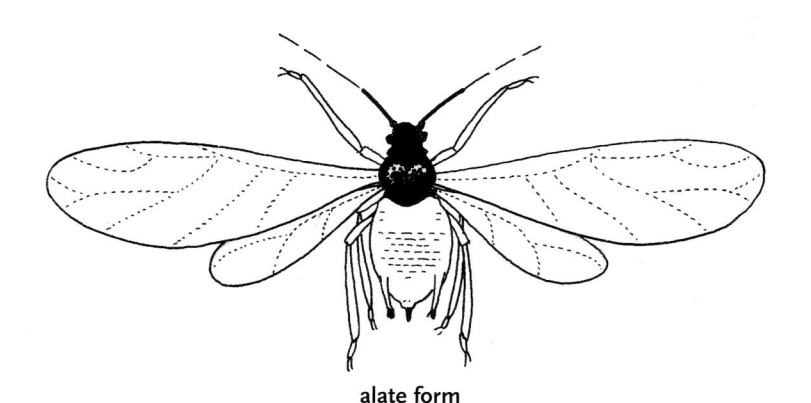

alate form

- Important insect pests
- Can appear the whole year round
- Polyphagous

Myzus persicae subsp. *persicae*
Myzus persicae subsp. *nicotianae*
Peach potato aphid and tobacco aphid

The peach potato aphid (*Myzus persicae* subsp. *persicae*) is an important insect pest in sweet pepper, tomato, cucumber and many other glasshouse crops.

The aphid may originate from Asia, where its winter-hardy host plant, the peach tree, is native, but it is now a pest with a world-wide distribution. Outdoors, various brassicas, potato, beet and green vegetables are colonized. *M. persicae* subsp. *persicae* is a particularly polyphagous aphid with summer host plants from more than 40 different families.

In around 1990 a number of new subspecies of aphid were identified, all closely related to the peach potato aphid. One of these is the tobacco aphid, *M. persicae* subsp. *nicotianae*. The tobacco aphid probably evolved in the Far East from the peach potato aphid and is a key pest of tobacco crops. Plagues of a red peach aphid on tobacco were described as early as the beginning of the last century, and today this subspecies is a serious pest on tobacco in both the United States and in South America. When red aphids looking very much like the peach potato aphid were first observed in The Netherlands in 1994 they were at first thought to be a red form of the latter species, but were soon identified as the tobacco aphid.

Both aphids are encountered in a range of crops in glasshouses, such as sweet pepper, aubergine, chrysanthemum, and various pot plants and cut flower crops.

Life-cycle and appearance

Adults are 1.2 - 2.1 mm in length and oval in form. Wingless aphids are generally smaller than winged individuals. In *M. persicae* subsp. *persicae* they may be green, white-green, a light yellow-green, grey-green, pink or red in colour, whereas wingless *M. persicae* subsp. *nicotianae* are always pink or red. They appear matt, never glossy. Winged individuals have a brown-black head and thorax and a yellow-green to green or even reddish abdomen. There is a dark brown spot on the abdomen and several transverse black bands across the body. These aphids have long antennae that reach as far as the cornicles. The cornicles are of average length and are somewhat thickened from halfway to their distal end; they are a light colour with a dark tip. Frontal tubercles are clearly present and turned toward each other, the middle tubercle being hardly developed (see overview on page 140). The cauda is rather small, elongated and finger-shaped. The aphids have relatively small legs. Nymphs from which winged adults develop are often pink or red in colour.

In temperate regions, peach potato aphid's eggs overwinter on the twigs of peach, plum or related species. If the winter is not too severe, female individuals may also overwinter on winter host plants. It is not known whether the tobacco aphid is also capable of overwintering outside and if so on which host plants. In the (sub)tropics and in glasshouses, where conditions remain favourable throughout the year, there may be no alternation of host plants and no winter eggs laid. Thus, in glasshouses the aphids can continue reproducing asexually throughout the year. The peach potato aphid and the tobacco aphid can only be distinguished under the microscope. The colour should not be used as a distinctive characteristic since pink-red forms are known in both subspecies. Further, the colour of an aphid has no relationship to its resistance to

7.11. Different forms of *Myzus persicae*

pesticides. In The Netherlands, it is well known that the commonly occurring *M. persicae* subsp. *nicotianae* is resistant to pirimicarb, whereas *M. persicae* subsp. *persicae* is usually not.
It would appear that the tobacco aphid develops more rapidly than the peach potato aphid.
Both aphids can be biologically controlled using broadly similar methods.

Damage
Of all the aphids, *M. persicae* subsp. *persicae* is the most important vector of viral diseases. It can transmit at least 100 different viruses and is thus feared by many growers. It is not known to what extent *M. persicae* subsp. *nicotianae* can also transmit viruses. *M. persicae* subsp. *persicae* and *M. persicae* subsp. *nicotianae* also cause the same kind of damage as other aphids, by extracting plant sap, secreting honeydew and introducing toxic substances.

Distribution and dispersal
Peach potato aphids can generally be found distributed over the entire plant, although their preference is for young leaves. They are usually found in closely packed groups, although they are also capable of dispersing through the crop by walking. Dense colonies can be found in flowers as well as on the leaves. The tobacco aphid tends to form denser colonies closer to the top of the plant than the peach potato aphid.

- Prefers higher temperatures
- Causes damage especially in cucumber and melon
- Very rapid population growth

Aphis gossypii
Cotton aphid

Among the many plant species affected by this species, the cotton aphid (*Aphis gossypii*) is an important pest in cotton, in other crops of the same family (the Malvaceae), in cucumber and other cucurbits and in several ornamental crops, such as kalanchoë. It is also found on many vegetable crops.
The cotton aphid is distributed throughout the world but prefers the warmer regions. In America and the Middle East the species has long plagued the cultivation of cotton and melons. It has been a pest in European glasshouses since the late 1980's, particularly in cucumber.

Life-cycle and appearance
Although the species shows no evidence of alternation of host plants in Europe, such alternation has been reported in North America, and it is closely related to European species of *Aphis* that also alternate host plants. Worldwide, the cotton aphid consists of a number of strains that differ in their host range and their resistance to insecticides. Some strains show a unique association with a particular species of host plant. The cotton aphid can be distinguished from other European aphids by the colour of the two cornicles (or siphunculi) that project from the fifth abdominal segment. These cornicles are always black, whatever the body colour, which can vary widely from light yellow to light green or even a green-black colour (see plate 7.12). This body colour depends on temperature, food source and the density of the population. Large individuals (up to 1.8 mm long) are usually dark green to black, while aphids produced in overcrowded colonies at a high temperature can be much smaller (from 0.9 mm) and yellow or cream-coloured. Cotton aphids have red eyes and relatively short antennae.

7.12. Different forms of *Aphis gossypii*

Population growth

The development time for nymphs on cucumber is 9.6 days at 16°C and 6.5 days at 28°C. Because females can produce 3 to 10 young aphids per day, population growth can be extremely rapid.

Under glass, cotton aphids may sometimes be found very early on if the temperature is high enough for efficient reproduction. An infestation can begin early in spring, and after some weeks the population can be enormous. However, most damage is done at the start of the summer. A population of *A. gossypii* can multiply fourfold (in aubergine) to 12-fold (in cucumber) in 7 days.

Despite the fact that there are enough host plants in northern Europe, the insect is only ever a serious problem in glasshouses. This is because the temperature and relative humidity under glass are highly suitable for the cotton aphid's generation and growth. The somewhat lower temperatures outdoors ensure that population growth is less rapid.

Damage

A. gossypii is an important vector of viral diseases. This aphid species can transmit more than 50 different viruses, including the cucumber mosaic virus. The cotton aphid also causes damage in the same manner as other aphids, by extracting plant sap, secreting honeydew and by introducing toxic substances. The species shows a preference for the underside of the leaf and for young shoots and leaves.

- **A pest particularly on tomato, aubergine and roses**
- **Especially active in summer, although sometimes earlier on tomato**
- **Thrives at relatively low glasshouse temperatures**

Macrosiphum euphorbiae
Potato aphid

Macrosiphum euphorbiae (the potato aphid) is of North American origin, but is these days distributed throughout the world. It is a highly polyphagous species with a preference for Solanaceae (particularly potato), but has also been described on more than 200 plant species from more than 20 families. Various vegetable and ornamental crops grown under glass are colonized, including tomato, aubergine, roses and alstroemeria.

Life-cycle and appearance

The adult potato aphid is a large, slender aphid with long, slightly convergent, green cornicles, long legs and a long cauda. Adults are usually green but may also be pink or red, depending on the food source. Even the winged forms can be red.

Wingless potato aphids are 1.7 - 3.6 mm long, elongated and with a dark longitudinal stripe running along the dorsal surface. In green individuals this stripe is dark green, while in pink specimens it is dark red. This longitudinal stripe is specifically characteristic of the potato aphid and is particularly clearly visible in the nymphs. Between the two antennae the head is slightly depressed by the presence of two divergent frontal tubercles. The antennae, which are longer than the body, have visible dark rings. The cornicles are extremely long, light brown in colour with a dark tip and are curved slightly outward. The cauda is also extremely long, narrow, either colourless or light brown and finger-shaped. The eyes are bright red. The aphids fall from the plant at once when touched. Winged females are slightly shorter than wingless forms. Pink alates have a yellowish head and a pinkish-yellow thorax.

Plate 7.13 shows different forms of *M. euphorbiae*.

Population growth

The potato aphid mainly appears in the summer and requires either a mild climate or a glasshouse before it can overwinter. The most favourable temperature for its development is in the range of 15 - 18°C. Each adult female is capable of producing around 30 offspring.

7.13. Different forms of *Macrosiphum euphorbiae*

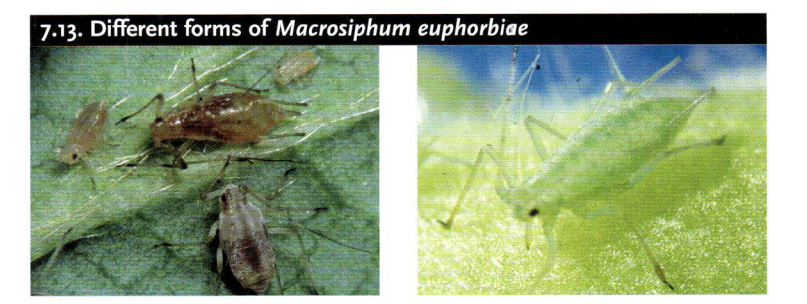

Damage

The potato aphid occurs on many crops, but is not always necessarily a big problem. In most crops the species gives rise to the same kind of damage as other aphids, and the contamination caused by the secretion of honeydew and shed cuticles causes a major problem in tomatoes. In roses the aphid immediately goes for the buds, thus causing damage very quickly.

Distribution and dispersal

It is mainly young parts of the plant that are affected. As well as the leaves, this aphid can often be found on the stalks of various crops. It is a highly mobile aphid and therefore rapidly spreads through the crop. Colonies mostly develop on the undersides of leaves. In cases of serious infestation, however, they also spread to the upper surfaces of the youngest growing parts of the plant. In tomato they are often first encountered on leaves lower down on the plant, on the main stem, and later over the entire plant. The colonies are easily disturbed by the approach of natural enemies, with the aphids immediately falling to the ground.

- ■ An increasing problem
- ■ Causes virus-like symptoms in sweet pepper
- ■ Causes considerable damage even at low population densities

Aulacorthum solani
Glasshouse potato aphid

Aulacorthum solani (the glasshouse potato aphid) is a native European species that in recent years has been reported with increasing frequency. Crops of sweet pepper in particular can be seriously affected. The species also occurs in begonia and other ornamental crops, as well as on potato, lettuce, beans, aubergine and sometimes on tomato.

Life-cycle and appearance

The wingless adult stage of the glasshouse potato aphid is rather squat, of medium size (1.8 - 3.0 mm long) and of oval-round form. It is a light yellow-green to brown-green colour, possesses long antennae marked by a number of darker bands and whose length exceeds that of the body. The legs are also long. The cornicles are of medium length, straight with bulging ends and light green with a darker tip. There are two dark green spots on the abdomen at the base of the cornicles. The cauda is green, finger-shaped and rather short. The frontal tubercles are parallel. Several specimens are shown in plate 7.14.

There are both light and dark forms of winged individuals. In the light forms the head and thorax are a light brown colour, while the dorsal surface is green, without pigmentation. A large dark spot is visible at the base of the tubercles. In the darker form the head and thorax are brown-black, the antennae are brown and the abdomen carries dark spots and transverse bands. The cauda is grey-brown.

The glasshouse potato aphid manifests both anholocyclic (without eggs being laid) and holocyclic (when winter eggs are produced) life cycles.

7.14. Apterous (top) and alate (bottom) forms of *Aulacorthum solani*

Knowing and recognizing **Aphids**

There is no alternation of hosts, and the two types of life-cycle occur on the same host plant. The aphid can overwinter not only in glasshouses but also outdoors in sheltered places.

Damage
When the glasshouse potato aphid pierces the leaves or fruit of sweet pepper it secretes a toxic substance which leads to deformations in the growing tip of the plant. Precisely where the plant is pierced is unimportant. Thus piercing the lower leaves in the crop can cause deformed growth above with lumpy leaves and curled leaf margins. The immediate effects of being pierced by the aphids are visible in lower leaves in the form of yellow patches. Heavy yellowing may even be followed by defoliation. In any case, the yellowing means loss of photosynthesis and as a result reduced growth. Black rings and spots may also appear on the fruit.
The glasshouse potato aphid causes yellow patches on the young leaves in chrysanthemums.
Such damage resembles the symptoms of viral infection and is evident even at low aphid population densities. The other symptoms in glasshouse crops are much the same as those caused by other aphid species.

Distribution and dispersal
The species is mainly found low down on plants. The colonies are easily disturbed by natural enemies, with the aphids immediately dropping to the ground which often leads to further dispersal to other plants. If repeated often, however, this may also lead to high mortality.

7.15. Damage caused by glasshouse potato aphids in sweet pepper

Leaf deformation

Yellow spots on leaves

Overview

Overview of important aphid species

There follows an overview of the more important species of aphid in glasshouse crops. For purposes of comparison with other species, those aphids already dealt with above are also included in this review. In addition, the two species used in the open culture of natural enemies of aphids are also dealt with. The specific characteristics of each species are described, and an apterous form shown in the accompanying plate. Drawings of the head of the apterous form are also given, showing the frontal tubercles of the different species. These frontal tubercles are too small to be visible in the field and can only be observed at a magnification of at least x50 after moving the antennae aside. However, they provide an important distinguishing feature for the identification of different aphid species. The following species are included in this overview:

- *Myzus persicae* subsp. *persicae* — peach potato aphid
- *Myzus persicae* subsp. *nicotianae* — tobacco aphid
- *Aphis gossypii* — cotton aphid
- *Macrosiphum euphorbiae* — potato aphid
- *Aulacorthum solani* — glasshouse potato aphid
- *Aphis fabae* — black bean aphid
- *Myzus ascalonicus* — shallot aphid
- *Rhodobium porosum* — yellow rose aphid
- *Nasonovia ribis-nigri* — currant lettuce aphid
- *Aphis nasturtii* — buckthorn potato aphid
- *Brachycaudus helichrysi* — leaf curling plum aphid
- *Aulacorthum circumflexum* — lily aphid
- *Macrosiphum rosae* — rose aphid
- *Rhopalosiphum padi* — bird cherry-oat aphid
- *Sitobion avenae* — grain aphid

Myzus persicae subsp. *persicae* - peach potato aphid
Myzus persicae subsp. *nicotianae* - tobacco aphid

7.16

apterous
colour: *M. persicae* subsp. *persicae* variable green, whitish yellow-green or grey-green, pink or red; *M. persicae* subsp. *nicotianae* always pink or red
length: 1.2 - 2.1 mm
antennae: approx. same as body length
cornicles: light green or light brown, 0.6 x body length, dark tip, somewhat swollen from roughly halfway to the end
cauda: rather small but longer than diameter of base, elongated and finger-shaped
legs: short
alate
colour: head and thorax brown-black to black, dark brown spot on yellow-green to green abdomen and with several black transverse bands
length: 1.2 - 2.1 mm
distribution and dispersal
in groups over the whole plant but especially on the upper surface of the topmost leaves
crops damaged
many, but notably Solanaceae and chrysanthemum
remarks
M. persicae subsp. *persicae* is the most important vector of viruses among the aphids

Aphis gossypii - cotton aphid

7.17

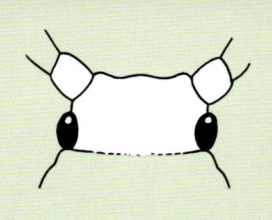

apterous
colour: varies from light yellow to light or even dark green
length: 0.9 - 1.8 mm
antennae: 0.7 x body length
cornicles: short (0.2 x body length), slightly convergent, black
cauda: short, lighter than the cornicles, tongue-shaped with 4 - 7 hairs
legs: short
alate
colour: head and thorax black, abdomen yellow to dark green
length: 1.1 - 1.8 mm
distribution and dispersal
show preference for the underside of the leaf and for young shoots and leaves
crops damaged
many, especially cucurbits
remarks
very rapid population growth; transmits more than 50 viruses

Macrosiphum euphorbiae - potato aphid

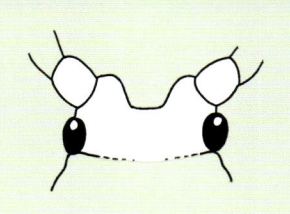

apterous
colour: light green, pink or red with a dark longitudinal stripe and bright red eyes
length: 1.7 - 3.6 mm
antennae: 1.2 x body length, clear, darker toward the tip
cornicles: light brown, 1.2 x body length, slightly bent outward, sometimes with darker tip which bears a netlike structure
cauda: very long, light coloured and finger-shaped
legs: very long
alate
colour: yellow-brown head and light wings, distinct brown bands on a light green or pink abdomen
length: 1.7 - 3.4 mm
distribution and dispersal
mainly in the young parts of the plant. In tomato, often on the lower parts of the plant, mainly the leaves but also on the main stem
crops damaged
many, particularly Solanaceae

Aphis fabae - black bean aphid

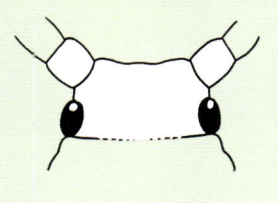

apterous
colour: dark olive green to black with black bands on the abdomen
length: 1.5 - 3.1 mm
antennae: 0.7 x body length, light yellow, darker toward the end
cornicles: short, but longer than the cauda, usually darker than the body colour
cauda: dark, with more than 10 hairs, tongue-shaped
legs: light yellow, darker toward distal extremities
alate
colour: dark olive green to black with black bands on the abdomen
length: 1.6 - 2.6 mm
distribution and dispersal
infestations begin in the young shoots but spread over the whole plant
crops damaged
under cover, beans and hedera especially, outdoors in many crops
remarks
this aphid is very common outdoors

Aulacorthum solani - glasshouse potato aphid

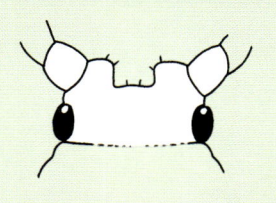

apterous
colour: clear yellow-green to brown-green, dark green spots at the base of the cornicles
length: 1.8 - 3.0 mm
antennae: 1.3-1.5 x body length, with a number of dark transverse bands
cornicles: 0.2 x body length, straight, light green in colour with dark spots, bulging toward the end
cauda: short, green and finger-shaped
legs: long with dark banding
alate
colour: - light form: head and thorax yellow-green to light brown. Evenly green abdomen with a large dark spot at the base of the cornicles
- dark form: head and thorax light brown, brown-black pattern of spots and lines on the abdomen
length: 1.8 - 3.0 mm
distribution and dispersal
mainly lower down on the plant, in colonies
crops damaged
many, especially Solanaceae
remarks
causes virus-like symptoms in sweet pepper

Myzus ascalonicus - shallot aphid

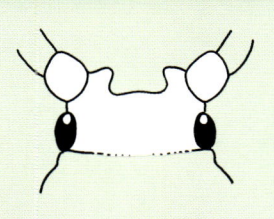

apterous
colour: pale yellow, yellowish to greenish brown, similar to *M. persicae*, but the abdomen is plumper and glossier and in colour paler and browner
length: 1.1 - 2.2 mm
antennae: light with darker tips
cornicles: short and thin, swollen at the ends, light coloured
cauda: blunt, tongue-shaped, hardly visible in dorsal view
legs: dark extremities
alate
colour: head and thorax dark, abdomen straw-coloured
length: 1.3 - 2.4 mm
crops damaged
occurs in flower bulbs, in strawberry and many other crops

Overview of important aphid species

Rhodobium porosum - yellow rose aphid

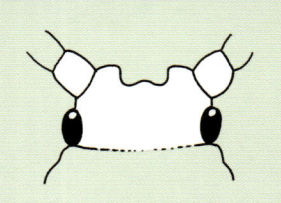

apterous
colour: light to dark green body with yellow-brown head and black eyes
length: 1.2 - 2.5 mm
antennae: somewhat longer than the body
cornicles: extremely long with dark tips, slightly broader at the base
cauda: tongue-shaped
alate
colour: clear green
length: 1.4 - 2.2 mm
distribution and dispersal
mainly in young shoots
crops damaged
roses, strawberry

Aphis nasturtii - buckthorn potato aphid

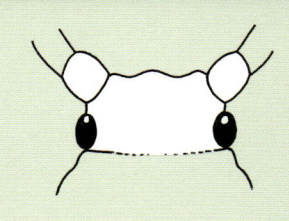

apterous
colour: lemon to light green or brown-yellow
length: 1.3 - 2.0 mm
antennae: less than half body length
cornicles: short, rather slender, black toward the tip
cauda: distinctly lighter than the cornicles, tongue-shaped, with hairs
alate
colour: head and thorax brown-black, yellow to yellowish green abdomen with several dark spots
length: 1.2 - 2.0 mm
distribution and dispersal
on the undersides of lower leaves, immobile
crops damaged
sweet peppers, tomato, aubergine and gerbera among others

Nasonovia ribis-nigri - currant lettuce aphid

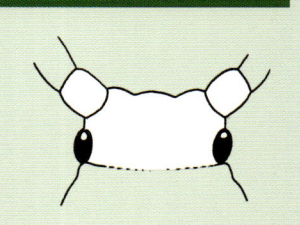

apterous
colour: light yellow to apple green, 7 pairs of dark spots on the abdomen
length: 1.3 - 2.7 mm
antennae: black tips
cornicles: mid length
cauda: elongated
legs: distal extremities black
alate
colour: head and thorax black, abdomen yellow-green to dark green
length: 1.5 - 2.5 mm
distribution and dispersal
colonies are found at the growing tips
crops damaged
mainly a problem in lettuce

Brachycaudus helichrysi - leaf curling plum aphid

apterous
colour: variable, white, yellow, green or brown
length: 0.9 - 2.0 mm
antennae: short, reaching to half the length of the body
cornicles: short, light coloured
cauda: light, short, helmet-shaped
alate
colour: usually brown, with a large irregularly shaped dark patch on the abdomen
length: 1.1 - 2.2 mm
distribution and dispersal
in the tops of plants, on the outer side of buds, between the ray florets of ornamental Compositae
crops damaged
gerbera, chrysanthemum and other Compositae especially
remarks
causes virus-like symptoms. In ornamental crops the threshold for damage is low

Aulacorthum circumflexum - lily aphid

apterous
colour: pale yellow-green with a black horseshoe-shaped marking and dark bands dorsally
length: 1.2 - 2.6 mm
antennae: long, longer than the body
cornicles: light coloured
cauda: light coloured and tongue-shaped
legs: light coloured
alate
length: 1.6 - 2.4 mm
crops damaged
many including in particular ornamental crops and sweet pepper

Rhopalosiphum padi - bird cherry-oat aphid

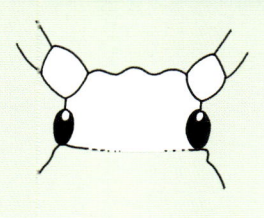

apterous
colour: brownish to olive green, single merged red-brown patch at the base of the cornicles
length: 1.2 - 2.4 mm
antennae: short, half body length
cornicles: 2x length of cauda, black with distinct constriction
cauda: light coloured, rather small
alate
colour: black head and thorax, abdomen light to dark green with a patch around and between cornicles
length: 1.2 - 2.4 mm
distribution and dispersal
only in the banker plant system
crops damaged
only causes damage in Graminae, and can therefore be introduced without danger into dicotyledonous crops
remarks
used in banker plant system

Macrosiphum rosae - rose aphid

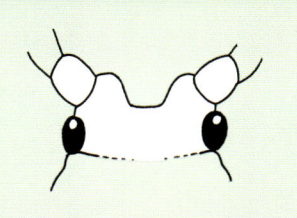

apterous
colour: glossy green, dark green, pink or red-brown
length: 1.7 - 3.6 mm
antennae: yellow and black, long
cornicles: black, long
cauda: light yellow
legs: yellow and black, long
alate
colour: abdomen is light green or pink, black head
length: 2.2 - 3.4 mm
distribution and dispersal
highly mobile
crops damaged
roses

Sitobion avenae - grain aphid

apterous
colour: variable, yellow-green, green, red-brown, black and all intermediate colours, usually with dorsal spots
length: 1.3 - 3.3 mm
antennae: brown, slightly shorter than body length
cornicles: black with broad base
cauda: slightly shorter than cornicles
legs: brownish, distal extremities dark brown to black
alate
colour: dark brown head and thorax, abdomen has broad transverse bands and brown spots laterally
length: 1.6 - 2.9 mm
distribution and dispersal
only in the banker plant system
crops damaged
only causes damage in Graminae, can therefore be used without danger in dicotyledonous crops
remarks
used in banker plant system, and sometimes known as *Macrosiphum avenae*

Natural enemies of aphids

Important natural enemies of aphids in glasshouse crops include:

The gall midge *Aphidoletes aphidimyza*
The ladybird *Adalia bipunctata*
The lacewing *Chrysoperla carnea*
The hoverfly *Episyrphus balteatus*
The parasitic wasp *Aphidius colemani*
The parasitic wasp *Aphidius ervi*
The parasitic wasp *Aphelinus abdominalis*
Entomophthorales fungi
The fungus *Verticilium lecanii*

7.30. *Aphidoletes aphidimyza*

7.31. *Adalia bipunctata*

7.32. *Chrysoperla carnea*

7.33. *Episyrphus balteatus*

7.34. *Aphidius colemani*

7.35. *Aphidius ervi*

7.36. *Aphelinus abdominalis*

7.37. *Erynia neoaphidis* op *Sitobion avenae*

7.38. *Verticilium lecanii* op *Aphis gossypii*

There are various species of predators, parasitic wasps and fungi which contribute to the biological control of aphids. Precisely which is the best strategy to apply in any given situation will depend on the species of aphid, and the duration of the crop concerned.

The parasitic wasps *Aphidius colemani*, *Aphidius ervi* and *Aphelinus abdominalis* are fairly specific, and the *Aphidius* species are especially suitable for release at the beginning of an infestation. Hyperparasites, however, can impair the efficacy of these natural enemies.

The predators are more suitable in situations where aphid populations are already at high densities. Their mode of operation is less specialized and they can often clear up large numbers of aphids quickly. However, the ladybirds tend to fly out of the glasshouse when they reach adulthood without having consumed all the prey. The efficacy of lacewings and hoverflies mainly depends on the crop in which the aphids are established.

Under favourable conditions, naturally occurring fungi can have a major impact on an aphid population. They mainly appear in the late summer and autumn when the climate is more humid.

Naturally occurring parasitic wasps and predators can also contribute significantly to the biological control of aphids, as can *Orius* spp. released to combat thrips.

- Commercially introduced, but can also occur naturally
- A predator of all aphid species commonly occurring in glasshouses
- Aphids die after attack, even if not eaten

Aphidoletes aphidimyza

The gall midge *Aphidoletes aphidimyza* is a predator of aphids in glasshouses. Like other gall midges, it is a member of the Cecidomyiidae, within the order Diptera (the true flies). The larvae of most gall midges live on plants, often causing galls and sometimes harming the plant. However, there are also gall midge species that prey on aphids, scale insects, whitefly and other insects or mites.

There are five species of gall midge that feed on aphids: four species of the genus *Aphidoletes* and one of the genus *Monobremia*. *A. aphidimyza* is the most common species and will prey on all aphids of the family Aphididae.

An account of *Aphidoletes aphidimyza* published in 1847 (when it was known as *Cecidomyia aphidimyza*) described the larval habit of feeding on aphids. At the time it was assumed that all gall midges were restricted to a single food source, as is the case in the gall-forming species, with the result that many different names were given to *A. aphidimyza*. Only after the Second World War was it demonstrated that *A. aphidimyza* was capable of developing on many different aphid species, leading to a clarification of the nomenclature. Since 1989, *A. aphidimyza* has been deliberately introduced to control aphids, although it can appear naturally in glasshouses.

Population growth

The development time of the gall midge depends mainly on temperature, the species of prey, prey density, and relative humidity. The adult gall midges are nocturnal, with mating and egg-laying mainly occurring between sunset and sunrise. Mating is necessary before a female can lay eggs, and each female produces either male or female offspring. The species of aphid consumed has little influence on fecundity. There is a preference for the lower leaves, where conditions are darker and more humid. Most eggs are deposited when night temperatures are above 16°C and when atmospheric humidity is high. The number of eggs laid on a plant is highly dependent on the density of prey. No eggs are laid on plants without aphids, whereas many eggs are laid in large aphid colonies.

The number of eggs laid per female depends on the climate, the quantity of food consumed during the larval stage and the quantity of honeydew ingested as an adult. Without the ingestion of honeydew, egg-laying can be severely reduced. Most eggs are laid in the first two or three days after emergence of the female.

The absence of honeydew also has a negative influence on the lifespan of the gall midge.

Relative humidity appears to have a major effect on mortality among gall midges, which have a shorter lifespan under dry conditions. Table 7.2 provides data relating to the development of *A. aphidimyza*.

Overwintering

In nature, the first gall midges appear in spring, and in the summer they can be observed on a range of crops. At the end of the summer the full-grown larvae retreat to the ground where they overwinter in a cocoon approximately 2 cm below the surface. The adults emerge from these cocoons the following year.

Table 7.2. Development time (in days) of different stages of *Aphidoletes aphidimyza* at different temperatures (after Havelka, 1980)

stage	temperatuur (°C)			
	15	20	21	25
egg	5.1	2.5	2.3	1.7
larva	10.4	7.1	6.6	5.4
pupa	16.6	10.8	10.1	8.0
total development time (days)	32.1	20.4	19.0	15.1

Feeding behaviour

When a gall midge larva attacks an aphid, it injects a paralyzing toxin that also starts pre-digestion of the body contents. The larva then sucks out the body fluids, leaving the dead aphid hanging by its proboscis from the leaf to turn brown or black, and eventually decay.

The quantity of aphids consumed by a gall midge larva depends on temperature, relative humidity, age and size of the aphids and the aphid species. In total, between 10 - 100 aphids are consumed per gall midge, with 50% of these being consumed during the last larval stage.

With a sufficiently large number of aphids, the gall midge larva can kill more aphids than it needs to satisfy its nutritional requirements. The larger the aphid population, the more aphids are killed without being eaten. Once attacked, the paralysis toxin with which they are injected means that the aphids will die whether they are eaten or not.

The time required to consume an aphid can vary from a few minutes to several hours, depending on the age and nutritional condition of the predator as well as the size of prey.

A. aphidimyza cannot survive on scale insects or mealy bugs, although it is cosmopolitan as far as aphids are concerned, with over 60 species of suitable aphid described, including all those that occur commonly in glasshouses. It seems likely that this species of gall midge can consume any member of the Aphididae.

Searching behaviour and dispersal
The searching behaviour of adult female gall midges is extremely efficient. For example, they can find a plant infested by aphids amid many other non-infested plants, ensuring rapid establishment of midges through the crop.
For egg laying, a female gall midge prefers large aphid colonies. For example, five times as many eggs will be deposited on a single leaf with 60 aphids than the total on five leaves each with 12 aphids.
Newly hatched larvae can move about 6 cm without feeding and can detect aphids within a radius of about 2.5 cm. Since eggs are generally laid within an aphid colony, finding prey is not normally difficult.

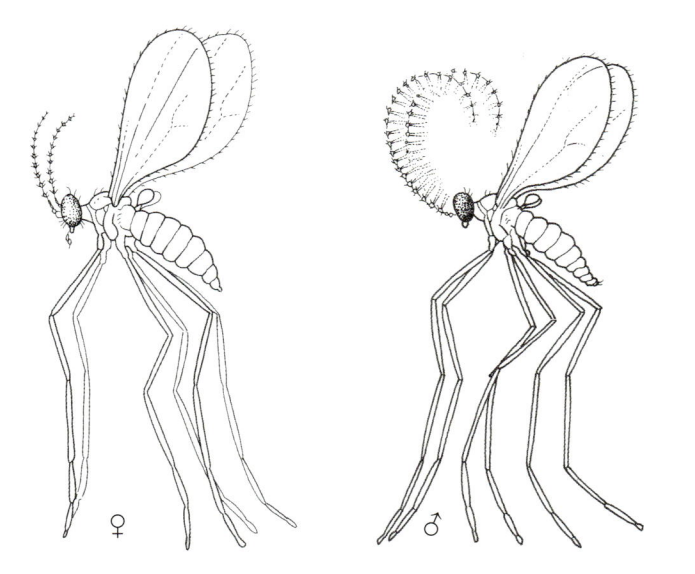

7.39. Yellow (left) and orange (right) larva with prey

7.40. Cocoons covered with sand grains with abandoned pupal skins after emergence of adults

7.41. Adult male

Life-cycle and appearance
Aphidoletes aphidimyza

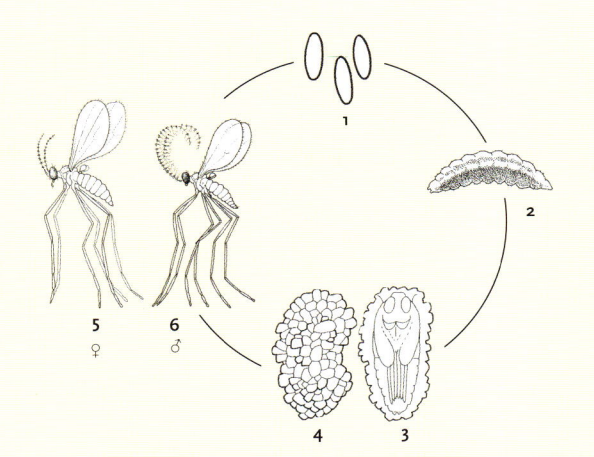

1 eggs
2 larva
3 pupa
4 cocoon with sand grains
5 adult female
6 adult male

The adult midges are active at night and during the crepuscular periods of dawn and dusk. During the day they remain motionless in protected places on the plant, often hanging in spiders webs close to the ground, and when disturbed will fly up and search for another resting place. Mating occurs in the webs, and both mating and egg-laying take place from dusk onwards. The eggs, which are laid beside or sometimes under aphids, are oval in shape and have a glossy, orange-red appearance. Their small size (about 0.3 x 0.1 mm) means that they are difficult to detect in an aphid colony.

Newly hatched larvae are about 0.3 mm long, elongated in form and transparent orange in colour. Like the eggs, the larvae are difficult to detect among the aphids at first on account of their small size. Newly hatched larvae sometimes feed on honeydew, but if they are to avoid dehydration they must quickly find aphid prey. Depending on the body contents of their prey, larvae may later change colour to yellow, orange, red, brown or even grey.

As the larvae become larger (up to approximately 2.5 mm in length) they become easier to detect. When full-grown they migrate from the leaves and pupate in the top layer (up to 1 cm deep) of moist soil, where they form an oval silk cocoon constructed from long sticky threads. This cocoon is covered with small particles of soil, frass etc. and is about 2 mm long.

These cocoons can sometimes be found in the vicinity of the plant. If the ground is covered with plastic sheeting and *A. aphidimyza* is unable to reach the soil, mortality will be high during pupation. When the adult emerges, the moulted pupal skin is left hanging on the outside of the cocoon.

The adults emerge during a period beginning shortly before sunset and lasting for several hours thereafter. The delicate adults are roughly 2.5 mm long and the female has a wing length of 2.5 - 3.5 mm. The legs are long and thin. The males possess long, backward bending antennae covered with long hairs, whilst those of the female are shorter.

7.42. Eggs

7.43. Yellow larva with prey

7.44. Cocoons covered with sand grains

7.45. Adults hanging in a spiders web

- Often present naturally, but some species are also introduced
- Larvae and adults combat all stages of aphids
- Capable of eradicating an aphid colony very quickly

Coccinellidae
Ladybirds

Ladybirds belong to the order Coleoptera (beetles) and the family Coccinellidae, which contains roughly 5,000 species. Apart from a small number of species that feed on plants or fungi, all other members of this family, both larvae and adults, feed on insects and mites. The vast majority eat aphids.

One group has the familiar orange or red wing cases with black dots, whilst other smaller black species, which although less conspicuous, are highly important natural predators. These insect-eating ladybirds have played an important role in the development of biological control. One well-known example is the release of *Rodolia* sp. in California in 1888 against the scale insect *Icerya purchasi* in citrus orchards.

Because ladybirds often require a minimum prey population before they can be effective, and because they tend to escape from the glasshouse when adult, their introduction is often only worthwhile in cases of high prey density or where there are concentrated infestations which they can control very quickly. Most species are not suitable for keeping pest populations at a low level. Many species occur naturally and can provide a welcome supplement in biological control.

Several important species are described here and are also included in the following overview, which deals with those ladybirds that are significant in the biological control of various glasshouse pests.

Population growth

Many species of ladybird can spend long periods in a resting state, usually in the adult instar, which enables them to withstand times of drought and food shortage. The profundity of this resting state varies according to species and the situation. In northern Europe the beetles enter diapause in the autumn with the onset of colder weather and re-emerge in May. In spring, after the overwintering stage, the beetles fly off to areas where food is abundant. As soon as prey are found and the females have fed on them they lay their eggs in the colonies. In the absence of prey no new eggs are produced, and those that have already been produced are immediately consumed.

After several days the larvae hatch from the eggs. Instar-specific survival, development time and adult weight are all strongly influenced by the species of prey. However, the growth of a population of ladybirds does not depend entirely on the prey, but also on the host plant and the microclimate. Adult ladybirds have a lifespan of three months to a year.

Feeding behaviour

When given a choice, adult and larval ladybirds of a particular species always feed on the same food. Since the beetles lay their eggs where they feed, the food for the next generation is effectively chosen by the beetles. Ladybirds grasp their prey with their jaws and suck them empty. The older larvae and the adult beetles will completely consume prey that is not too large. First and second instar larvae eat relatively little, but food consumption increases rapidly in the third and fourth instars. Adult beetles generally eat less than fourth instar larvae. Young larvae are cannibalistic: because eggs do not hatch simultaneously, and because some eggs are not fertilized, a larva will exploit other eggs that it encounters in its search for prey as its first source of food. A considerable percentage of eggs can be lost in this manner, although when prey are relatively scarce this practice can greatly enhance chances of larval survival.

In addition to aphids, many species feed on pollen, nectar and honeydew, but reproduction usually depends on the consumption of live prey.

Ladybirds are highly specific in their diet. Species that live on spider mites can usually not develop to full maturity on any other food source, whilst species that feed on aphids can only develop properly on aphids. Some ladybirds are even species-specific in their choice of prey, while others will eat several species. Often the species of prey can have a considerable influence on the rate of development, mortality and reproductive capacity. The quantity of food ingested varies enormously and depends as much on the species of ladybird as the species of prey. The effect of ladybirds on prey populations can sometimes be disappointing. In many cases, the growth of the prey population is inadequately stemmed, for one or other of the following reasons (in no particular order):

- the prey colonies are protected by ants,
- the influence of parasites and predation on ladybird eggs and larvae can be considerable, including the effect of larval cannibalism (only 4% reach adult maturity),

7.46. Pupa of *Harmonia axyridis*

Life-cycle and appearance
Ladybird

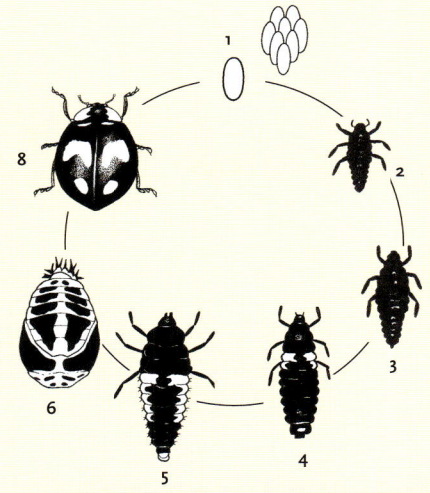

1 egg (single and cluster)
2 first instar larva
3 second instar larva
4 third instar larva
5 fourth instar larva
6 pupa
7 adult

7.47. The commonly occurring form of *Coccinella septempunctata* and aberrant colour forms (Gunst, 1978)

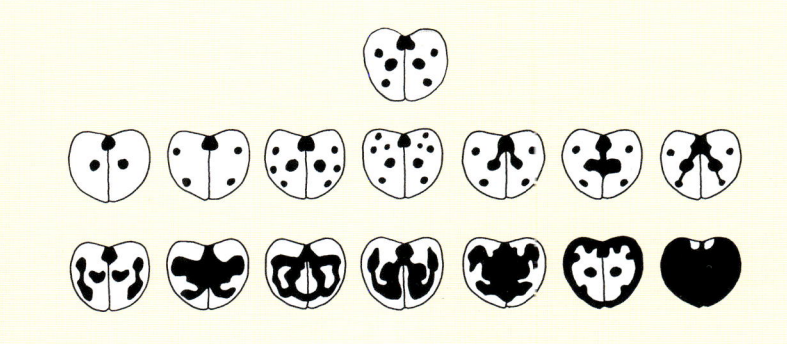

7.48. Eggs of a ladybird

7.49. Larva of *Coccinella septempunctata*

As adults, many species are bright red, orange or yellow with black dots. Unfortunately these are not good characteristics for identification purposes since the number of dots can vary, and there may also be different colour forms within a particular species. The unusual forms are known as 'aberrations'. This is mainly due to genetic variation, but the sex, food source, season and geographical location may also be relevant. The bright colours serve as a warning to predators that they have a very bitter taste. The length of an adult is 1.5 - 8 mm and its breadth roughly ²/₃ of this. The beetles are hemispherical and most have rather short legs, a small head with large eyes and short antennae.

The eggs are orange-yellow and stand upright in groups of 10 to 100 packed closely against each other on the underside of leaves, usually close to their prey. Eggs are cylindrical in form and about 2 mm long.

For the first 24 hours, newly hatched larvae remain sitting on or in the vicinity of the egg batch. They first consume the empty egg shell and often other eggs with their un-hatched larval contents, after which they go in search of other prey. In marked contrast to the adults, the larvae have the appearance of voracious predators. They are grey-black with yellow to red spots or stripes with three pairs of relatively long legs. There are usually four (sometimes five) larval instars. At the rear the larva has a special sucker which ensures that it does not fall from the leaf, and which also serves to anchor the larva during pupation. The body is slightly flattened and gradually tapers to a point. Depending on the species, the dorsal surface may be covered with few or many branched projections and sometimes, as in the larvae of *Scymnus* spp. and *Cryptolaemus montrouzieri*, with a waxy secretion.

Pupation takes place on the underside of the leaves. The pupa never occupies a cocoon but is fixed to the substrate by the posterior point of the abdomen. A newly emerged beetle, apart from the black eyes and sometimes the marking on the thoracic shield, is entirely pale yellow to almost white in colour. The pattern of spots and the basic colour of the elytra begin to develop slowly during the hours that follow.

Coccinellidae possess a number of different methods of defense against attack from other animals. The first mechanism is to withdraw the legs and antennae beneath the body, feigning death, a behavioural ploy known as 'thanatosis'. Few animals will attack or eat insects that show no movement. A second mechanism possessed by both adults and larvae is the reflex excretion of a bitter yellow fluid that contains substances poisonous to the attacker. Of course, there are a few attackers that are resistant to these toxic substances.

- the ladybirds do not search efficiently,
- ladybirds often leave the aphid colonies while living aphids still remain.

Searching behaviour
For a long time it was assumed that ladybirds search for prey randomly, and that prey were only discovered when physical contact was made. It seems, however, that adults in particular can detect their prey from a short distance both by sight and scent. Once prey have been detected, they begin to search more intensively. As prey are not evenly distributed but aggregated in colonies, this is the most productive strategy. If, after some time no prey have been found, the search activity becomes gradually more extensive. The fact that they sometimes appear to walk over their prey without eating them may be the result of a temporary preoccupation with finding a partner or a place to deposit eggs. In any case, it is clear that ladybirds cannot locate prey from any great distance.

Adalia bipunctata
Two spotted ladybird

Adalia bipunctata occurs naturally throughout Europe, Central Asia and North America,
and is commonplace on a wide variety of plants. It is an important predator in orchards and soft fruit beds, and may also enter glasshouses. It has been released commercially for the biological control of aphids in European glasshouses since 2000.

Life-cycle and appearance
Adults are 3.5 – 5.5 mm in length, with males often smaller than females. The head and thorax is mainly black with small, white markings, and the legs and underside of the body are black to reddish brown. The adult is easily identified by its bright red to orange-yellow elytra (or wing cases), each of which has a single black spot. Occasionally, the colouration may be reversed, giving black elytra with red spots. Various other colour forms also occur.
The eggs are 1.0 - 1.5 mm long, orange-yellow in colour, and are deposited on the underside of leaves close to a potential food source. When the larvae emerge, they start searching for prey immediately. The larvae are brown or blackish grey with black and yellow or orange spots, and are 5 - 6 mm long when fully grown. The pupae are dark grey to black with yellow dots, and are 3.0 - 3.5 mm long. They are often found hanging from leaves.

Population growth
Adults hibernate throughout the winter, often congregating in large numbers in suitable refugia. Eggs are laid in small upright batches on leaves, usually close to aphid colonies. Under optimal conditions a female can lay up to 1,500 eggs during her lifetime, laying on average 20 eggs per day.
The number of eggs in each batch can vary from 10 to 50. Larval emergence is dependant on climatic conditions, but normally takes 4 to 8 days, with development from egg to adult taking several weeks. There is only one generation each season, although adults are long-lived (1 - 3 months during the growing season), and all stages occur simultaneously throughout the summer months.

7.50. Two common forms of *Adalia bipunctata*

Knowing and recognizing Aphids

Feeding behaviour

Adults and larvae will eat more than 50 species of aphid, including all those that are important in vegetable crops. Adults can consume several thousand aphids during their lifetime. Fourth instar larvae consume per day as many aphids as adult ladybirds, if not more. Each larva may consume several hundred aphids during its development. If food is scarce, they will also attack other small organisms such as spider mites and lepidopteran eggs.

For biological control the larvae are used, as adults tend to fly away in sunny conditions.

Distribution and dispersal

A. bipunctata searches for food in two distinct ways: the initial search is rapid and superficial, becoming slower and more intensive once suitable prey is located. This search method is characteristic of predators whose prey are found in groups, as the chances of more prey being encountered in the vicinity are high.

The adults are able to disperse and fly out of the glasshouse.

7.51. Larva of *Adalia bipunctata*

There follows an overview of the characteristics of important ladybird species in glasshouse crops. Species already mentioned in other chapters are included for comparative purposes.

Overview of important ladybird species

Harmonia axyridis

7.53  7.54

egg
colour: orange-yellow
size: 1 mm
larva
colour: first black, later with two orange-yellow longitudinal stripes
size: 8 mm
pupa
colour: brown with black spots
adult
colour: highly variable from light yellow or orange-red to black with 0 to 21 dots
size: 5 - 7 mm
prey
polyphagous: aphids, scale insects and other insects with soft cuticles, lepidopteran eggs
remarks
is introduced in some countries

Adalia bipunctata

7.57  7.58

egg
colour: orange-yellow
size: 1.0 - 1.5 mm
larva
colour: brown or blackish grey with black and yellow or orange spots
size: 5 - 6 mm
pupa
colour: dark grey to black
size: 3.0 - 3.5 mm
adult
colour: thorax mainly black with small white spots, elytra bright red to orange-yellow with two black spots or black with red spots
size: 3.5 - 5.5 mm
prey
adults and larvae will eat more than 50 species of aphid, including all those that are important in vegetable crops; also eggs of Lepidoptera
remarks
occurs naturally, particularly in fruit crops, and may also be present or introduced in glasshouses

Coccinella septempunctata

7.55  7.56

egg
colour: orange-yellow
size: 1.5 - 2.0 mm
larva
colour: bluish grey with black and orange spots
size: 8 - 10 mm
pupa
colour: yellow and black
size: 5 - 7 mm
adult
colour: thorax black with white patch to left and right, elytra red or orange with seven black spots
size: 6 - 8 mm
prey
various species of aphid, including the most important species found under glass. Can survive for some time on pollen
remarks
as a result of introductions, now has worldwide distribution. In northern Europe one generation per year, very common in the wild

Adalia decempunctata

7.59  7.60

egg
colour: orange-yellow
size: 1.0 - 1.5 mm
larva
colour: light grey with yellow and white spots
size: 5 - 6 mm
pupa
colour: yellow and black
adult
colour: bright red to orange-yellow with 10 black spots
size: 3.5 - 5.0 mm
prey
adults and larvae will eat more than 50 species of aphid including all those that are important in vegetable crops
remarks
occurs naturally, particularly in fruit crops, and may also be present in glasshouses

Scymnus spp.

egg
colour: transparent white
size: 0.35 - 0.40 mm
larva
colour: white with downy wax secretion
size: up to 13 mm
pupa
colour: covered with white, downy wax secretion
adult
colour: dark brown
size: 1.8 - 2.3 mm
prey
aphids and scales are the most important prey, depending on species
remarks
can appear spontaneously in glasshouses. Most active at temperatures between 20 and 30°C, enters diapause below 16°C

Stethorus punctillum

egg
colour: pale white
size: 0.4 x 0.22 mm
larva
colour: grey-black with many long, branched hairs and black spots; as the larva ages the colour changes to reddish, beginning laterally
Size: 2.5 - 3.0 mm
pupa
colour: black
adult
colour: glossy black with fine yellow-white hairs
size: 1.2 - 1.5 mm
prey
all stages of spider mite
remarks
see Chapter 2

Cryptolaemus montrouzieri

egg
colour: white
size: ± 1.5 mm
larva
colour: white with waxy appendages
size: 13 mm
pupa
colour: white with waxy appendages
adult
colour: head, anterior thorax tips of the elytra and the abdomen orange to red-brown, the rest dark brown
size: ± 4 mm
prey
mealy bugs
remarks
originally from Australia but now released world-wide. Does not occur naturally in northern Europe, see Chapter 9

- Very common
- *Chrysoperla carnea* is the most familiar species
- Aphids are often eaten, but other insects and mites can also serve as prey

Lacewings

Lacewings belong to the order Neuroptera and sub-order Planipennia. There are two families of lacewings, the Chrysopidae (the green lacewings) and the Hemerobiidae (the brown lacewings). Adult Chrysopidae are green and feed on pollen and honeydew. The eggs are laid at the end of mucus threads so that each egg has a stalk, probably as protection against predators. When the larva is ready to hatch the egg bends toward the leaf surface. The small larvae are formidable predators that will even attack their own species. The Hemerobiidae are distinguished from the Chrysopidae by their brown colour, their stalk-less eggs and by the predatory feeding habit of the adults and the larvae.

Lacewings pass through 7 stages: the egg, three larval instars, the prepupal instar, pupa and adult. The larvae will eat almost anything whose contents they are capable of sucking up with their mouthparts, but there is usually a preference for a particular type of food. If there is an excess food supply, more prey items are killed, although each may only be partially consumed. The last larval instar spins a hairy, white cocoon in which it pupates. *Chrysoperla carnea* is the most common European species.
There are numerous species of lacewings distributed throughout the world, with an estimated 50 species in Europe.

Chrysopidae
Green lacewings

Among the Chrysopidae, *Chrysoperla carnea* (= *Chrysopa carnea*) is the most common and familiar species. The descriptions and data in this section mainly refer to this species although they may be valid for other green lacewings. *Chrysoperla carnea* prefers aphids as prey, but will also feed on other insects and mites. It is distributed worldwide except for the Australian continent.

Population growth
Population growth is dependent on temperature, the species of prey and on atmospheric humidity. Adult lacewings do not like extremes of temperature, and in warm periods they leave the glasshouse.
The development from egg to adult takes on average 69 days at 16°C, 35 days at 21°C and 25 days at 28°C. At constant temperatures below 10°C development is not completed. When low temperatures (even sub-zero temperatures) alternate with higher temperatures however, development can take place to maturity. Temperatures above 35°C are lethal.
Adults feed on pollen, honeydew and nectar. Fertility is particularly affected by the food intake of the adult lacewing. Mating takes place at dusk and after dark immediately following the emergence of the adults. A female can lay between 400 and 500 eggs.
After 1 - 2 weeks the males die, but females have a longer life span. The adult stage overwinters. Overwintering adults acquire a yellow colour. Diapause is triggered by shortened day-length and is broken in the spring by rising temperatures.
In mid Europe there are usually 2 generations per year, but in southern Europe there are 3 - 4 generations. More generations are possible in the glasshouse. The larvae are fairly resistant to insecticides.

Feeding behaviour
Lacewing larvae search for prey randomly on the plant. Their efficiency is seriously impaired by the presence of leaf hairs. The larvae are not easily observed and are mainly active at night, sheltering under the plant during daylight hours.
The larvae grasp their prey from below and lift them in their jaws. They are injected with a salivary fluid that digests the body contents which are then sucked out. The prey can sometimes be larger than the larva itself. On average a larva will consume 300 - 400 aphids of various sizes during its development, although the actual quantity depends on prey species and temperature. Roughly 75% of the entire lifetime consumption of a lacewing will be eaten during the last larval instar. Female larvae eat more than the male larvae, and the remains of consumed aphids are so shriveled that they are difficult to detect in the crop.
Lacewings prefer to feed on aphids, but will also prey on whitefly, spider mites and thrips. If present, useful insects may also be consumed. Lepidopteran eggs and mealy bugs are also eaten. When prey is scarce, the larvae can resort to cannibalism, with the older larvae eating the younger. Eggs are often ignored because they stand on stalks. The duration of development and the body weight of the lacewing depend on the nutritional value of the prey consumed.
Adults make no contribution to biological control. They feed exclusively on pollen, nectar and honeydew and most will fly out of the glasshouse. It has been found that the larvae of *C. carnea* do not establish well in tall growing crops. They establish better in lower-growing crops since they easily fall from the leaves, and in tall crops are unable to reach the growing tips again where their prey congregate.

Life-cycle and appearance
lacewings

1 eggs
2 larva
3 pupa
4 adult

Adult green lacewings are 23 - 30 mm long, slender, yellow-green in colour with a yellow-white dorsal stripe and golden eyes. The anterior thorax has black markings. They have large, finely-veined wings that are a transparent green with green veins. The wings cannot be folded and at rest are held tent-like above the abdomen. The hind-wings are slightly smaller than the fore-wings. Adult brown lacewings resemble green lacewings apart from colour and size: they are brown and about half the size. A pale yellow stripe runs over the thorax with dark brown spots on either side. The wings are light yellow with yellow-grey or light brown spots. Like green lacewings, they are active fliers in the evening and after dark. The adult female green lacewing lays oval eggs on transparent stalks attached to either the upper or undersurface of leaves. The length of these stalks is highly variable, but averages 3.5 mm. The eggs sit singly or in groups of about 10, although sometimes up to 40, often in the vicinity of aphid colonies. Sometimes however, the eggs can be found in strange places, such as on aphids, on stems or on structural parts of the glasshouse. The eggs are 0.9 mm long, 0.4 mm wide and a transparent whitish green colour when newly laid. After a few days the green disappears and the egg turns a whitish grey with reddish strips. On the upper side where the egg is slightly flattened there is a lid, while the underside tapers to a point that is barely wider than the stalk. The egg shell has a fine network of veins. The function of the stalk is in the first place to protect the un-hatched larvae from their cannibalistic siblings, while also offering protection from other natural predators. Female brown lacewings can lay several hundred oval eggs. These eggs, which have no stalks, are 0.7 mm long, cream to beige-coloured and are spread over the leaves. After a few days they turn a pink or mauve colour.

The larvae of Chrysopidae are 2 - 10 mm long. They have large, forwardly projecting jaws and well developed legs. The larva is colourless and transparent when newly hatched, later becoming cream-coloured to light brown, although the colour can vary depending on the prey. The head is a light grey with black eyes and dark mouthparts and antennae, with two converging dark brown stripes dorsally that become broader towards the rear. Two chocolate-brown or red-brown bands run over the body. The larvae of Hemerobiidae are at most 7 - 8 mm long, cream-coloured with red-brown spots and in both appearance and habit resemble the larvae of green lacewings. The legs, antennae and mouthparts are dark. Green lacewing pupae develop in a bullet-shaped cocoon that is hard but flexible and enclosed in numerous white threads, by which it is also attached to the leaf. The pupae are usually separate from each other, but sometimes appear in groups when the population density is particularly high. The development of the pupa is visible externally. The outer cocoon is 8 mm long and 4 mm thick. The inner, more closely spun cocoon is off-white and egg-shaped, about 3 mm long and 2.5 mm thick. The space between the inner and outer cocoon is filled with fine loose webbing. Cocoons can be found attached to the upper or lower surface of leaves, to various other sheltered places, and even on the ground. Brown lacewing pupae are cream-coloured and surrounded by an elliptical, woven brown cocoon and can be found in sheltered sites under bark or between leaves.

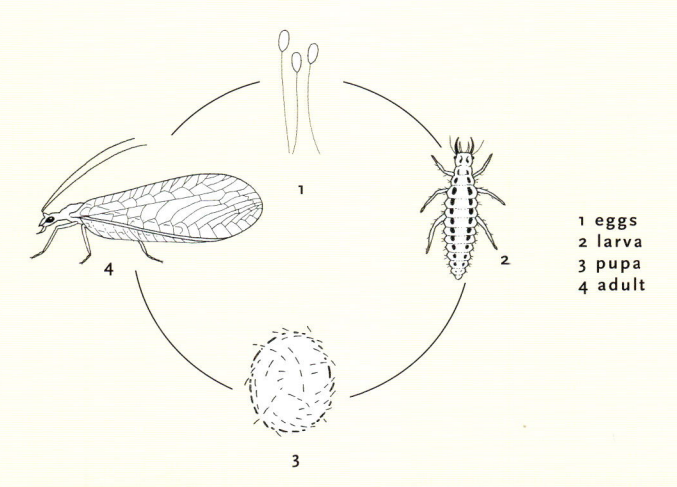

7.67. Egg of *Hemerobius humulinus*

7.68. Predatory larva of *Chrysoperla carnea*

7.69. Adult of *Chrysoperla carnea*

7.70. Adult of *Hemerobius humulinus*

Hemerobiidae
Brown lacewings

Brown lacewings (Hemerobiidae) survive better than green lacewings outside the glasshouse because the minimum temperature for growth and development is lower (around 8°C). They can, however, also appear in the glasshouse where they prey on spider mite and aphid colonies. Two familiar species are *Hemerobius humulinus* and *Micromus variegatus*.

Population growth
Brown lacewings overwinter as full-grown larvae in cocoons spun between dead leaves or in other places offering protection. Pupation then takes place in the spring. The adults emerge and lay eggs on both sides of leaves, beside the main veins. These eggs hatch after about a week. Larvae spend two or three weeks feeding on their prey before they begin to spin a cocoon in which to pupate. Outdoors, different generations appear each year and adults can be found until autumn. Development from egg to adult takes about 4 to 6 weeks in the summer.

Feeding behaviour
Both larvae and adults are predatory. The larvae suck their prey empty, whereas the adults have biting mouthparts.

7.71. Larva of *Hemerobius humulinus*

- Very common outdoors
- *Episyrphus balteatus* is released in glasshouses but can also appear naturally
- Larvae can be predatory

Hoverflies

Hoverflies belong to the order Diptera (the true flies) and constitute the family Syrphidae. Worldwide approximately 5,000 species of hoverfly are known. The adult flies can be identified on the basis of their wing venation. The different species vary markedly in size and shape, but often have clear yellow and black markings, which sometimes resembles those of wasps, bees or bumblebees. However, unlike these hymenopterans, hoverflies cannot sting. This mimicry protects hoverflies against attack by hostile predators such as birds. Hoverflies have only a single pair of wings whereas bees and wasps have two pairs, although these may not always be clearly visible. The wings of hoverflies are held more widely spread, often almost horizontal. In addition they are usually smaller and have short antennae, whereas bees and wasps often have long, thread-like antennae. Finally they may be recognized by their flight behaviour, alternating between periods of motionless hovering in the air with sudden, darting flight. The males in particular are capable of hovering for long periods. The females are usually seen hovering in the air above an aphid colony while they assess its suitability for egg-laying, or over a flower as a potential source of food.

Not all hoverflies feed on aphids. Some live on compost or vegetable material and there are even species whose larvae are aquatic and feed on submerged organic matter. There are also species that occupy the nests of flies, wasps or bees and consume the host larvae, or feed on other insects such as caterpillars, beetles or thrips.

Adult hoverflies require pollen and nectar as food, especially for egg production. Many species are thus efficient pollinators.

Male hoverflies can be distinguished from the females by their larger eyes, which occupy almost the entire head.

In those hoverflies whose larvae feed on aphids the eggs are often laid separately in or by an aphid colony. The eggs are elongated and glossy white, but their colour darkens with age. The larvae are spindle-shaped, with the anterior end always smaller than the posterior. They are usually a transparent white to light orange with black or orange-brown stripes. Like other dipterans, the larvae of hoverflies have no legs. They feed mostly at night, sucking larger aphids completely empty, while young

aphids may be ingested whole. Because the larvae can consume huge numbers of aphids, the aphid population can show a rapid decline once hoverflies appear in the glasshouse.
After moulting several times the larvae pupate. The pupae are teardrop-shaped or barrel-shaped and are green or brown in colour. Among other places they can be found attached to leaves or the ground-litter. Larvae, pupae and adults can all overwinter.

Episyrphus balteatus
Marmalade hoverfly

Episyrphus balteatus occurs naturally in northern Europe and since 1999 has been released against aphids in crops of sweet pepper. In the wild this species of hoverfly appears in considerable numbers during the summer. There are 3 - 5 generations per year. They can be released directly into the glasshouse or reared on banker plant systems.

Population growth
At temperatures around 20°C a female lays on average 500 eggs in the vicinity of aphid colonies, although in some cases up to 1,000 may be laid. The number of eggs laid is particularly dependent on the density of the aphid population. There is a high rate of mortality among the eggs. The time for development from egg to adult depends on temperature (see table 7.3).
At day-time temperatures of 15°C or less reproduction ceases, although development from egg through to the adult stage can occur at temperatures as low as 10°C. At 7.5°C, however, development is no longer possible.
Females live longer than males.

Feeding behaviour
The larvae search for aphids specifically on glabrous plants such as roses and sweet pepper and not on plants with leaf and stem hairs such as cucumber and tomato. Larvae, feeding mainly at night, consume around 300 - 500 aphids during their larval lifespan although this depends on the aphid species and on the density of the aphid population. Larval food uptake increases with the age of the larvae up to the end of larval development, at which point feeding ceases abruptly and the gut is purged, leaving characteristic black specks on the leaves. Adult hoverflies consume pollen and nectar, which is essential for the females' egg production.
In the case of food shortage, hoverfly larvae cease searching and wait for prey. Such passive behaviour contributes little to controlling the aphid infestation. Hoverfly larvae feed during the night.

Table 7.3. Development time (in days) of different stages of *Episyrphus balteatus* at different temperatures (Hart et al., 1997)

stage	temperature (°C)			
	10	15	17	22
egg	10.4	4.4	4.0	2.3
larva	36.2	13.3	10.3	7.7
pupa	34.5	12.1	10.6	6.9
egg-adult	81	30	25	17

Life-cycle and appearance
Episyrphus balteatus

1 egg
2 larva
3 pupa
4 adult

An adult is 10 - 20 mm in length. The thorax is black with a yellow-brown shield, and the abdomen is conspicuously yellow with irregularly broad and narrower transverse black stripes.

The eggs, which are white, elongated and easily seen with the naked eye, are laid close to aphid colonies. Females are attracted to these colonies by the scent of honeydew and the excretions from the cornicles of the aphids.

The larvae are 10 - 20 mm long and a dirty white colour, with the intestine visible through the epidermis. There are 3 larval instars.

Pupation takes place beneath the leaves. The pupa is 7.5 mm long, pear-shaped and of orange-brown colour with dark, wave-like bands.

These hoverflies are excellent fliers and can travel large distances, sometimes flying in swarms. Part of the population flies to Southern Europe or North Africa in the autumn, returning to Northern Europe in the spring. However, there is probably also a small proportion of the population that overwinters in the adult female stage. These females have already mated before entering diapause.

Long leaf hairs injure the larvae, as a result of which they die. In crops such as cucumber, aubergine and tomato, therefore, it is not possible to build up a good population.

7.72. Egg

7.73. Larva

7.74. Pupa

7.75. Adult

- Commercially introduced but can also occur naturally
- *Aphidius* spp. and *Aphelinus abdominalis* are the most important species
- The parasitic wasps have different preferences for aphid species

Parasitic wasps

There are various species of parasitic wasps that parasitize aphids, the most important belonging to the genus *Aphidius*. This genus belongs to the order Hymenoptera and the family Braconidae (braconid wasps). The sub-order Aphidiinae, to which *Aphidius* belongs, consists entirely of aphid parasites.

The three *Aphidius* species of importance are *A. colemani*, *A. ervi* and *A. matricariae*. All these species can appear spontaneously, although the first two are now commercially released in glasshouses. *A. colemani* is an efficient parasite of *Aphis* and *Myzus* species, while *A. matricariae* parasitizes *Myzus persicae*, but not *A. gossypii*. *Aphidius ervi* parasitizes precisely those aphid species that are not parasitized by *A. colemani*, such as *Aulacorthum solani* and *Macrosiphum euphorbiae*.

Aphelinus abdominalis is an endoparasite of various aphid species, including *Macrosiphum*, *Aulacorthum* and *Myzus* species, but prefers the potato aphid. This aphelinid wasp is mainly released to combat the potato aphid and glasshouse potato aphid.

Aphidius colemani, *A. ervi* and *Aphelinus abdominalis* are discussed below.

Aphidius colemani

Aphidius colemani, a braconid wasp that can parasitize around 40 species of aphid, originates from the Near East, but has been introduced in many areas and is now naturalized in The Netherlands. The species is widely used in the biological control of the cotton aphid, peach potato aphid and tobacco aphid. *A. colemani* also parasitizes *Rhopalosiphum padi*, the aphid used in the banker plant rearing system.

Population growth

The development of a population of *A. colemani* can be very rapid, depending on temperature (see table 7.4).

Mating generally takes places within a day after the emergence of the adult from the mummy. Females mate only once, whereas males are capable of multiple matings. Mated females lay both fertilized and unfertilized eggs, the latter producing male offspring, the fertilized eggs producing females. A single female can lay more than 300 eggs. Most eggs are laid during the first three days after emerging as adults (see table 7.5.). The ♀:♂ sex ratio is approximately 2:1.

The life span of *A. colemani* is about 10 days at 18 - 22°. In the summer (and especially the late summer), the population growth of *A. colemani* can be slowed by the appearance of hyperparasites (specifically *Dendrocerus carpenteri*). These are parasitic wasps that parasitize *A. colemani* and other parasitic wasps. They are discussed below.

Parasitization

A. colemani has very effective searching behaviour, which enables the wasps to track down and parasitize aphids even when they are at low density. The behaviour of the female during egg-laying is typical of the Aphidiidae. Having made contact with a prospective host, the female stands on outstretched legs and bends her abdomen forward beneath the thorax and between her legs. By moving her abdomen forward she pierces the aphid with her ovipositor and deposits an egg. All this happens in less than half a second.

A. colemani is an efficient parasite of *Aphis* and *Myzus* species. Large aphids that can stand high on their legs, such as *Macrosiphum euphorbiae*

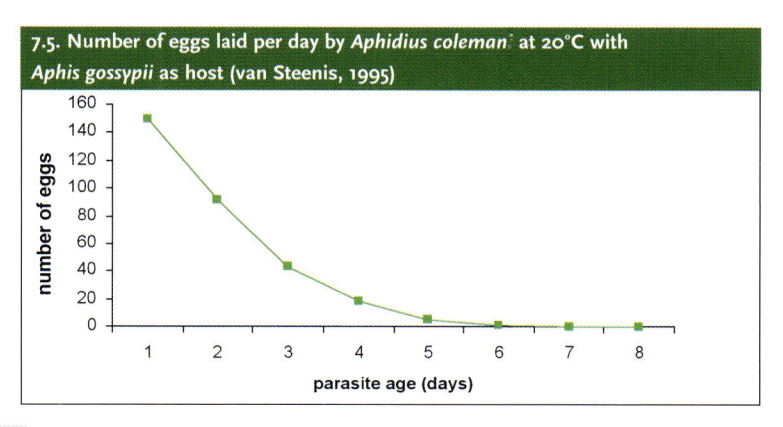

Table 7.4. Development time (in days) of *Aphidius colemani* on *Aphis gossypii* at 20°C and 25°C (van Steenis, 1995) and at 18°C, 22°C and 26°C (van Schelt, 1994)

temperature (°C)	egg - mummy	egg - adult
18	11.2	
20		13
22	8.3	
25		10
26	6.5	

7.5. Number of eggs laid per day by *Aphidius colemani* at 20°C with *Aphis gossypii* as host (van Steenis, 1995)

and *Aulacorthum solani*, are not suitable hosts. Within the space of two days, a single female can parasitize hundreds of aphids. The aphids do not die at once but usually continue to feed and secrete honeydew. They also remain capable of transmitting viruses. Parasitized adult aphids can even produce offspring for two more days, but if an aphid is parasitized in a young stage there will be no reproduction. Winged individuals can also be parasitized.

Apart from the parasitization itself the disturbance of the aphid population by the presence of the parasitic wasp also has an effect; the alarmed aphids secrete a warning substance which effects the entire population, often causing aphids to drop from the leaf and fall to the ground. Although many aphids die as a result, many will survive and migrate to other plants to begin new colonies.

In order to avoid becoming covered in their secretions, the parasite will not establish in very dense aphid colonies. At high temperatures (>30°C) the parasite is less effective.

7.76. Parasitizing *Aphidius colemani*

Aphidius ervi

Aphidius ervi is a good supplement to *Aphidius colemani* because it controls aphids that *A. colemani* does not parasitize (in particular the potato aphid *Macrosiphum euphorbiae* and the glasshouse potato aphid *Aulacorthum solani*). This species develops faster than *Aphelinus abdominalis*, which can parasitize the same species. *A. ervi* also parasitizes *Sitobion avenae*, the aphid used in the banker plant system. *A. ervi* is a native European species and better adapted to cooler conditions.

Population growth
The growth of a population of *A. ervi* is comparable to that of *A. colemani*. Both species multiply rapidly and disperse efficiently. This *Aphidius* species is also susceptible to being hyper-parasitized. On emergence as an adult wasp, a female already has around 100 mature eggs in her abdomen and can begin parasitizing aphid hosts at once, with most eggs being laid during the first three days after adult emergence.

Parasitization
The pattern of parasitization is much the same as for *A. colemani* (see plate 7.77).
After approximately two weeks the parasitized aphid is completely consumed and only a light brown mummy remains, from which a new wasp emerges through a hole which it gnaws in the cuticle.

Searching behaviour and dispersal
The flight behaviour is highly efficient and wasps actively search for their aphid hosts. Even at a temperature of 10°C the wasps still fly.

7.77. Parasitizing *Aphidius ervi*

Life-cycle and appearance
Parasitic wasp

1 adult
2 egg
3 aphid with egg
4 aphid with larva
5 aphid mummy
6 aphid mummy with pupa
7 empty aphid mummy

Female *Aphidius* spp. deposit an egg in an aphid (see diagram of life-cycle) in which there subsequently develop four larval instars. Once larval development is complete the larva spins a cocoon inside the aphid cuticle, causing it to swell. The cuticle turns to a hard, leathery brown to golden yellow casing known as a mummy.

An adult braconid wasp leaves this mummy via a round hole. The colour and form of the mummy do not, however, provide a reliable key to the species of wasp within: many braconid wasps that parasitize aphids produce a mummy similar to that of *Aphidius*.

Although the aphid continues to grow and feed after being parasitized, the size of the full-grown wasp is highly dependent not only on the aphid species but also on the size of the host at the time of parasitization.

A male wasp has slightly longer antennae, a rounded abdomen that is shorter than the wings, and is black with dark brown legs. A female has a more pointed abdomen that is as long as the wings, is equipped with an ovipositor, and is black with light brown legs.

A. colemani and *A. ervi* closely resemble each other, although *A. ervi* is darker, particularly the waist, and roughly twice the size.

An adult *Aphelinus abdominalis* is 2.5 to 3 mm in length, with short legs and short antennae. Females have a black thorax and yellow abdomen. Males are slightly smaller and their abdomen is a slightly darker colour.

A parasitized aphid hardens to become a leathery, black mummy. The adult wasp appears through an irregularly serrated hole at the posterior end of the mummy, on which the aphid's antennae and cornicles can still be recognized. The *Aphelinus* mummy may be distinguished from the mummies of *Aphidius* species by their less swollen, more elongated form and their black colour.

7.78. Mummy of *Aphidius colemani*

7.79. *Aphidius ervi* (left) and *Aphidius colemani* (right)

7.80. *Aphelinus abdominalis*

7.81. Mummies of potato aphids parasitized by *Aphelinus abdominalis*

Aphelinus abdominalis

Aphelinus abdominalis (Hymenoptera: Aphelinidae) is an endoparasite of different species of aphids, including *Macrosiphum*, *Aulacorthum*, and *Myzus* species, but with a preference for potato aphids. The adult parasitic wasp also feeds on aphids, particularly small nymphs, which are pierced and sucked empty. Nymphs of species that are too small to parasitize can also be consumed. Since 1993 this species has been used to control glasshouse potato aphid and potato aphid. *A. abdominalis* also parasitizes *Sitobion avenae*, an aphid used in banker plant systems.

Population growth
It takes 16 days for the larva to develop to the adult stage at 24°C, with the first mummies turning black after 7 days.
During the first 3 weeks of adult life the female lays 5 - 10 eggs per day. In total 250 eggs may be laid, although this depends strongly on the host. The sex ratio is 1:1. All host stages can be parasitized including the winged adult.
A. abdominalis is better able to withstand high temperatures than *Aphidius* species.
The average lifespan of an adult female is roughly 30 days. This parasitic wasp thus has a slow rate of development but a long lifespan, and although it lays few eggs per day it does so over a longer period, and is slower to migrate.

Parasitization
Once the female detects an aphid, she turns herself around, curls up the tips of the wings and repeatedly stabs her ovipositor into one of the aphids legs, probably to paralyze it. She subsequently pierces the underside of the aphid with the ovipositor to inject her egg (see plate 7.82). This can take several minutes.
Host feeding plays an important role in reducing pest numbers, with the adult wasp piercing the aphid with her ovipositor and sucking it empty without laying an egg (see plate 7.83).
The small aphids are used for host feeding while the larger individuals are parasitized, and the control of the aphids is due to host feeding and parasitization in roughly equal measure. Superparasitization (multiple eggs in a single aphid) rarely occurs, which would indicate that females are able to distinguish between aphids that have already been parasitized and those that have not.
Hyperparasitization (especially by *Asaphes* species) is seen, but in practice this is less of a problem than with *Aphidius* spp.

7.82. Parasitizing *Aphelinus abdominalis*

7.83. Host feeding *Aphelinus abdominalis*

- Lay their eggs in parasitized aphids
- Cause decline in the population of aphid-parasitic wasps
- Can seriously disrupt the biological control of aphids by parasitic wasps

Hyperparasites of aphid-parasitic wasps

Hyperparasitic wasps are parasites of parasitic wasps. Because these hyperparasites appear in nature in large numbers in the summer and spontaneously enter the glasshouse they can constitute a serious threat to the biological control of aphids.

Most aphid parasites, with their rapid development and short life-span, are adapted to a growing aphid population. But as soon as a large percentage of aphids are already parasitized they have trouble maintaining their population level. There is thus no equilibrium between the population of wasps and aphids. In this sort of situation, despite their slower development, the hyperparasites have the advantage. Only the use of gall midges or a regular re-introduction of parasites can then give the desired result.

There are various species of hyperparasites. The most common species is *Dendrocerus carpenteri*, although *Alloxysta* species are often also encountered. These are discussed briefly below.

Aphidius species are particularly susceptible to parasitization by hyperparasites. Occasionally, however, *Aphelinus abdominalis* is also parasitized, by *Asaphes* spp. and *Alloxysta* spp. among others.

Externally, the mummy gives no evidence of whether it contains a hyperparasite or not. Only when a parasitic wasp emerges from the mummy can the distinction be made; a parasitic wasp often leaves a lid which covers the exit hole it has made, whereas a hyperparasite gnaws a round hole with an irregular edge and never leaves a lid (see plate 7.84). In addition, the exit openings of hyperparasites can be found anywhere on their mummies whereas *Aphidius* always emerges from the posterior end.

7.84. Mummy from which a hyperparasite has emerged

Dendrocerus carpenteri

Dendrocerus carpenteri belongs to the order Hymenoptera, the superfamily Ceraphronoidea and the family Megaspilidae. *D. carpenteri* is a parasite of Aphidiinae, the subfamily to which belong the genera *Aphidius, Praon, Ephedrus* and *Lysiphlebus*. *Aphelinus*, however, belongs to a different family and is less affected by this hyperparasite. *Dendrocerus carpenteri* is distributed almost worldwide and is found on wasps parasitic on a very wide range of aphids. Parasitized *Aphis, Myzus, Macrosiphum* and *Acyrthosiphon* are among the most frequently observed hosts.

The adult wasps are 1.5 mm long and a glossy black. They are almost as large as *Aphidius*, but of somewhat stockier build (see plate 7.85). With good magnification one can see the characteristic wing venation and the clear 'elbow' in the antennae. It is possible to identify these wasps with the naked eye if one observes their behaviour; the hyperparasites run in a zigzag fashion over the leaf, drumming agitatedly with their antennae, whereas *Aphidius* moves in a much calmer manner over the leaf and holds its antennae fairly still.

The hyperparasitic wasp parasitizes full-grown larvae or young pupae of Aphidiinae, laying its egg in the aphid, beside the larva of the aphid-parasite which is then subsequently eaten from the outside in. The young hyperparasite pupates within the cocoon of its host wasp. After approximately 2 weeks the adult hyperparasite emerges from the mummy. Because the aphid-parasitic wasp is parasitized in a late developmental stage, the hyperparasites always emerge later than *Aphidius*. The females can live for more than 10 days. Overwintering occurs within the host mummy, although sometimes adults can also overwinter.

Figure 7.86 shows a drawing of an ectoparasitic hyperparasite.

7.85. Dendrocerus carpenteri

7.86. Hyperparasite laying an egg against the parasitic wasp larva in an aphid

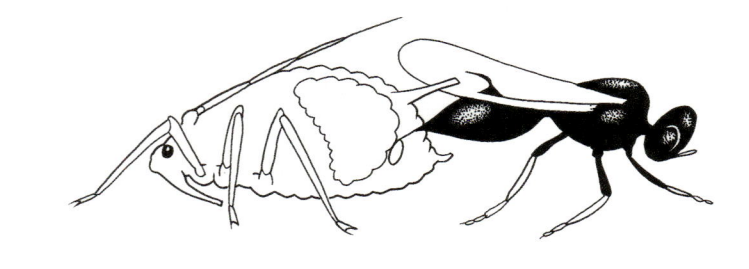

Alloxysta spp.

Hyperparasites of the genus *Alloxysta* belong to the order Hymenoptera, the superfamily Cynipoidea and the family Alloxystidae. They are obligate hyperparasites. Unlike *D. carpenteri* they lay their eggs *inside* the larvae of the aphid-parasitic wasp occupying the mummy of the aphid, and are therefore referred to as endoparasites. *Alloxysta* can attack many wasps parasitic on various different aphid species. Figure 7.87 shows a drawing of an endoparasitic hyperparasite.

7.87. Hyperparasite laying an egg in the larva of a parasitic wasp in an aphid

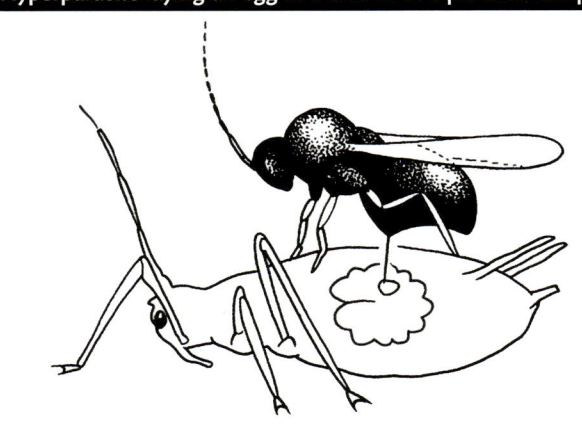

- Occur naturally
- Appear in late summer and autumn
- Can sometimes eliminate an entire aphid population

Entomopathogenic fungi

Under favourable circumstances, such as the warm humid weather of late summer and autumn, fungal moulds can significantly reduce aphid populations. Several Entomophthorales species are specialist parasites of particular species of aphid.

The susceptibility of aphids to Entomophthorales moulds differs from species to species. An aphid infected by the fungus can be distinguished from a healthy individual by its unusual colour: affected green species betray the presence of the mould by their grey-white appearance and a creamy or white fungal coating, whilst red-coloured individuals usually assume a pale pink colour with a grey-white covering. The different species of Entomophthorales cause characteristically different discolourations, which are often grey-green, but sometimes violet or red-brown. At a later stage the affected aphid can always be recognized by the external covering of hyphal threads. Plate 7.88 shows an aphid infected by an Entomophthorales mould.

More information on Entomophthorales moulds can be found in chapter 5. So far it has proved impossible to produce these fungi commercially. Strains of the entomopathogenic fungus *Verticillium lecanii* are also known that preferentially infect aphids. This fungus has been described in detail in chapters 4 and 5. This species can appear spontaneously, with affected aphids showing a white fungal "fuzz" (see plates 7.89 and 7.90).

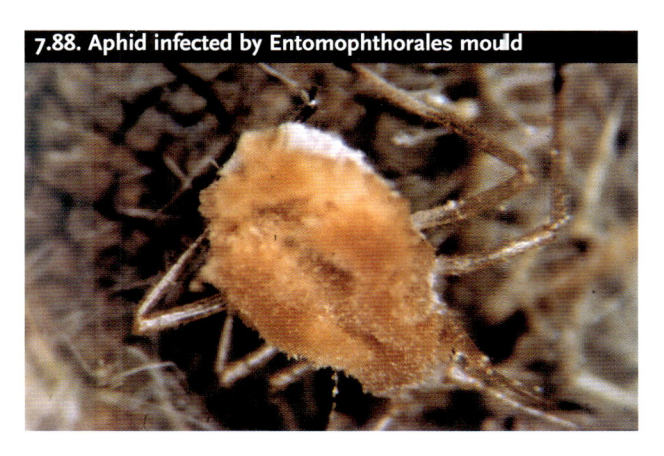

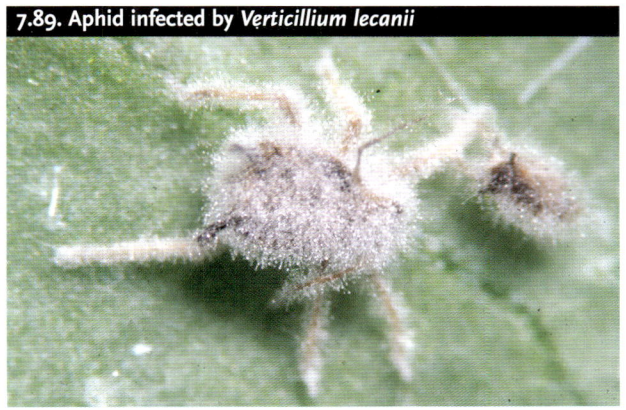

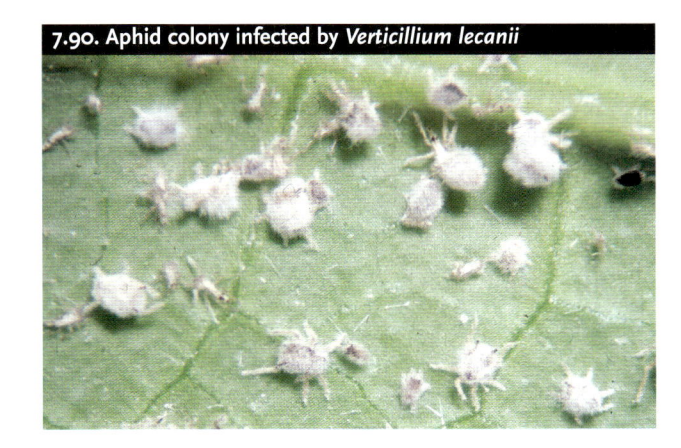

7.88. Aphid infected by Entomophthorales mould

7.89. Aphid infected by *Verticillium lecanii*

7.90. Aphid colony infected by *Verticillium lecanii*

Overview of important aphid parasitic wasp species

characteristics	*Aphidius colemani*	*Aphidius ervi*	*Aphidius matricariae*	*Aphelinus abdominalis*
taxonomy				
superfamily	Ichneumonoidea	Ichneumonoidea	Ichneumonoidea	Chalcidoidea
family	Braconidae	Braconidae	Braconidae	Aphelinidae
subfamily	Aphidiinae	Aphidiinae	Aphidiinae	
hosts	many species small aphids standing low on their legs, notably melon cotton aphid, peach potato aphid and tobacco aphid	among others, potato aphid and glasshouse potato aphid	among others, peach potato aphid	different species of aphid, but especially potato aphid and glasshouse potato aphid
natural or introduced	natural and introduced	natural and introduced	natural	natural and introduced
adult appearance				
colour	♀: black with light brown legs ♂: black with dark brown legs	darker than *A. colemani*	black with brown legs	♀: black thorax and yellow abdomen ♂: black thorax, abdomen darker than in ♀
size	± 2 mm	- two times the size of *A. colemani*	± 2 mm	2.5-3 mm
tergite 1*	- laterally: small number of deep grooves - dorsally: transparent brown	- laterally: no grooves - dorsally: black	- laterally: large number of fine grooves - dorsally: transparent brown	
mummy	golden yellow	golden yellow as *A. colemani*	golden yellow as *A. colemani*	black

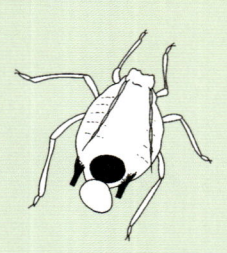

*tergite 1 = small plate between thorax and abdomen, view under 100 x magnification

characteristics	*Praon volucre*	*Dendrocerus carpenteri*	*Alloxysta* spp.
taxonomy			
superfamily	Ichneumonoidea	Ceraphronoidea	Cynipoidea
family	Braconidae	Megaspilidae	Alloxystidae
subfamily	Aphidiidae		
hosts			
	notably potato aphid	parasitic wasps of Aphidiidae	many aphid-parasitic wasps
natural or introduced			
	natural	not applicable	not applicable
adult appearance			
colour	glossy black and brown	glossy black, with 1 distinct wing vein	glossy black
size	2-4 mm	1.5 mm, stockier than *Aphidius*	≤ 2 mm
mummy			
	white transparent, pupates under the empty aphid cuticle	identical to host mummy	identical to host mummy, see *D. carpenteri*

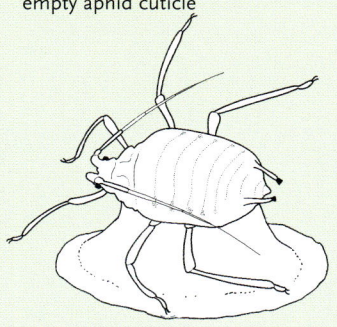

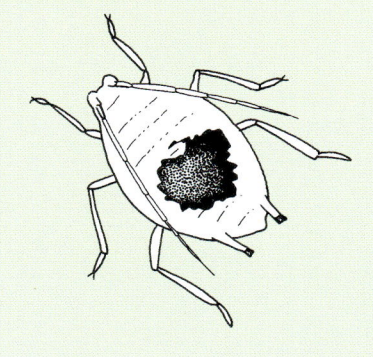

Biological control of aphids

OVERVIEW OF THE BIOLOGICAL CONTROL OF APHIDS				
product name	natural enemies	aphid pest stage controlled	stage that combats aphids	remarks
APHIPAR	*Aphidius colemani*	notably melon cotton taphid, tobacco aphid and peach potato aphid	adult female	recommended at start of infestation
ERVIPAR	*Aphidius ervi*	notably potato aphid and glasshouse potato aphid	adult female	recommended at start of infestation
CHRYSOPA	*Chysoperla carnea*	many species of aphid	larva	with high prey density on smooth-leaved, low-growing crops
SYRPHIDEND	*Episyrphus balteatus*	many species of aphid	larva	with high prey density, on smooth leaves, adults need honey and pollen
APHIDEND	*Aphidoletes aphidimyza*	many species of aphid	larva	use if colonies are about to form, sometimes preventively
ADALIA	*Adalia bipunctata*	many species of aphid	larva and adult	with high prey density, on smooth leaves
APHILIN	*Aphelinus abdominalis*	notably potato aphid, glasshouse potato aphid and *Myzus* spp.	adult female	slow, long-lasting effect
APHIBANK	*Aphidius colemani*	notably the melon cotton aphid, peach potato aphid	adult female	banker plant system (*R. padi*) used preventatively
ERVIBANK	*Aphidius ervi*	notably glasshouse potato aphid	adult female	banker plant system (*S. avenae*) used preventatively

Where aphids develop very rapidly, natural enemies are introduced as soon as the aphids first appear. However, preventive introductions are also possible. Combating aphids with natural enemies calls for a different strategic approach to each situation and each crop. The parasite *Aphidius colemani* is mainly recommended for situations with few, dispersed aphids, and is often used in a preventive manner.

If the aphids are discovered in colonies, and the infestation increases, gall midges and lacewings are recommended. These are also capable of killing many different species of aphid whereas parasitic wasps are much more specific. If the ground is covered with sheeting, the larvae of the gall midge *Aphidoletes aphidimyza* has problems with pupation and population growth is restricted. The material then has to be repeatedly introduced. The lacewing *Chrysoperla carnea* is usually applied in low-growing crops.

The parasitic wasp *Aphidius ervi* is mainly recommended for the early stages of infestations, particularly of the potato aphid *Macrosiphum euphorbiae* and glasshouse potato aphid *Aulacorthum solani*.

Aphelinus abdominalis is also mainly released against the potato aphid *Macrosiphum euphorbiae* and the glasshouse aphid *Aulacorthum solani*.

The ladybird *Adalia bipunctata* is recommended as a biological corrective agent in cases of increasing aphid infestation, where aphids are present in large colonies. An excess of ladybirds, however, can disrupt the build-up of a population of parasitic wasps.

The gall midges are supplied as cocoons, the parasitic wasps in the pupal form (in mummies) and the lacewings as larvae.

The international trade-names of the different products are as follows:
the gall midge product is APHIDEND;
the product containing the parasitic wasp *Aphidius colemani* is APHIPAR;
the product containing the parasitic wasp *Aphidius ervi* is sold under the name ERVIPAR, while the parasitic wasp *Aphelinus abdominalis* is sold under the name APHILIN;
the lacewing is sold under the name CHRYSOPA.

To be able to keep an aphid infestation to a very low level in its early stages, one can also use a banker plant system where natural enemies are reared in the glasshouse on an aphid species which is incapable of attacking the crop. In this manner, a considerable population of natural enemies is built up before the aphid infestation occurs. These will then disperse further in the glasshouse as soon as aphids become established. Two banker plant systems have been developed, one rearing *Aphidius colemani* and the other for rearing *Aphidius ervi* or *Aphelinus abdominalis* using winter wheat as the host plant. The grain aphids *Rhopalosiphum padi* and *Sitobion avenae* are used on each respectively. These two systems cannot be used simultaneously because *R. padi* will always over-run the culture of *S. avenae*.

The international trade-name of the products is APHIBANK for the *Aphidius colemani* rearing system and ERVIBANK for the *Aphidius ervi* or *Aphelinus abdominalis* system. The products consist of boxes of growing wheat plants and aphid hosts onto which ERVIPAR of APHIPAR has to be released.

Aphidoletes aphidimyza and *Aphelinus abdominalis* can also be introduced in the same way. It is advised that the gall midge or the *Aphelinus* parasitic wasp should be introduced as soon as hyperparasitization is observed. These will then take over the work of the *Aphidius* spp. When hyperparasites appear the efficacy of *Aphidius* spp. is so seriously disrupted that there is no further point in using APHIBANK cultures. A better strategy is to use ERVIBANK for the culture of *Aphelinus abdominalis* and of the gall midge. In sweet pepper the hoverfly *Episyrphus balteatus* can be introduced as pupae on cards. The product name for this is SYRPHIDEND.

Orius spp. (especially *Orius majusculus*) can also contribute to the control of aphids, although these predatory bugs are not specifically used for this purpose.

Knowing and recognizing **Butterflies and moths**

8. Butterflies and moths and their natural enemies

There are various species of butterflies and moths, the larvae (or caterpillars) of which cause damage to horticultural crops. Caterpillars were a serious problem until broad-spectrum pesticides became widely used in 1940, with substrate sterilization killing many of the species that pupated in the soil. However, the increasing use of selective control methods, plus a reduction in the use of substrate sterilization has meant that caterpillars have become increasingly important pests in glasshouse crops in recent years. Common species include the tomato looper, *Chrysodeixis chalcites*, the tomato moth, *Lacanobia oleracea*, the cabbage moth, *Mamestra brassicae*, the silver-Y moth, *Autographa gamma*, the beet armyworm, *Spodoptera exigua*, the Egyptian cotton leafworm *Spodoptera littoralis* and the tomato fruitworm *Helicoverpa armigera* all of which belong to the family Noctuidae (the owl or noctuid moths). The most common, but not always the most harmful of the cutworms (*Agrotis* spp.) also belong to this family. Two other problem species worthy of mention are the cabbage leafroller, *Clepsis spectrana*, and the carnation leafroller, *Cacoecimorpha pronubana*, both of which belong to the Tortricidae. In recent years there have also been reports of several new species, such as *Duponchelia fovevalis*, and the banana moth *Opogona sacchari*.

The bacterium *Bacillus thuringiensis* was first found in caterpillars at the beginning of the last century. In Japan this bacterium was a serious problem in silkworm culture, whereas German scientists saw the possibility of exploiting the bacterium as a biological control agent. Over subsequent years, many strains of *B. thuringiensis* have been looked at, and a number of commercial products have been developed. At present, the *kurstaki* strain is the most widely used in biological control. In some recent products, the *aizawai* strain has also been used.

Knowing and recognizing **Butterflies and moths**

Egg parasites of the genus *Trichogramma* are also available, and several species such as *Trichogramma brassicae* have been used throughout the world.

Apart from these organisms, the predatory bugs *Macrolophus caliginosus* and *Orius* spp., together with spontaneously occurring parasites such as *Cotesia plutellae*, *Eulophus* spp., *Pimpla instigator*, and birds such as the rusty-capped fulvetta, *Alcippe brunnea* can all contribute to the control of lepidopteran larvae. A virus product is also available that is effective against *Spodoptera exigua*.

8.1. *Chrysodeixis chalcites*

8.2. *Lacanobia oleracea*

8.3. *Mamestra brassicae*

8.4. *Autographa gamma*

Knowing and recognizing **Butterflies and moths**

Butterflies and moths

8.5. *Spodoptera exigua*

8.8. *Duponchelia fovealis*

8.6. *Clepsis spectrana*

8.9. *Opogona sacchari*

8.7. *Cacoecimorpha pronubana*

8.10. *Agrotis* spp.

The most important caterpillar species in glasshouse crops are:

Chrysodeixis chalcites (tomato looper)
Lacanobia oleracea (tomato moth)
Mamestra brassicae (cabbage moth)
Autographa gamma (silver-Y moth)
Spodoptera exigua (beet armworm)
Spodoptera littoralis (Egyptian cotton leafworm)
Helicoverpa armigera (tomato fruitworm)
Clepsis spectrana (cabbage leafroller)
Cacoecimorpha pronubana (carnation leafroller)
Duponchelia fovealis
Opogona sacchari (banana moth)
Agrotis-spp. (cutworms)

Butterflies and moths form the order Lepidoptera, a very large group with more than 100,000 species that are easily distinguishable from other insects. Despite the number of lepidopteran species, it is a highly uniform group, both in appearance and life habit.

The most harmful caterpillars that appear in glasshouses belong to the Noctuidae (the noctuid or owl moths), which is the largest of all lepidopteran families. Most adults of this family are nocturnal, and apart from a few strikingly conspicuous species, most are rather drab. The fore-wings are mostly various shades of brown and grey, whilst the hind-wings are often strikingly coloured. The caterpillars usually have a sparse covering of hairs and mostly pupate in the ground. The pupae, which are remarkably similar throughout the family, are an inconspicuous brown. The most important species for glasshouse crops are *Chrysodeixis chalcites* (the tomato looper), *Lacanobia oleracea* (the tomato moth), *Mamestra brassicae* (the cabbage moth), *Autographa gamma* (the silver-Y moth), *Spodoptera exigua* (the beet armyworm) *Spodoptera littoralis* (Egyptian cotton leafworm), *Helicoverpa armigera* (tomato fruitworm) and various *Agrotis* species (cutworms).

Another family containing many species that damage glasshouse crops is the leafroller family, the Tortricidae. The name 'leafroller' is due to the fact that the caterpillars often live in rolled-up leaves, although some species live between leaves or flowers that have been spun together, in holes in stems, flowers or fruit, or in the bark of trees. Another diagnostic feature is that, when touched, leafrollers will wriggle and dangle on the end of a thread. Although some species are restricted to a single crop, most are polyphagous. Most species are brownish or mottled grey in colour. Leafroller species encountered in crops under glass include *Clepsis spectrana* (the cabbage leafroller) and *Cacoecimorpha pronubana* (the carnation leafroller).

The increasingly common *Duponchelia fovealis* belongs to the family Pyralidae, whilst in the absence of agreement amongst taxonomists, the banana moth (*Opogona sacchari*) is currently assumed to belong to the family Tineidae.

Life-cycle and appearance
Butterflies and moths

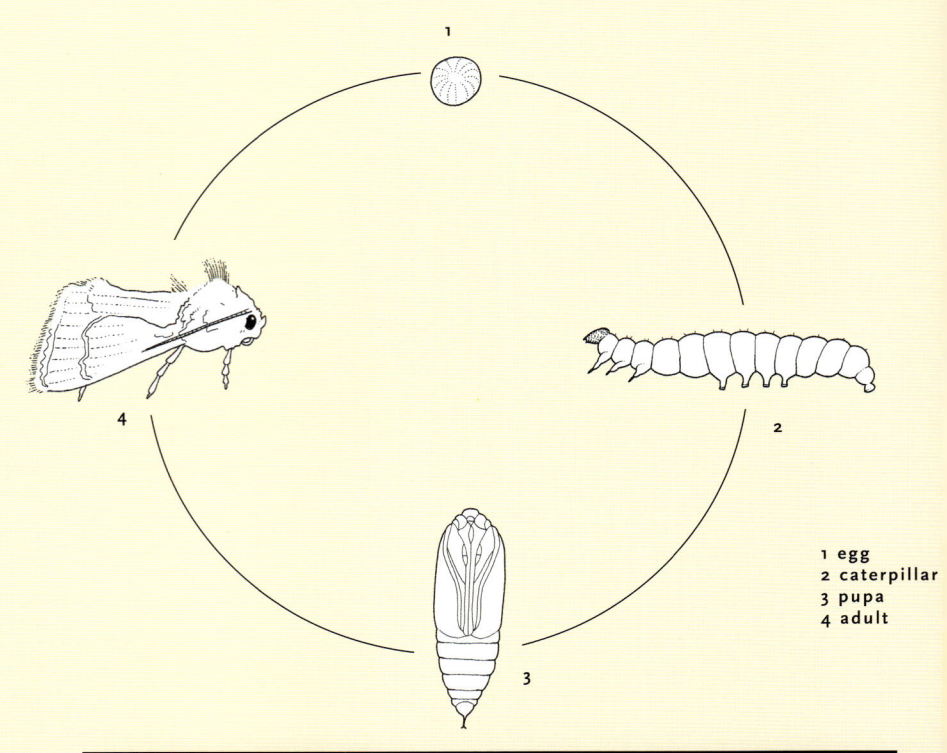

1 egg
2 caterpillar
3 pupa
4 adult

Butterflies and moths undergo four different stages in their development; the egg, caterpillar, pupa and adult.

Adults are of various sizes, but usually have two pairs of membranous wings that are covered, as is the body, with very small scales. These scales, which overlap each other like roof tiles, vary in size and form and are responsible for the vivid colours of some adults. The head is equipped with two large compound eyes and a pair of antennae which function as olfactory organs, and which vary widely in form. Most antennae are threadlike, with males having pinnate (feathery) antennae with which females can be detected over long distances. Females probably use their antennae to select the plant on which to lay their eggs.

Almost all butterflies and moths feed by means of a specialized suction tube, the proboscis, which when not in use is held rolled up beneath the head.

The eggs are often deposited in clusters, sometimes separately, on the leaf or even on glasshouse material. A cluster of eggs is often precisely arranged, either single-layered in regular lines or overlapping each other, and sometimes masses of eggs are deposited in a multi-layered fashion. The eggs of each species have a characteristic shape, colour and marking. The number of eggs laid by a single female can vary from a few dozen to more than 100 and in a few moths up to a few thousand.

The caterpillar has a well developed head equipped with tough jaws, three pairs of legs on the thorax, generally four pairs of false legs on abdominal segments 3 to 6 (the prolegs), and a single pair on the last segment (the clasper) at the end of the body. There are exceptions to this number of legs in the caterpillars of the geometers or inchworms (also known as loopers) of the family Geometridae, and the other loopers, or semi-loopers, of the family Noctuidae. These only have one or two pairs of anterior prolegs and a pair of terminal claspers behind the three pairs of thoracic true legs, enabling them to move in a characteristic looping fashion.

The head capsule of the small caterpillar can be made out through the shell of the egg just before it hatches. Once the caterpillar is sufficiently developed to leave the egg, it gnaws a hole in the shell just large enough for the head to emerge, after which it frees itself completely in one sudden movement. At the end of the prolegs there are tiny hooks by which a caterpillar can attach itself to the leaf very securely.

8.11. Caterpillar of *Mamestra brassicae*

Normal caterpillar (above) and (pseudo)looper caterpillar (below)

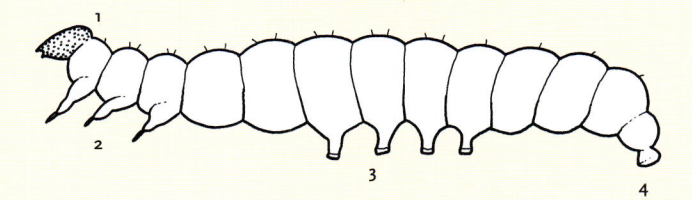

1 head capsule
2 true legs
3 prolegs or 'false legs'
4 clasper

8.12. Eggs of *Spodoptera exigua*

8.13. Caterpillar of *Autographa gamma*

8.14. Pupa of *Chrysodeixis chalcites*

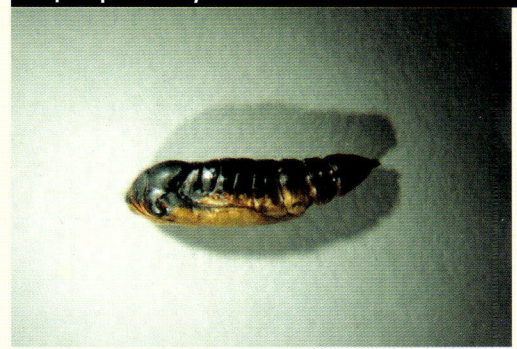

8.15. Adult of *Lacanobia oleracea*

Some caterpillars feed on flowers, developing seeds, stalks or roots. However, by far the majority feed on foliage, using their powerful jaws to nibble leaves back to the mid-rib. Caterpillars eat almost uninterruptedly, apart from when moulting. The number of moults varies from three to more than ten, with most species moulting four or five times. During their development, caterpillars grow in length roughly 10 to 20-fold, and their weight increases by a factor of two to three thousand.

Once fully grown, the caterpillar ceases feeding and goes in search of a suitable place to pupate. Although different species have different preferences, the chosen site must be suitable for pupation and for the adult to emerge. Thus, caterpillars must not pupate too deep in the ground. Once a suitable place has been found, silk threads are extruded from the silk glands. By moving to and fro the caterpillar spins a network of silk threads that completely enclose it inside a cocoon. Some species do not spin a cocoon, but pupate in a hole in the ground which they reinforce with saliva and silk threads. The pupa is shorter and blunter than the caterpillar, and has a wrinkled cuticle. This old larval cuticle soon splits to reveal the glossy pupal cuticle beneath. In butterflies that do not form a cocoon, this visible pupa is known as a 'chrysalis'. A pupa does not feed and does not move; only a small posterior part of the abdomen is capable of movement, whilst the anterior portion (where the head with its proboscis, and the thorax with wings and legs develop) is completely immobile. It is during the pupation process that the reorganization of the caterpillar into the adult takes place. This metamorphosis involves a breakdown of the larval tissues, and the re-development of the adult form from special, previously separated groups of adult primordial cells called the imaginal discs.

Eventually, the adult moth or butterfly emerges from the pupa. The newly emerged insect first searches for a suitable place in which to anchor itself by its fore-legs, and then unfurl its wings. This is followed by an almost imperceptible pumping action of the abdomen, which fills the wing veins with body fluid, thus inflating them until they reach full size. This process takes about 10 to 20 minutes, and after a further hour or two the wings harden and the adult flies away.

Most species overwinter as pupae, although overwintering in the egg, larval or adult stage is not unknown. The rate of development is much slower at lower temperatures.

Knowing and recognizing **Butterflies and moths**

Damage

The caterpillars of butterflies and moths can cause serious economic damage. The majority of pest species feed on leaves and young shoots, and their enormous feeding capacity can rapidly decimate a plant. Very small caterpillars are not able to chew, and so graze the underside of leaf. This causes a window effect where the epidermis of the upper leaf surface remains undamaged. Young caterpillars are often found in groups, spreading out over the plant as they grow larger. As their size increases, so too does the damage they inflict; leaves are often skeletonized, with only the mid-rib left intact. Flowers, fruits and growing tips can also be eaten by larger caterpillars. Some species bore holes into fruit or stems, making them difficult to detect and control.

In addition to consumption of the plant, caterpillars also spoil the plant with the large quantities of frass they leave behind. With fruiting vegetables (particularly sweet pepper) this means the fruit has to be polished after harvesting, adding to the cost of the operation.

Several examples of caterpillar damage are shown in plate 8.16.

8.16. Damage caused by caterpillars

Leaf damage (cucumber)

Leaf damage (sweet pepper)

Damage to fruit (sweet pepper)

General view of leaf damage (sweet pepper)

- The commonest lepidopteran species in glasshouses
- Found in glasshouses throughout the year, but mostly encountered in late summer and autumn
- Widely distributed through the crop

Chrysodeixis chalcites
Tomato looper

The tomato looper, *Chrysodeixis chalcites*, is native to tropical and sub-tropical regions of the world, and is also the most frequently encountered lepidopteran species in glasshouses. The caterpillars inflict damage on many crops, including cucumber, aubergine, sweet pepper, tomato, chrysanthemums and roses.

Originally an annual migrant from southern Europe, *C. chalcites* appeared in more northerly countries each year, and although unable to overwinter outside in northern Europe, it can survive throughout the year in glasshouses. The species belongs to the family Noctuidae and the sub-family Plusiinae. It is very similar to the silver-Y moth, *Autographa gamma*.

Life-cycle and appearance

The moth has a wing span of 32 - 37 mm. They are predominantly brown-gold (hence the name *Chrysodeixis*, which means "gold-shining") with two conspicuous droplet-shaped white marks on the fore-wing that sometimes overlap each other. The hind-wings are brown. The body is brown and hairy and, when at rest, the hairy dorsal plume is characteristically distinct. The antennae are long and thin. The moths do not fly during the day unless disturbed.

The eggs, which are dingy white in colour, round and ridged, are mostly deposited separately and distributed throughout the crop. They are often attached to the undersides of leaves.

The caterpillar has a green head capsule and a yellowish-green body with a clear yellow longitudinal stripe on either side, and several less conspicuous longitudinal stripes dorsally. The body is sparsely covered with a few stiff hairs. In older caterpillars, each segment has a clear black dot above the lateral yellow lines. There are two pairs of anterior prolegs and the caterpillar moves with the same looping motion as an inchworm. Last stage caterpillars measure about 40 mm in length. The caterpillars of this species are very aggressive, and encounters often involve bouts of pushing, sometimes culminating in death.

Pupation takes place in a web, often in a leaf that has been folded together, but can also occur in glasshouse materials. Unlike most Noctuidae, however, the pupa is not found in the soil. The pupa has a different appearance from that of other noctuids; it is 2 cm long, green to yellowish-brown with a darker dorsal side. Eggs, larva and adults of *C. chalcites* are shown in plate 8.17.

Population growth

Unmated females lay sterile eggs and thus produce no progeny. A mated female may lay more than 1,000 eggs which hatch four to six days after they are laid. The caterpillar stage lasts two to three weeks, and the pupal stage lasts a further one to two weeks. After this, the moth lives for another one to one and a half weeks. In the glasshouse, several overlapping generations may be present simultaneously, and thus all stages can be present throughout the year. However, they are most frequent in late summer and autumn.

Damage

The caterpillars consume both leaves and fruit, causing the familiar damage associated with lepidopteran larvae.

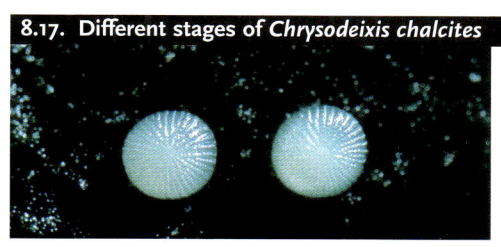

8.17. Different stages of *Chrysodeixis chalcites*

Eggs

Caterpillar

Moth with coalesced white marking on the forewing

Moth with 2 separate spots on the forewing

- A common species
- Mostly found in summer and autumn
- Can damage leaves, stems and fruit

Lacanobia oleracea
Tomato moth

Lacanobia oleracea, the tomato moth, is a particular pest in glasshouses but can also inflict damage on outdoor crops. It is a polyphagous species affecting many different crops – tomato, sweet pepper, lettuce, brassicas, cucumber, cut flowers (particularly chrysanthemums), apple and other perennial and woody shrubs and trees. *L. oleracea* belongs to the family Noctuidae and the sub-family Hadeninae. It is found throughout Europe, North Africa and Asia Minor.

Life-cycle and appearance
Adult *L. oleracea* have a wingspan of 30 - 40 mm. The fore-wings are a reddish-brown colour with a feint light brown, kidney-shaped marking, and a thin, toothed white line along the posterior edge of the fore-wing. As in most noctuid moths, the hind-wing is greyish and lighter than the fore-wing, with a darker shadow at the margin. The antennae are long and thin. Head and thorax are reddish-brown while the abdomen is a lighter grey-brown. The moths are nocturnal, hiding during the day in fissures in the ground, and are often only seen when they are startled, such as during crop watering.

The females deposit their spherical eggs on the underside of leaves in groups of 50 to 300, often arranged in layers. Apart from a smooth underside, the transparent shell is strongly ridged and bears a net-shaped brown pattern. These eggs are greenish to begin with, but this gradually changes to a light yellow or almost white.

The caterpillar has four pairs of abdominal prolegs and can grow to a length of 50 mm. The head is a pale green colour in the first two larval instars, later becoming a white to grey-brown with mottled markings. The pattern of markings on the caterpillar varies according to the host-plant and developmental stage. At first, the body is a glossy green. Later, a conspicuous yellow longitudinal stripe with black dots above it appears laterally, on either side. Older caterpillars may vary from light green to light brown, or even a reddish colour with three dark grey dorsal stripes partly overshadowed by the underlying grey-black stripes. Yet another grey stripe runs just above the yellow lateral stripes. Each segment has several darker stripes both laterally and dorsally. When touched, the caterpillar rolls itself up, or drops from a silk thread and hangs motionless. The young caterpillars remain together, beginning to move out over the plant after the second moult.

When about to pupate, the caterpillars seek a protected place just below the ground surface or under rockwool matting. A light, loose silk cocoon is spun and two or three days later a red-brown pupa is formed, which gradually turns a glossy black.

The tomato moth is a strictly nocturnal moth, and remains motionless throughout daylight hours. The caterpillar too feeds almost exclusively at night.

Eggs, a larva and an adult *L. oleracea* are shown in plate 8.18.

Population growth
Although in temperate regions there is a single generation per year outdoors, there are two generations in glasshouses, and a partial third generation in parts of southern Europe. Because the females have a long lifespan, these generations overlap each other, so that all stages can be found in the glasshouse throughout the year.

At 20°C, eggs hatch after seven days. The larval instars, of which there can be five to seven, are completed after another 33 - 39 days. The pupal stage can last from two to seven weeks.

Mating takes place immediately after the wings of the female are unfolded, and two to three days after mating the female begins to lay eggs. A single male can mate with several females, with each female laying about 1,000 eggs in irregular clusters of 50 to 300. In temperate climates outdoors, the eggs and adults can be found in June and July, the caterpillars from July to August and the pupae from October to the following May.

Damage
Larger caterpillars eat large areas of leaves and can also attack stems, rendering them fragile and liable to break. Young plants may be stripped completely, and the frass deposited on fruit can cause significant problems. In fruiting vegetables, large holes can be formed in both green and ripening fruit. The second generation of caterpillars often cause greater loss of fruit than the first.

8.18. Eggs, caterpillar and adult of *Lacanobia oleracea*

Eggs

Caterpillar

Adult

- **Commonly occurring species**
- **Found mainly in summer and autumn**
- **Causes damage particularly in sweet pepper**

Mamestra brassicae
Cabbage moth

Mamestra brassicae, the cabbage moth, can cause serious damage in sweet pepper, but also in tomato, lettuce and sometimes in chrysanthemums and carnations. Outdoors, the moth lays its eggs on many host plants, with cabbage being the most common. The species belongs to the family Noctuidae and the sub-family Hadeninae. *M. brassicae* is found in Europe, Japan and sub-tropical Asia, including India.

Life-cycle and appearance
The cabbage moth has grey-brown to black fore-wings with a wingspan of 40 - 50 mm, with a kidney-shaped white-rimmed marking in the middle. The hind-wings are light brown with a rather inconspicuous spot in the middle. Head and thorax are grey-brown with white spots, and the abdomen is a pale, greyish brown. The moth flies only at dawn and dusk, hiding in the crop during the day.

Several days after emerging from their pupae the females lay their first eggs, either on the underside of the leaves of the host plant, or elsewhere within the glasshouse structure. The eggs can be laid separately or in groups of up to 100, although they are usually deposited beside each other in masses of around 20 to 30. They are initially light and translucent, becoming steadily darker until they are a brown-black or sometimes even a purple colour. They are ridged and have a light network of markings.

After hatching, the larvae remain together, eating the edge of the leaves on which they were laid. From the third instar onwards, they swarm out over the whole plant. The caterpillar stage consists of six instars; in the youngest they are a transparent yellow to grey-green colour with a characteristic brown-black head capsule. With the first moult, the head capsule becomes yellow and the green contents of the gut are clearly visible. After the third moult, the caterpillar is green with a dark back and thick, yellow, lateral longitudinal stripe. The full-grown caterpillar is about 40-50 mm in length, and its colour varies from green or brown to black. In younger instars, the relatively large head capsule is conspicuous. In older instars the lighter colour of the bands between the body segments gives them a characteristic ringed appearance. Young caterpillars fall from the leaf if disturbed, whereas older caterpillars (which react much more slowly) curl up *in situ*. The caterpillars are nocturnal.

Pupation takes place in the ground. The pupae are approximately 2 cm long and a glossy brown colour, turning black shortly before the adult moth emerges. Plate 8.19 shows eggs, caterpillars and the adult of *M. brassicae*.

Population growth
Outdoors, the moths are active from mid spring onward. They are hardly seen during the day as they are crepuscular, only flying during the hours of dawn and dusk. Five days after emerging from the pupa, the females lay their first eggs on the underside of the leaves of the host plant or on other glasshouse material. The optimal temperature for egg-laying is around 20°C, at which temperature anything from 400 to a maximum of 1,000 eggs may be laid by each female. At 20°C the eggs hatch after 6 - 10 days. The six larval instars are completed after about 30 days, while the pupal stage lasts about 3 weeks. The sex ratio is 1:1.

Outdoors in temperate climates there are generally two generations produced each year, the first during May / June, the second from mid July onwards. The second generation is thus much larger than the first. Under glasshouse conditions there are more generations, which overlap. Overwintering usually occurs in the pupal stage, within the ground, although the caterpillars can also often overwinter.

Damage
Large caterpillars feed mainly at the top of the plant, eating the youngest leaves.

8.19. Eggs, caterpillars and adult of *Mamestra brassicae*

Eggs

Adult

Different colour forms of caterpillars

- Appears throughout the year, but mainly in summer and autumn
- Highly polyphagous, can seriously damage fruit, flowers and buds
- Resistant to many pesticides

Spodoptera exigua
Beet armyworm

Spodoptera exigua, the beet armyworm, originates from the tropics and sub-tropics, but can now be found in warm and temperate regions throughout most of the world. Geographically, its original habitat was South East Asia. It is an active flier, and can thus occur in regions where it is not able to overwinter. This species has become a regular problem for horticulturalists due to its resistance to nearly all insecticides, and its ability to develop outside on wild vegetation during the summer. *Spodoptera exigua* belongs to the family Noctuidae and the sub-family Amphipyrinae.

It is highly polyphagous and occurs in many greenhouse crops where it can cause serious problems, especially in ornamental crops such as chrysanthemum, gerbera, roses and pot plants. Among the various vegetable crops in which this species causes problems, sweet pepper is particularly susceptible to large infestations. Before it was recognized as a pest, the species was seen several times in Northern Europe as a migrant from the Mediterranean region. These migrants do not create problems as they are controlled by the chemical treatments applied against other species. However, the beet armyworm moths currently present in glasshouses show varying resistance to conventional controls, including those based on the bacterium *Bacillus thuringiensis*. This situation corresponds with that found in Florida, and it may thus be inferred that glasshouse populations in Europe are derived from a Florida population.

Life-cycle and appearance
The beet armyworm is a small, nondescript moth which hides during the day and is active only at night. Its wingspan is 17 - 30 mm and its body length approximately 15 mm. The fore-wings are grey-brown suffused with a dark brown or black. The head and thorax are brown and the abdomen grey-brown. The forewings have a yellowish, kidney-shaped mark. The white hind-wings have clearly outlined brown venation. Because the moths show such a strong aversion to light, they are not readily identified within the crop.

The eggs are laid in packets, sometimes in several layers, and on top of a mass of white, cottony hairs and scales from the moth's body. This makes them less likely to become dislodged, and less likely to dry out. The eggs, which are usually grey but sometimes greenish or pinkish, are laid at night, preferably low down in the crop on the underside of leaves, in groups of 10 to 250. Occasionally eggs may be laid on parts of the glasshouse structure.

The eggs hatch after several days and small, drab yellow-green caterpillars appear, their colour gradually changing to yellow, green, brown or even black. The colour and pattern of the caterpillars is highly variable and depends partly on the host plant, on the stage of development and also partly on the climate. As a result, the caterpillars are easily confused with those of other species. A full-grown beet armyworm caterpillar can be 25 - 38 mm in length, with dark, wrinkled stripes dorsally and a yellow band running the length of the body on either side, above which there is a black dot on each segment. The caterpillar has four pairs of prolegs and closely resembles the larval tomato moth. The two species can be distinguished, however, by the behaviour of the caterpillars when touched: the tomato moth rolls up, the beet armyworm does not. The young caterpillars live in groups and spin loose webs over the leaves under which they remain until the third or fourth instar, with older caterpillars spreading out over the crop. The beet armyworm feeds mainly at night and conceals itself during the day, although the youngest instars will feed in the daytime. Full-grown caterpillars move to the ground in preparation for pupation. Although pupation sometimes occurs in the crop, it usually takes place on the surface of the ground in a loosely spun cocoon, consisting of soil particles glued together with a sticky secretion. Pupae are brown and 15 to 20 mm long, and resemble those of other noctuid moths. After 5 - 10 days the adult moth emerges from the pupa.

An adult, a caterpillar and eggs of *S. exigua* are shown in plate 8.20.

Population growth
The beet armyworm can develop quite rapidly, especially at higher temperatures. Combined with the great egg-laying capacity of the adults, this can lead to a population explosion in the crop within a very short time. An adult female moth can lay 10 - 250 eggs in a single night, and because adults live for 10 - 20 days, a female moth will lay some 500 - 600 eggs during her life, and in some cases maybe as many as 1,500. In a warm climate, there are 5 - 6 generations per year, with a total development time of 23 - 25 days per generation. In a temperate climate they cannot survive the winter outdoors, but under glass conditions they can survive and reproduce throughout the whole year. Data concerning the population growth of the beet armyworm are given in table 8.1.

Table 8.1. The population growth of *Spodoptera exigua* (Fye & McAda, 1972)

	temperature (°C)		
	20	25	30
development time (days)			
egg	5.6	2.9	2.0
caterpillar	18.7	13	9.9
pupa	10.4	7.7	5.1
total egg-adult	34.7	23.6	17.0

Damage

Older caterpillars migrate to the top of the plant where they feed mainly on the growing tips. They produce large holes in the leaf, sometimes stripping the leaf down to the mid-rib. Flowers and buds are also attacked. This large-scale damage has considerable consequences for the growth of the crop. With a substantial population of caterpillars the damage can spread to the stems, and in the worst cases to the fruit. This often leads to the entire contents being consumed. In ornamental crops the worst damage is caused by attacks on flowers and buds.

Distribution and dispersal

The caterpillars are active walkers and can cover considerable distances, thus caterpillars from the same cluster of eggs can cause damage at various locations within the crop.

8.20. Eggs, caterpillar and adult moth of *Spodoptera exigua*

Eggs

Caterpillar

Adult

- Migrant moth
- Appears in crops in summer and autumn
- Can attack sweet pepper, chrysanthemums and various other crops under glass

Autographa gamma
Silver-Y moth

Autographa gamma, the silver-Y moth, belongs to the family Noctuidae and to the sub-family Plusiinae, and is similar to the tomato looper, *Chrysodeixis chalcites*. The species occurs in Europe, North Africa and Asia. *A. gamma* is a migrant moth that appears in northern parts of Europe until November, after which it migrates back to southern Europe; the moths can fly 2,000 to 2,500 km. In temperate climates, the species develops one or perhaps two overlapping generations, of which it is the second, from August onwards, that inflicts much damage to agricultural crops, particularly brassicas. They are sometimes also encountered in glasshouse crops such as sweet pepper, lettuce, beans, chrysanthemums and various other ornamental crops where they can cause considerable damage. Damage occurs in summer and autumn, but there are no indications that the species overwinters in the glasshouse.

Life-cycle and appearance
The wingspan of the silver-Y moth is 30 - 45 mm. The fore-wings have varied markings, from different shades of brown to grey, but the most distinctive feature is the irregular white mark resembling the Greek letter gamma (hence the Latin name), or less convincingly a letter Y (hence the common name). The hind-wings are brown with a broad, dark band along the margin. In the resting position, the hairy plume along the back is also distinctive. The body is light brown and hairy, the antennae are long and thin. The moths fly during the day as well as at night, often in large swarms. They fly rapidly from flower to flower, often resting with their wings folded roof-like above the body.
The female lays up to 2,000 eggs on the underside of the leaves of their host plant, usually separately but sometimes in small groups. These eggs are a greyish white suffused with a blue-grey in the middle and have irregular surface ridging.
The caterpillars have a small head that is grey-green to brownish green, sometimes with a black lateral line. The body, which has stiff hairs, is of variable colour, but is usually greenish with a light lateral line. Below the lateral line there is a dark dot on each segment. Dorsally there are alternating light and dark stripes. The caterpillar, which is up to 40 to 45 mm in length, has only two pairs of prolegs rather than the standard four, and behaves like a looper. A characteristic feature of this caterpillar is the oscillatory swaying movement of the anterior end. The caterpillars are nocturnal. When the caterpillar is full-grown, it will form a greenish to black pupa, either in leaves which are woven together with silk threads, or in a suitable hiding place spun somewhere in glasshouse material.
Plate 8.21 shows an egg, a caterpillar and an adult of *A. gamma*.

Population growth
In northern Europe the moths are found outdoors during May to June and August to September, which is also when they lay their eggs. The eggs hatch after 12 days, so that caterpillars can be found from June to July and September to October. Pupae can be found outdoors in July and August. In southern Europe, there are probably three generations annually, but because the species migrates it is difficult to distinguish between migrant and indigenous generations. Caterpillars can survive low temperatures down to 2°C, but the second generation of caterpillars does not survive the winter outdoors.
In the glasshouse, populations develop faster than they do outdoors, resulting in overlapping generations. At 20°C, eggs hatch after only 6 - 10 days.

Damage
The larvae feed on leaves, causing typical caterpillar damage.

8.21. Egg, caterpillar and adult moth of *Autographa gamma*

Egg

Caterpillar

Adult

- **Found in Africa, the Middle East and parts of Mediterranean Europe**
- **Highly polyphagous**
- **In warm climates generations develop continuously**

Spodoptera littoralis
Egyptian cotton leafworm

Spodoptera littoralis is a highly polyphagous species that originates from Egypt, and is currently found in Africa, the Middle East and parts of Mediterranean Europe, including Spain.

Life-cycle and appearance
The life-cycle of *S. littoralis* consists of an egg stage, six larval stages, a pupal stage and the adult stage.
Each adult female can lay up to 3,000 eggs. Each egg is roughly spherical, about 0.6 mm long and whitish yellow. They are laid in clusters of 20 to 500 on leaf surfaces, often on the lower parts of the plant. The female covers the eggs with brownish-yellow hairs from her abdomen to make them less conspicuous and to protect them from desiccation.
Young caterpillars are pale green with a brownish head. In the last larval stage they are 35 to 45 mm long. Colouration varies from grey to reddish or yellowish, with a median dorsal line bordered on either side by two yellowish-red or greyish stripes and small yellow dots on each segment. The underside of the caterpillar is greyish-red or yellowish.
The caterpillars are easily distinguished from other *Spodoptera* species by 4 black triangular dots on their body. The appearance of the caterpillar is similar to that of *S. litura*, (the Asian cotton leafworm) which occurs naturally in other regions.
Before pupation, caterpillars crawl to the ground and settle in the surface layers. The pupa is 15 to 20 mm long, and reddish brown in colour.
The adult is also easily recognized (although it too resembles adults of *S. litura*). It has a wingspan of 35 to 40 mm. The fore wings are brownish with bluish overtones and straw yellow along the median vein. The ocellus is marked by two or three oblique whitish stripes. The front of the wing tip has a blackish marking, which is more pronounced in the male. The hind wings are whitish, with a brown leading edge. Adults are only active at night.

Population development
Outside, adults appear in early spring. The moths emerge at or just before nightfall. The eggs, which are laid in clusters, take 3 to 4 days to develop at 25 - 28°C. The caterpillars are gregarious at first, becoming solitary in the 4th larval stage. In common with the adults, they are mainly nocturnal, sheltering in the soil during the day. Larval development takes about 2 weeks.
Under optimal conditions (30°C) the total development from egg to adult takes 3 weeks, increasing to around 34 days at 25°C, 50 days at 20°C and almost 150 days at 15°C. In warm climates generations develop continuously, with up to ten generations per year. Overwintering takes place in the soil during the pupal stage.

Damage
Many crops can be affected, including tomato and sweet pepper. The majority of damage is caused by the older caterpillars.

8.21.a. Caterpillar and moth of *Spodoptera littoralis*

Caterpillar

Moth

- Occurs mainly in tropical and sub-tropical areas, including Spain and southern France
- In glasshouses most abundant in summer and autumn
- Polyphagous, but under glass tomato is particularly severely attacked

Helicoverpa armigera
Tomato fruitworm

The tomato fruitworm (*Helicoverpa armigera* or *Heliothis armigera*), causes damage mainly in the tropics and sub-tropics. Caterpillars of *H. armigera* are polyphagous, feeding on crops such as cotton, tomato, sweet corn, tobacco, sunflowers and various fruit crops. In glasshouse crops, tomato is particularly severely damaged.

Life-cycle and appearance
H. armigera normally undergoes six larval stages, although under certain circumstances five or seven stages may occur.

Eggs are deposited separately on young shoots near to buds, flowers, fruits or on leafy parts. They are almost spherical with a flattened base and a diameter of about 0.5 mm. They are a shiny yellowish white at first, although their colour changes to brown just before the larvae emerge.

Young caterpillars are yellowish-white to reddish-brown with dark spots. In later instars the head is mottled, and the body is marked with three prominent longitudinal dark bands with numerous lighter colored wavy lines. The colour of the caterpillar is extremely variable, and may be green, straw yellow, black, or pinkish / reddish brown. Full grown larvae are 30-40 mm long. The first instars are the most mobile, and move like loopers. The caterpillars are often aggressive and cannibalistic. Pupation occurs in the soil, with pupae being 14 to 18 mm long, with a brown, smooth body surface.

Adult female *H. armigera* are well developed, brownish orange moths approximately 18-19 mm in length and with a wing-span of about 40 mm. The grey-green males are smaller. The fore wings are edged with a line of black spots, and the hind wings in both sexes are cream with a dark brown band around their outer margin.

Population growth
One to four days after the adults emerge, mating and egg-laying takes place. Oviposition continues for two to five days, with a single female laying between 500 and 3,000 eggs. At 25°C it takes approximately 4 days before eggs emerge. At 22°C larval development takes approximately 18 days, increasing to about 50 days at 17°C. The pupal stage usually lasts from 11 to 17 days. At the end of summer the pupae enter diapause, resuming activity when the soil temperature rises above 18°C. In warmer areas the species may remain active throughout the year, with multiple overlapping generations.

The longevity of the adult moths is strongly dependent on the availability and quality of nectar or equivalent food. Females live longer than males (13.5 days and 8.7 days respectively). The moths are active at dusk and throughout the night, when they feed on nectar or water from their host plants.

Although all stages can occur throughout the year under glass, the species is most abundant in summer and autumn.

Damage
First instar larvae feed on soft leaves, creating small holes in them. When they reach the second instar, they can penetrate fruit through a small hole, often bored near the stalk. All fruits can be damaged, although the preference is for smaller specimens. During development the caterpillars damage most fruits by mining. The fruits stop growing, mature rapidly, and drop off. In legumes, flowers are attacked, and seed pods may be pierced.

8.21.b. Caterpillar and moth of *Helicoverpa armigera*

Caterpillar

Moth

- A common species
- Mainly found in ornamental crops, but also in sweet pepper
- Can bore into stalks and fruit

Clepsis spectrana
Cabbage leafroller

Clepsis spectrana, the cabbage leafroller, is a native species of North-West Europe which belongs to the family of leafrollers (Tortricidae).
C. spectrana has many host plants and causes damage in outdoor cabbage, apple and pear. Because fewer broad-spectrum pesticides are used these days, the cabbage leafroller has become a pest in crops grown under glass. Great damage can be inflicted on ornamental crops (in particular gerbera), but also in roses, alstroemeria, azalea, cyclamen, and kalanchoë. In chrysanthemum too, this species is an increasing problem, and has caused problems in sweet pepper since the mid 1980's. In protected soft fruit cultivation, strawberry, currants, blackberry and raspberry can all serve as host plants.
This family of moths derives its name from the fact that the caterpillars spin themselves into leaves which roll up as a consequence. This concealed, protected habit makes control very difficult.

Life-cycle and appearance
The adult moth has a wingspan of 16 - 22 mm. An adult female is 12 mm long and is ochre-yellow with darker markings. The male is slightly smaller and slimmer, more clearly marked and lighter in colour. Both have two brown marks on each wing, the more proximal pair almost forming a "V" when at rest. At rest, the wings are held folded almost horizontally over the body. The antennae are thin and of medium length. The moths fly when it is dark.
The yellow eggs are deposited tile-fashion on the underside of leaves and sometimes stalks, in batches of 10 to 90.
The caterpillar has a dark green to black head capsule and prothoracic shield. The body is a drab brown with lateral stripes that are sometimes light in colour. The ventral surface is lighter than the dorsal, with hairs located in warty spots that are lighter than the general body colour. Overall the caterpillar has a rather dull appearance. In the fruit of sweet pepper, however, the caterpillars are much lighter and have dark lateral stripes. Further, the caterpillars are relatively thin (up to 25 mm in length), are slightly hairy and have four pairs of prolegs. They either make webs in which they hide, or live inside fruit or flowers. It is these webs that cause the leaves to curl. Many of these webs are empty, and it is thought that the caterpillars probably make a series of webs during their development. When disturbed, the caterpillar makes a contorted wriggling movement and drops on the end of a silk thread.
The pupa is about 1 cm long and brown, with pupation taking place within the webbing.
An egg, a caterpillar and an adult moth are shown in plate 8.22.

Population growth
The cabbage leafroller only develops at temperatures between 10 and 35°C. In glasshouse crops, this leafroller can be encountered throughout the year, in all stages, with 8 - 12 overlapping generations per year. In addition, during summer months, moths may enter the crop through the vents and lay eggs. A female lays on average 400 - 500 eggs.
Outdoors the moth generally produces only two generations per year,

8.22. Egg, caterpillar and adult moth of *Clepsis spectrana*

Egg

Caterpillar

Adult

probably with a winter resting period. The caterpillar overwinters in a dense web behind bark or other rough part of the food plant.
An infestation of cabbage leafrollers does not develop especially rapidly, and tends to go unnoticed. The eggs and young caterpillars are rarely seen in the crop, and often damage only comes to light when the fruit is being sorted.

Damage
Caterpillars feed mainly on leaves, flowers and buds. Older caterpillars loosely spin newly opened leaves together such that subsequent growth leads to an untidy roll of leaves with large feeding holes. Caterpillars can also bore into the plant; young caterpillars eat a path into an opening bud to the growing tip, while older caterpillars eat a hole in the shoot and hollow it out, causing the shoot or growing point to wilt and die. In sweet pepper, the caterpillar lives under the sepals for a while before eating into the fruit itself, leaving a hole of about 2 mm diameter at the edge of the sepals. The caterpillar subsequently feeds unnoticed on the inside of the fruit. The cabbage leafroller never inflicts visible damage to a sweet pepper crop, and it is difficult to detect affected fruit while it is still growing. The small holes near the sepals are the first sign, with closer inspection revealing fine threads of webbing and dry, sand-size granules of frass. When the fruit is cut in half, the inside appears rotten, and there may be a caterpillar or an adult moth inside. The damage can be considerable and is limited mainly to the fruit. This may be the result of selection, since the larvae that manage to find their way into the fruit evade natural enemies and pesticides.
In ornamental crops, leaves, shoots and flower buds can all be seriously damaged. In crops where the leaf damage remains visible for longer (e.g. bromeliads or kalanchoë), even a moderate attack can cause considerable cosmetic damage.

Distribution and dispersal
The adults are good fliers and lay their eggs in several different locations within the crop. Dispersal to other glasshouses often happens on warm nights, when the vents are open for a long time.

- Becoming increasingly common in glasshouses
- Mainly found in ornamental crops
- Difficult to combat with chemical pesticides due to its cryptic habit

Duponchelia fovealis

Duponchelia fovealis belongs to the family Pyralidae, and originates from the Mediterranean region and the Canary Islands. The caterpillars can attack many different crops and cause considerable damage in ornamental plants grown under glass. The secretive habit of the caterpillars of this species makes them very difficult to control with chemical pesticides.
D. fovealis was first observed outside its original geographical range in 1984 in Finland, since when there have been further reports of damage from other parts of Europe. Chemical control is only possible with broad-spectrum pesticides, and even then the secretive habit of the caterpillars means that very little of the chemical comes into contact with them.

Life-cycle and appearance
Adults of *Duponchelia fovealis* are light to dark brown with a clearly visible white crinkled line on the fore-wing. The wingspan is 9 to 12 mm. The end of the remarkably long abdomen bends upward almost vertically, with the abdomen of males often being longer than that of females. Another distinctive feature is the triangle formed by the head and fore-wings. Moths mainly fly at night, but can sometimes be seen in the daytime.
Eggs are laid in an overlapping tile-like fashion in clusters of 5 - 10 on the underside of leaves, often close to the veins. The females often lay their eggs low down in the crop, at the base of the stalks. The pinkish red eggs are about 0.5 mm in diameter and become steadily darker. Just before hatching, which occurs about a week later, the young larva is visible within the egg.
The full-grown caterpillars are 20 to 30 mm long and a creamy white colour. Along the dorsal surface of the body there are pairs of dark dots. The dark head is also conspicuous, as are the four dark dots behind the head capsule. There are 4 pairs of prolegs. Although in appearance the caterpillars are difficult to distinguish from certain other species, they have a very different life habit. However, on the basis of their rapid movement and appearance alone, they could easily be confused with the banana borer. Caterpillars are found in webs and have a preference for damp, humid places and are thus mainly found low down in the crop,

often in decaying organic material on the ground or substrate. The caterpillars often eat into the stalk and remain concealed within, thus making chemical control so difficult.

The caterpillars pupate in an earthen cocoon, concealed either within the crop or in another suitable spot in the vicinity.

Plate 8.23 shows eggs, a caterpillar and an adult of *D. fovealis*.

Population growth

Under normal glasshouse conditions, the caterpillars emerge from the eggs after eight days. They are full-grown after about 4 weeks, after which they pupate. The pupal stage lasts about one to two weeks, and the lifespan of the adult moths is also about one to two weeks.

A female will lay about 200 eggs during her lifetime, mainly in damp places such as the moist ground near the root of the plant. The eggs are deposited either singly or in small groups of up to ten.

Damage

The caterpillars often stay well down and in the middle of the crop, sometimes attacking the stem of the plant as well as the leaves. In gerbera, the flower buds may also be eaten. It has even been known for the caterpillars to inflict damage to the roots of crops cultivated in an ebb-and-flow irrigation system.

The damage threshold varies considerably according to the crop: in begonia and kalanchoë a relatively small population can cause a great deal of damage, whereas roses appear to tolerate greater numbers.

Plate 8.24 shows damage caused by *D. fovealis*.

Distribution and dispersal

The caterpillars prefer damp, humid conditions and are therefore mainly found concealed deep in the crop, close to the substrate. In gerbera, for example, they can often be found on the lowest leaves closest to the damp ground. In crops that have dense exposed roots they are often found on the lowest leaves, just above ground level. In bushy crops they are found higher up the plant, often inside protective webs. The caterpillars may also live as stem-borers in poinsettias, having bored down through the stem from their oviposition site in the top of the plant.

The adults can be disturbed very easily during the day, whereupon they fly off over the crop in search of an alternative resting place. The moths are good fliers, with most flight activity occurring primarily during the night.

8.23. Eggs, caterpillar and adult of *Duponchelia fovealis*

Eggs

Caterpillar

Adult

8.24. Damage in eustoma caused by *Duponchelia fovealis*

- A quarantine organism
- Occurs specifically in imported plant material
- Can inflict disastrous damage without being noticed

Opogona sacchari
Banana moth

It is not entirely clear to what natural family the banana moth, or banana borer, *Opogona sacchari* (also known as *Opogona subcervinella*), belongs. It is usually assigned to the family Tineidae, but sometimes also to the family Lyonetiidae or Hieroxestidae. The genus *Opogona* contains some 225 species worldwide of which only a few are economically significant. The moth originates from the humid tropical and sub-tropical areas of Africa, and outside these climatic zones *O. sacchari* can only become established in glasshouses. At present, the species is found on various African islands, in West-Africa (Nigeria), on the Canary Islands and Madeira, in Mid and South America, in Europe and in Florida. The insect has been repeatedly encountered in different European countries and measures are still in force to control any infestations. However, since 1970 the insect has become permanently established in glasshouses in Southern Italy.

The banana moth is an important pest in banana and can also attack tropical crops such as pineapple, bamboo, maize, potato and sugarcane. The caterpillars are polyphagous and prodigious feeders, attacking a large number of ornamental plants of tropical or sub-tropical origin that are now cultivated in European glasshouses. Dracaena, yucca, strelitzia and cacti are hosts, and also occasionally dieffenbachia, euphorbia, bromeliads and ficus. It would seem that any plant species that contains sufficient food and into which the caterpillars are able to bore can be affected. Once the caterpillars have bored inside the plant they are extremely difficult to combat.

Life-cycle and appearance

Adults are almost uniformly a clear, yellowish-brown colour. The moth is about 1 cm long with a wingspan of 18 - 26 mm. The fore-wings are brown suffused with a golden glow and with 2 small black dots, and may display dark brown bands. The hind-wings are paler and clearer.
O. sacchari is a typical nocturnal moth, but unlike other nocturnal moths it holds its antennae straight out in front when at rest.

The caterpillars are at most about 3.5 cm long, are highly mobile, photophobic and gluttonous feeders. They withdraw very rapidly into affected plant material. They are a dirty white to grey-brown colour, and partially transparent such that the internal organs can be seen. They have a bright red-brown head with clearly visible brownish plates on the thorax and abdomen. The larvae normally live in the crown of banana, in the stems of ornamental plants (*e.g.* cacti and dracaena), or in leaves or stalks.

The pupae are brown and usually less than 1 cm long. The cocoon is spun at one end of a borehole and is usually 1.5 cm long. Before the moth emerges, the pupa works itself partially out of the plant tissue by means of two small hooks. Beside these hooks, the pupa also has conspicuous spines on each segment. The most striking feature of this insect is that the empty pupal case is left hanging out of the passage in the plant tissue. Occasionally pupation may take place in pots. Plate 8.25 shows a caterpillar and an adult moth of *O. sacchari*.

Population growth

At 15°C the complete life-cycle of the banana moth takes roughly 3 months, but under optimal (warmer) conditions the life-cycle is completed in 6 weeks. At 23 - 25°C under laboratory conditions the egg stage lasts 4 days, the larval stage 58 days and the pupal stage 16 days. Adults live for 3 to 4 weeks.
The adult female lays her eggs in splits and crevices in the plant tissue, in batches of about 5 eggs. A single female can lay 250 - 300 eggs.
This species is incapable of overwintering outdoors in northern Europe.

Damage

Caterpillars make passages and holes in both woody and succulent crops. In ornamental crops the larvae are found mainly in the stalks, and sometimes in leaves and cotyledons. The presence of yellow white dust left in the openings of bore passages is characteristic of an attack of the banana borer. Seedlings can be very badly affected. The attack is not usually recognized in its earliest stages. The caterpillars normally begin to eat into the phloem and woody tissues and leave the bark standing. Because the flow of sap is interrupted, the plant begins to wilt. In yucca, for example, the bark is flaccid to the touch and can be dented with the finger in the affected area. After some time, a caterpillar can hollow out the interior of the plant completely: in cacti, yuccas and dracaena, for example, the stems may be completely hollowed out. Plants whose roots or stem base have been attacked first lose a few leaves and then suddenly collapse (*e.g.* sanseveria and palms).
In addition to the direct damage they inflict the indirect damage should also not be underestimated. Damaged areas can be invaded by moulds and bacteria and begin to rot as a result. This can exacerbate the damage and produce unpleasant odours.

8.25. Caterpillar of *Opogona sacchari*

8.25.a Adult of *Opogona sacchari*

- Often mistaken for *Clepsis spectrana*
- Males fly during the daytime, females are fairly inactive
- Can cause damage in ornamental crops, sweet pepper and strawberry in glasshouses

Cacoecimorpha pronubana
Carnation leafroller

The carnation leafroller *Cacoecimorpha pronubana* is a Mediterranean species that causes serious damage in the culture of carnations throughout the Mediterranean region, particularly in Israel, Italy, France and Spain. This leafroller is also well known to carnation growers in southern England and Wales. Like the cabbage leafroller (*Clepsis spectrana*), the species belongs to the family Tortricidae. *C. pronubana* has many wild and cultivated host plants but is most notable for the damage it inflicts on carnations and strawberries under glass. For some time it has been clear that the damage caused by leafrollers to the fruit of sweet pepper has been incorrectly attributed to the cabbage leafroller *Clepsis spectrana*; a great deal of this damage is in fact probably due to *C. pronubana*.

Life-cycle and appearance
Male carnation leafrollers are good fliers and are active diurnally, unlike most leafrollers which mainly fly at dusk and after dark. The females have a limited flight capacity and only cover short distances. The moth is rather small, with a wingspan of 15 - 25 mm. At rest, the wings are folded almost horizontally over the body. The contrasting colours between the yellow to purple fore-wings (with a dark brown band in the females and two red bands in the males) and the hind-wings (bright orange) ensure that the moths of the species are easily recognized. This colour pattern is most easily observed in the males during flight, and thus the moth is easily distinguished from the cabbage leafroller.

Each female is capable of laying up to 700 eggs, deposited in batches of 10 to 200 on the upper surface of leaves or on other glasshouse materials. At first, the eggs are a light green colour, later becoming yellow. They measure 1 x 0.6 mm and are a flattened oval to round shape.

Caterpillars have an olive-green to bright or yellow green colour with a black head that later turns yellowish brown with dark spots. They can grow up to 20 mm in length. If touched or disturbed they move

backward with powerful, wriggling movements and readily drop from the end of a silk thread. The caterpillars typically hide in sturdy rolls of leaves, but also in bunches of growing tips, flowers and fruit that have been spun together with silk.
Plate 8.26 shows a caterpillar and an adult moth of *C. pronubana*.

Population growth
The carnation leafroller overwinters as a caterpillar, but as it cannot survive low temperatures and is not very tolerant of high rainfall, the species survives winters outdoors in temperate climates very badly. However, overwintering in glasshouses is possible, with the first moths appearing from April, after which there may be more than 5 successive, overlapping generations. At 15°C the life-cycle takes about 135 days, decreasing to about 35 days at 30°C. Below 15°C, development almost ceases. A relative humidity of 40 - 70% is optimal, with higher humidities proving deleterious for both caterpillars and pupae.

Damage
The feeding caterpillars damage flower buds, fruit and leaves. The species is particularly troublesome in the culture of carnations and strawberries under glass; the damage inflicted to strawberries and other crops grown outdoors is less serious. For several years there have been problems with leafrollers attacking the fruit of sweet pepper. Female moths lay their eggs on or in the vicinity of the fruit, and as soon as the caterpillars emerge they eat their way into the interior of the fruit, almost always through or near the calyx. As a result, it is not immediately obvious that the fruit has been attacked. The caterpillars only feed on the green calyx and the white seed-bearing tissue inside the fruit. They often pupate within the fruit, in which case the moth will eventually emerge from the fruit. It has been clear for some time that this kind of damage, first attributed to the cabbage leafroller *Clepsis spectrana*, is in fact mainly caused by *C. pronubana*.

8.26. Caterpillar and adult of *Cacoecimorpha pronubana*

Caterpillar

Adult

There follows below an overview of several different characteristics of the most important species of Lepidoptera in glasshouses. All the species dealt with above are included, in addition to the cutworms (*Agrotis* spp.) and the azalea moth (*Caloptilia azaleella*).

Overview of important lepidopteran species

Tomato looper - *Chrysodeixis chalcites* - NOCTUIDAE

eggs

- dingy white
- laid on underside of the leaf
- usually deposited separately

caterpillar

- light yellowish-green
- several less conspicuous light dorsal stripes
- on either side of the body there is a thicker yellow stripe
- in older caterpillars each segment has a clear dark dot above the yellow lateral line
- 2 pairs of abdominal prolegs (semi-looper)
- maximum length 4 cm

moth

- golden brown
- 2 conspicuous droplet-shaped white markings on each fore-wing, sometimes merged to form a single marking
- hairy plume or comb dorsally
- wingspan 3 - 4 cm

damage

- polyphagous
- damages both leaves and fruit
- most commonly occurring moth

Tomato moth - *Lacanobia oleracea* - NOCTUIDAE

eggs

- bright green, turning light yellow or almost white, with brown net-like marking
- laid on the underside of the leaf
- deposited in batches of at least 50 eggs, sometimes in several layers

caterpillar

- variable colour, mostly light green with a light brown head
- older caterpillars have a yellow stripe on either side of the body
- all segments spotted with several black dots and tiny white stippling
- 4 pairs of abdominal prolegs
- maximum length 5 cm

moth

- red-brown
- conspicuous, kidney-shaped, light brown marking on fore-wings
- wingspan 3 - 4 cm

damage

- polyphagous
- damages leaves, stalks and fruit

Overview of important lepidopteran species

Cabbage moth - *Mamestra brassicae* - NOCTUIDAE

8.31 8.32

eggs

- initially transparent, turning brown-black to purple
- laid on the underside of the leaf
- usually deposited in batches of around 30

caterpillar

- young caterpillars are green, older caterpillars of variable colour, often with lighter inter-segmental bands giving a ringed appearance
- broad yellow stripe on either side
- 4 pairs of abdominal prolegs
- maximum length 4 - 5 cm

moth

- dark brown
- obvious kidney-shaped, white rimmed marking on the fore-wings
- light brown hind-wings with a spot in the middle
- relatively large moth, wingspan 4 - 5 cm

damage

- damages sweet pepper, tomato and sometimes carnations and chrysanthemums
- larger caterpillars mainly feed at the tops of plants

Beet armyworm - *Spodoptera exigua* - NOCTUIDAE

8.33 8.34

eggs

- grey, sometimes greenish or pinkish
- laid on the underside of the leaf
- deposited in batches of 10 - 250 eggs, sometimes in several layers
- egg batches are covered with fine hairs

caterpillar

- variable in colour, mostly gree
- yellow stripe on either side
- a dark dot on each segment
- 4 pairs of abdominal prolegs
- maximum length 2.5 - 3.5 cm

moth

- grey-brown
- kidney-shaped, yellow or pink marking on the forewings
- wingspan 1.5 - 3 cm

damage

- causes damage in vegetable crops, but especially in ornamental crops
- damages leaves, fruit, flowers and buds
- larger caterpillars mainly eat growing tips

Silver-Y moth - *Autographa gamma* - NOCTUIDAE

8.35 　8.36

eggs

- white with blue-grey in the middle
- laid on the underside of the leaf
- usually deposited separately, sometimes in small batches

caterpillar

- variable in colour but usually greenish
- light stripe on either side
- 2 pairs of abdominal prolegs (semi-looper)
- maximum length 40 - 45 mm
- has a covering of stiff hairs
- anterior end makes a typical, side-to-side swaying movement

moth

- dark brown to grey
- a white Y-shaped marking on the fore-wings
- diurnally and nocturnally active
- wingspan 30 - 45 mm

damage

- causes damage in sweet pepper, chrysanthemums and various other crops
- destroys the leaves

Cabbage leafroller - *Clepsis spectrana* - TORTRICIDAE

8.37 　8.38

eggs

- yellow
- laid on the underside of the leaf
- deposited in batches of 10 - 90 eggs

caterpillar

- drab appearance, variable in colour, mostly brownish with a darker head capsule
- body tapered at both ends: a typical leafroller
- found in concealed places, within webs in leaves, fruit or flowers
- 4 pairs of abdominal prolegs
- maximum length 2.5 cm
- can drop on a silk thread

moth

- ochre-yellow
- 2 dark markings on each wing
- wingspan 1.5 - 2 cm

damage

- particularly damaging in ornamental crops and in sweet pepper
- caterpillars eat leaves and can also chew into fruit and stalks

Overview of important lepidopteran species

Egyptian cotton leafworm - *Spodoptera littoralis* - NOCTUIDAE

8.38 a   8.38 b

eggs
- whitish yellow
- laid on leaf surfaces, often on the lower parts of the plant
- deposited in clusters of 20-500
- clusters are covered with fine hairs

caterpillar
- variable in colour
- underside greyish-red or yellow
- 4 black triangular dots

moth
- ocellus marked by two or three oblique whitish stripes
- blackish marking on the front of the wing tip
- wingspan 35-40 mm

damage
- causes damage in many crops, including tomato and sweet pepper

Tomato fruitworm - *Helicoverpa armigera* - NOCTUIDAE

8.38 c   8.38 d

eggs
- yellowish white at first, changing to brown just before emerging of the larvae
- laid on young shoots near to buds, flowers, fruits or leafy parts
- deposited seperately

caterpillar
- extremely variable in colour
- full grown larva are 30-40 mm long
- the head of later instars is mottled and the body has three dark bands with numerous lighter colored wavy lines

moth
- female brownish orange, male greyish green
- fore wings edged with a line of black spots, hind wings cream with a dark brown band
- wingspan about 40 mm

damage
- damage fruits by mining
- flowers are attacked and seed pods may be pierced

Duponchelia fovealis - PYRALIDAE

8.39 8.40

eggs
- salmon pink
- laid on the underside of the leaf in damp places, often beside or near the leaf veins
- in batches of 5 - 10 eggs

caterpillar
- creamy white and glossy with dark dots
- black head capsule with 4 conspicuous black dots just behind the head
- 4 pairs of abdominal prolegs
- maximum length 3 cm
- in concealed, damp places, in the heart of the plant and at the base of the stem, often in rotting plant material

moth
- light to dark brown
- a wrinkled, white line visible on the forewings
- wingspan 9 - 12 mm
- long, often upturned abdomen

damage
- damages ornamental crops particularly
- weak plants are attacked first

Banana moth - Opogona sacchari - TINEIDAE

8.41 8.42

eggs
- laid in various places; in palms usually at the base of stalks

caterpillar
- dingy white to grey-brown and partially transparent
- bright red-brown head capsule
- 4 pairs of abdominal prolegs
- maximum length 3.5 cm
- bores its way into plant tissues

moth
- yellowish brown
- fore-wings are brown / gold with 2 small black dots
- wingspan 20 - 30 mm

damage
- mainly damages ornamental crops
- caterpillar makes holes and passages in both woody and succulent crops
- secondary damage caused by bacterial and fungal infections exacerbates the original damage
- weak plants are attacked first

Overview of important lepidopteran species

Carnation leafroller - *Cacoecimorpha pronubana* - TORTRICIDAE

8.43　8.44

eggs

- light green at first, turning yellow later
- laid in large batches

caterpillar

- olive green to bright green with a yellow-brown head with darker spots
- powerful backward wriggling when disturbed; may drop on the end of a silk thread
- body tapered at both ends: a typical leafroller
- can grow up to 20 mm in length
- lives hidden, typically in strongly rolled leaves, but also in bunches of growing tips, flowers and fruit that have been woven together

moth

- fore-wings yellow to purple and hind-wings bright orange
- males have red bands on the fore-wings, females dark brown
- wingspan 15 - 25 mm

damage

- many wild and cultivated species serve as host plants, but mainly damages carnation and strawberry under glass
- Inflicts damage mainly on the leaves, but also on flower buds and fruit.

Cutworms - *Agrotis* spp. - NOCTUIDAE

8.45　8.46

eggs

- white at first, rapidly turning darker
- laid separately or in batches on the ground, or on the plant close to ground level

caterpillar

- on hatching more or less yellow with a dark head capsule and prothoracic shield, later earth-brown to dingy grey.
- on hatching about 1.5 mm long, full-grown length 4 cm, sometimes longer.
- 3 pairs of true legs and usually 4 pairs of abdominal prolegs. In the first instar, the first two pairs of prolegs are still undeveloped; in the second instar, they are evident but not fully formed. During these first two instars, locomotion is by the same looping pattern as in inchworms. Only from the third instar are the 4 pairs of prolegs fully functional, and from this stage on the caterpillars live in the ground.
- may be encountered in the soil in a characteristic, circular position

moth

- pale grey to grey-brown fore-wings with a faint kidney-shaped marking and a red spot
- broad hind-wings, evenly light, dark grey or orange-yellow with a broad dark marginal stripe
- 3 - 4 mm
- dorsal thorax covered with close-lying hairs

damage

- chews at the roots and other plant parts close to the ground. The first two instars eat small round holes in the leaves

species

- turnip moth (*A. segetum*), heart and dart moth (*A. exclamationis*), black cut worm (*A. ipsilon*) and the large yellow underwing (*A. pronuba*).

Azalea moth - *Caloptilia azaleella* – GRACILLARIIDAE

8.47 8.48

eggs

- glossy milk white
- laid on the underside of the leaf along the veins
- mostly deposited separately, but sometimes in small batches

caterpillar

- at first transparent, turning yellow-green with a yellow-brown head
- young caterpillars do not have clearly visible legs, but older caterpillars do

moth

- fore-wings bright ochre-yellow to brown, hind-wings dark grey. Head, thorax and abdomen glossy grey-brown
- length approximately 10 mm

damage

- damages azaleas and rhododendrons
- young caterpillars form crooked passage mines in leaves that later transform to blister mines
- older caterpillars bend down the tops of leaves and weave them in this position, creating a sleeve inside which the leaf is eaten away by the feeding caterpillar, leaving the upper epidermis intact
- damaged leaves become brown, shrivel and fall off

Knowing and recognizing **Butterflies and moths**

Natural enemies of butterflies and moths

The most important natural enemies of caterpillars in glasshouse crops are:

The bacterium *Bacillus thuringiensis*
The parasitic wasp *Trichogramma brassicae*
Naturally occurring parasitic wasps such as the caterpillar parasite *Eulophus* sp.

8.49. *Bacillus thuringiensis*

8.50. *Trichogramma brassicae*

8.51. *Eulophus* sp.

It has been known since the beginning of the last century that the bacterium *Bacillus thuringiensis* kills caterpillars. The *kurstaki* strain proved particularly effective, especially against young caterpillars, and is used extensively as a biological insecticide.

Parasitic wasps have long been known as natural enemies of moths. Used widely in outdoor horticulture, particularly against *Ostrinia nubilalis* (the maize borer), *Trichogramma brassicae* has also been used in glasshouses since 1993.

Finally the predatory bugs *Macrolophus caliginosus*, *Orius* spp. and spontaneously occurring parasites such as *Cotesia plutellae*, *Eulophus* spp., *Pimpla instigator* as well as birds, can sometimes make an important contribution to the biological control of lepidopteran pests.

- Entomopathogenic bacterium
- Applied artificially, but also occurs naturally
- Mainly kills young caterpillars

Bacillus thuringiensis

Bacillus thuringiensis is a naturally occurring bacterium of the family Bacillaceae. Different varieties and strains of this bacterium kill different insects; *B. thuringiensis kurstaki* and *B. thuringiensis aizawai* are highly effective against lepidopteran larvae, whilst *B. thuringiensis tenebrionis* is effective against the larvae of some coleopteran species (see chapter 10). Within a particular variety, several strains can be distinguished on the basis of the crystalline proteins they produce (see below). *B. thuringiensis kurstaki* has been used against many species of caterpillar, in many parts of the world, on countless different crops and under wide ranging climatic conditions for thirty years.

B. thuringiensis was first discovered in diseased silk moth caterpillars in 1902 in Japan. In about 1940 *B. thuringiensis* was first used as a commercial preparation in France against the maize borer (*Ostrinia nubilalis*), since when many new strains of *B. thuringiensis* have been discovered that have insecticidal properties, including *B. thuringiensis kurstaki* and *B. thuringiensis aizawai*. *B. thuringiensis* was first registered as a biological insecticide in the early 1960's, since when a considerable amount of knowledge has been amassed concerning both the effect of the bacterium and its mode of action.

Action

B. thuringiensis is only effective when it is ingested by the targeted organism. Young larvae are thus better controlled than their larger counterparts, as they have to ingest less material. The bacterium has no effect on eggs or adult moths. It is therefore important to be familiar with the life-cycle and development time of the caterpillar concerned in order to time the application correctly. Timing is particularly important in species with a concealed habit, as only the first instar larvae that often wander on the plant unprotected will be susceptible. A few hours after the caterpillars have ingested the bacterium, feeding (and thus plant damage) ceases.

The bacterium produce spores and protein crystals, the latter being responsible for the death of the caterpillar. These are broken down by enzyme action in the alkaline (pH > 9) medium of the caterpillars gut. The smaller proteins damage the gut wall, causing holes to form through which the bacterial spores enter the body cavity of the caterpillar and multiply. The metabolism of the larva is disturbed and the jaws are paralysed, as a result of which feeding ceases within a few hours and the spread of damage is prevented. The caterpillar dies within 2 to 5 days of ingesting the bacterium. Paralysed and dead specimens hang by their claspers from the leaves and slowly decay, subsequently falling to the ground.

Action spectrum

The breakdown of the crystalline protein occurs only under highly specific chemical and biological conditions, which are generally found only in the guts of caterpillars. As a result of this specificity, *B. thuringiensis* is only active against caterpillars and is not dangerous to other animals, humans, or plants. The bacteria eventually die off and decay, breaking down to natural products that are taken up in the organic cycle. There are no toxic residues, and thus no accumulation of toxic substances in the food chain.

The Lepidoptera is a very large order of insects, and the efficacy of the bacterium is known only for relatively few species. Different species certainly show different susceptibilities, with many species of the large families such as the Geometridae (inchworms) and Noctuidae (owl-moths) being particularly sensitive. In contrast, the beet armyworm is insensitive to the bacterium. The activity of the product is strongly dependent on several factors, such as the behavioural habits of the caterpillars, their biotope (or ecological community), climatic conditions, the method of treatment, and the dose. In general, the 'unprotected', leaf-feeders are the easiest to control, whereas caterpillars with more secretive, concealed habits, such as stem-borers and leafrollers, and caterpillars that live within webs are more difficult to combat.

Life-cycle in nature

Bacillus thuringiensis

In nature, *Bacillus thuringiensis* is mainly a soil-living organism, although it may also inhabit the leaf canopy of trees, where it lives on waste products produced by the leaves.

The life-cycle of the bacterium consists of 2 phases: a phase of cell division and a phase of sporulation, or spore formation. During this latter phase the bacterium produces specific crystalline proteins which, in conjunction with the spores, enable it to kill caterpillars. The bacterium then feeds on and multiplies within the dead insect. The bacterium can also live saprophytically on dead organic matter.

The spores enable the bacterium to survive unfavourable conditions (such as food shortages), and can be found in many places, not only in insects and in the soil but also in water and on the leaves of trees. The bacteria are disseminated above ground mainly by the wind, but also by non-target organisms that are insensitive to its effects. Fish, birds and mammals may all ingest the bacterium along with their food and deposit it elsewhere in their faeces. However, reproduction and dispersal in nature does not appear to reach epidemic proportions.

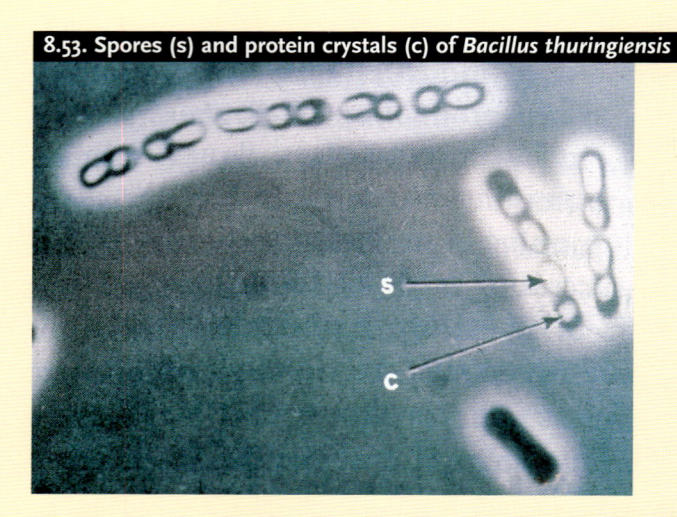

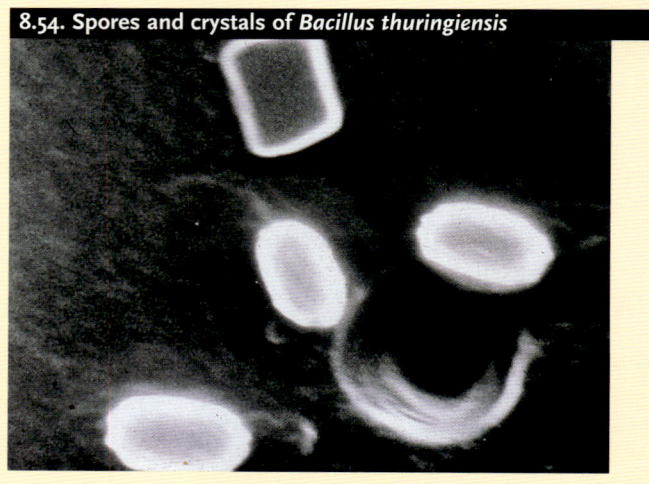

8.53. Spores (s) and protein crystals (c) of *Bacillus thuringiensis*

8.54. Spores and crystals of *Bacillus thuringiensis*

8.55. Caterpillar killed by *Bacillus thuringiensis*

- Parasitic wasp
- Females parasitize the eggs butterflies and moths
- More than one parasite egg can be laid in each host egg

Trichogramma brassicae

In the early 20th century, parasitic wasps of the genus *Trichogramma* (order Hymenoptera, family Trichogrammatidae) were being used as biological control agents in both the United States and Russia. However, from 1930 onward, the widespread use of cheap insecticides meant that biological control received less attention. Only in countries such as China and Russia, where the chemicals were not then available, was biological control still pursued. In the 1980's, investigations into the use of *Trichogramma* were resumed in the West under the weight of environmental pressure. *Trichogramma* is widely used in outdoor crops, particularly in maize against *Ostrinia nubilalis* (the maize borer). Wasps are released into the crop repeatedly during the cultivation period. There are some problems with the taxonomy of *Trichogramma*, in particular because within a single species there are often different races that may differ with regard to host range. The problem is compounded by the fact that the parasites are very small and of almost identical appearance. At present, the species used in glasshouses is *Trichogramma brassicae*.

Population growth

At 25°C, development from egg to adult takes about 10 days, increasing slightly to 12 days at 23°C. Males hatch slightly earlier than females. The development time of *Trichogramma* in the host egg depends on temperature, and to a lesser extent on the host. The maximum temperature at which development can occur is around 38°C. The species of host has a significant influence on the quality of the parasites that develop within. The size of the egg is particularly important; if they are too small, there will be no parasitization, whilst larger eggs are more prone to super-parasitization (the parasitization of the same egg by more than one female). The eggs of Noctuidae are generally ideally suited to the development of *Trichogramma* species. The availability of food (nectar and water) also has considerable influence on the development time and subsequent fertility of adult females. Female *Trichogramma* adults emerge from their pupae with a number of mature eggs that can be laid immediately, and a further number that are still maturing. The actual number of eggs is particularly dependent on the species and size of the host egg, and the number that are laid depends on the prevailing environmental conditions. At a temperature of 20°C, a female lays on average 73 eggs in the eggs of its host, *Mamestra brassicae*.

In most cases, the majority of these eggs are laid during the first two days after emergence. The sex ratio is usually around 7♀:3♂, but this is dependent on several factors such as host, the species and race of *Trichogramma*, climatic conditions and the number of eggs laid in a single host egg.

Parasitization

Because all the eggs of an adult female are already present in the abdomen on its emergence from the pupa, a female can begin parasitizing hosts very shortly after mating. After touching an egg with her antennae and accepting it as a host, the female wasp climbs on to it. By drumming on the egg with her antennae while turning herself around, the female 'measures' the egg and decides how many eggs she will lay in it. The larger the egg, the more eggs will be laid, and more young wasps will hatch per host egg. The female then begins to bore through the egg wall with her ovipositor and lays one or more eggs. The development of male or female offspring depends on whether or not the egg was fertilized; the female wasp is able to determine this herself, as she stores the sperm separately. Both sexes may develop together in the same moth egg.

After being parasitized, this host egg is 'marked' both inside and out, so that it will not be parasitized again. However, a female is only capable of recognizing this marking once she has laid an egg. Parasitized eggs will turn an even black colour after about 4 days at 24°C.

The eggs of most species of moth can be parasitized. Problems only arise if moths of a particular species cover their eggs with hairs or wing scales (such as *Spodoptera exigua*), or deposit the eggs in multiple layers (such as *Lacanobia oleracea*). In addition, very small eggs or eggs secreted in the axils of leaves and other awkward places are not easily parasitized. On the assumption that batches of eggs are more likely to be discovered than eggs deposited separately, the parasitization of species such as *Chrysodeixis chalcites* (which lays its eggs separately) will be less efficient. However, the larger the egg batch, the less likely it is that all the eggs will be parasitized.

The parasitic wasp searches through the plant from the bottom up, and narrows its search area whenever a suitable egg is found.

Life-cycle and appearance
Trichogramma brassicae

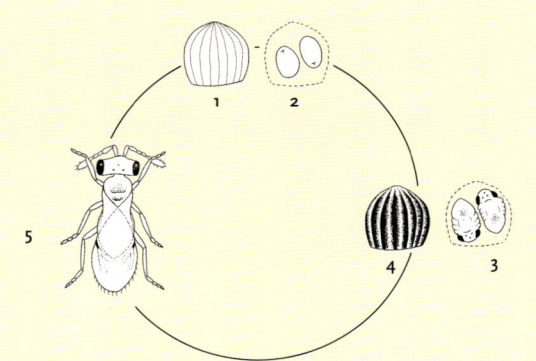

1. moth egg
2. egg with parasitic wasp larvae
3. egg with parasitic wasp pupae
4. blackened moth egg
5. adult

The parasitic wasp lays its eggs inside the eggs of moths. Within the host egg, one or more parasitic wasps develop. The number depends on the size of the host egg, and can vary from one to as many as 30 per host. Two or three eggs of *Trichogramma* are usually laid in the eggs of Noctuidae.
A *Trichogramma* egg is translucent, 0.14 mm long and 0.04 mm in diameter. The larva that emerges from the egg after 24 hours has no clear external characteristics apart from its sickle-shaped jaws. This larva consumes the entire interior contents of the moth egg. Larvae of *Trichogramma* only develop properly in freshly laid host eggs; the older the host egg, the less likely it is to be accepted by the parasite, and the higher the mortality of the parasite larvae.

After four larval instars the wasp forms a pupa inside the moth egg, turning it black.

A *Trichogramma* pupa has characteristic crystals of urea (from uric acid, the insect's excretory product) in the abdomen. This pupa is a light colour at first, and the developing compound eyes are clearly visible. Shortly before emergence, the stripes on the abdomen of the adult can also be clearly seen.

The adult wasps always emerge early in the morning. Female adults of *Trichogramma brassicae* are approximately 0.6 mm long, with a black head, a black thorax and a yellow striped abdomen. Adults have haltere-shaped fore-wings and very small hind-wings. Males and females can be distinguished by the construction of their antennae: females have an elbowed antenna with a knobbed end, males have a more curved antenna with long hairs. *Trichogramma* wasps are active diurnally

8.61. Eggs parasitized by *Trichogramma brassicae*

8.62. *Trichogramma brassicae*

Biological control of butterflies and moths

OVERVIEW OF THE BIOLOGICAL CONTROL OF LEPIDOPTERA					
Product name	natural enemy	controlling stage	stage of Lepidoptera controlled	moth species controlled	crop
TRICHO-STRIP	Trichogramma brassicae	adult	egg	particularly species that lay eggs uncovered in a single layer	all crops except tomato
DIPEL	Bacillus thuringiensis	spore and crystalline protein	caterpillar	particularly those species that live freely in the crop. S. exigua poorly controlled	all crops
MIRICAL	Macrolophus caliginosus	nymph and adult	egg	particularly those species that leave their eggs uncovered	all crops

The bacterium *Bacillus thuringiensis* is most effective against young caterpillars and should therefore be applied as soon as the first caterpillars are seen. If necessary, spraying should be repeated every ten days. The quantity of active material needed depends on the crop and the size of the caterpillars. It is critically important that the underside of the foliage should be adequately covered, since that is where the youngest larval instars feed; the upper epidermis is left intact. Only as caterpillars reach later instars, when they consume the entire leaf, will they come into contact with the active material on the upper leaf surface. However, the bacterium is less effective against these later instars.

The active ingredient in the preparation consists of spores and crystals of the bacterium *Bacillus thuringiensis*. Various products are supplied either as a powder or liquid spray. These can be used in many crops for the control of caterpillars, but the instructions on the label should be followed carefully.

The parasitic wasp *Trichogramma brassicae* is used against *Mamestra brassicae*, *Lacanobia oleracea*, *Chrysodeixis chalcites* and *Autographa gamma*, especially in sweet pepper. *T. brassicae* is supplied by Koppert as pupae on cards. The international trade-name of the product is TRICHO-STRIP.

Finally, the predatory bug *Macrolophus caliginosus* can consume moth eggs, and is therefore introduced in various situations under the name MIRICAL.

Knowing and recognizing mealy **Butterfliess and moths**

9. Mealy bugs and scale insects and their natural enemies

Mealy bugs (family Pseudococcidae), soft scales (family Coccidae) and armoured scales (family Diaspididae) form three important families within the superfamily Coccoidea. This superfamily belongs to the order Hemiptera (the true bugs) and to the sub-order Homoptera (cicadas, leafhoppers and plant lice). The Coccoidea, along with aphids and whiteflies, belong to the division Sternorrhynchae (commonly referred to as plant lice), which is one of the two divisions of the Homoptera. The other division of Homoptera, the Auchenorrhynchae, comprises the cicadas and leafhoppers.

Coccoidea are at first sight barely recognizable as insects at all. The females are wingless and immobile, and covered with a hard scale (armoured scales and soft scales) or with waxy threads (mealy bugs). They suck the sap of plants and are mostly host plant specific. In most soft scales and armoured scales the eggs are laid under the scale, and in mealy bugs in an egg sac. The nymphs that hatch from these eggs are responsible for the dispersal of the insects. Sometimes the females are viviparous that is eggs hatch whilst still in the female such that the young are born as nymphs. Generally, the females die after laying their eggs. Males are generally rare and are small, with wings and legs but with atrophied mouthparts. Populations can reach a high density. Apart from the damage they cause by sucking plant sap, mealy bugs and soft scales also produce honeydew (on which moulds grow) and this too is a serious economic problem, resulting in considerable damage in ornamental and fruit crops.

The concealed habit and the protective covering (whether wax-covered scale, armoured scale or threads) of these insects means that they are very well protected against natural enemies and also against chemicals. Because Coccoidea can be such a serious problem, it was one of the first groups of insects against which biological control was implemented in the classic form (*i.e.* the introduction of a natural enemy with the aim that it should become permanently established). As early as 1888, an Australian ladybird, *Rodolia cardinalis*, was introduced into California and successfully released against the cottony cushion scale *Icerya purchasi*, which had become a huge problem in citrus cultivation.

In this chapter, a description is first given of mealy bugs, the soft scales and the armoured scales, following which the mealy bug predator *Cryptolaemus montrouzieri* is described. Finally the parasitic wasp *Leptomastix dactylopii*, used against the citrus mealy bug, is described.

Overview of mealy bugs and scale insects

	mealy bugs Pseudococcidae	soft scales Coccidae	armoured scales Diaspididae
important species	*Planococcus citri* *Pseudococcus* spp.	*Coccus hesperidum* *Saissetia coffeae* *Parthenolecanium corni*	*Diaspis boisduvalii* *Aspidiotus nerii* *Pinnaspis* spp.
life-cycle female	egg nymph 1 nymph 2 nymph 3 sometimes nymph 4 adult	egg nymph 1 nymph 2 usually nymph 3 adult	egg nymph 1 nymph 2 adult
life-cycle male	egg nymph 1 nymph 2 prepupa pupa adult	egg nymph 1 nymph 2 prepupa pupa adult	egg nymph 1 nymph 2 prepupa pupa adult
external appearance	- body of females from 3rd nymphal instar covered with white, waxy material in the form of powder, threads, spiky projections or platelets - only after 2nd nymphal instar are males and females clearly distinct - adult male small and winged, rather rare	- scale is connected to th body and not easily removed from the insect - males and females easily distinguished from the 2nd nymphal instar, male scale is smaller, less round, flatter and lighter - adult male small and winged, rather rare	- scale is not connected to the body and easily removed from the insect - males and females often clearly distinct from the 1st nymphal instar, male scale is smaller, less round, flatter and lighter - adult male small and winged, rather rare
presence of legs	all instars have legs, but first nymphal instar and adult males in particular are mobile	all stages have legs, but first nymphal instar and adult males in particular are mobile	only first instar nymphs and adult male; as soon as scale begins to form the scale insects lose their legs
damage	- direct damage through ingestion of plantsap - indirect damage caused by exuded honeydew	- direct damage through ingestion of plantsap - indirect damage caused by exuded honeydew	- direct damage through ingestion of sap - no honeydew exuded - can inject toxic substances

Mealy bugs

The most important mealy bugs in glasshouse crops are:

Planococcus spp., especially *P. citri*
Pseudococcus spp.

9.1. *Planococcus citri*

9.2. *Pseudococcus affinis*

Mealy bugs constitute the family Pseudococcidae. There are about 15 species that have at some time been seen in glasshouses, the most important of which belong to the genera *Planococcus* and *Pseudococcus*. Of the former, *Planococcus citri* (the citrus mealy bug) is most important, while different species of *Pseudococcus* turn up from time to time in glasshouses. Mealy bugs appear frequently in ornamental crops, but can also generate problems in tomato, and to a lesser extent in cucumber, melon and aubergine.

Mealy bugs were one of the first insects against which biological control was implemented. Biological control programmes were already running in California by the beginning of the twentieth century against mealy bugs in citrus crops, in which the citrus mealy bug was the foremost pest. The first control insect introduced was the ladybird *Cryptolaemus montrouzieri*. At first, the results produced by this beetle were very variable and therefore various other predators and parasitic wasps were introduced. Later, however, better results were obtained.

The use of chemical pesticides against these insects has also created an increasingly difficult problem in glasshouses, because they can only be effectively controlled through regular, systematic application. The need for biological control has therefore become more pressing as they can only be controlled by the regular application of systemic insecticides.

Life-cycle and appearance
Mealy bug

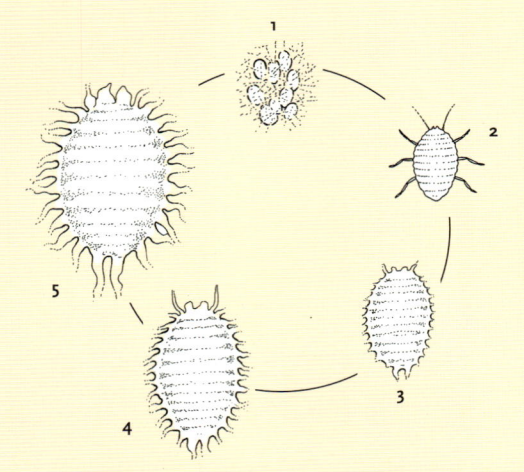

1 egg sac with eggs
2 first instar nymph
3 second instar nymph
4 third instar nymph
5 adult female

Mealy bugs get their name from the fact that the body of the females from the third nymphal instar onwards are covered with a white waxy material in the form of powder, threads, spiky projections or platelets. Unlike most other members of the superfamily Coccoidea, all nymphal instars and the adult males and females possess legs.

The females are wingless and may be up to 5 mm long. The body of these females consists of a single unit of fused head, thorax and abdomen with a white, powdery wax layer. They are often pink to yellow in colour. Their body form is rather reminiscent of a woodlouse.

Adult male mealy bugs are totally unlike the females. They are mostly no longer than 1 mm, possess wings, but lack mouthparts and are therefore incapable of feeding. They have a brief lifespan during which they are wholly engaged in seeking females in order to fertilize them. Eggs are laid in a sticky, foamy mass of wax threads, called an egg sac, which temporarily sticks to hands and clothing, falling off later. Once the batch of eggs is laid, the female dies.

First instar nymphs are yellow-brown, not yet covered with wax, about 0.6 mm long and 0.2 mm wide. They are actively mobile and known as 'crawlers'. At this stage the male and female nymphs are indistinguishable.

Second instar nymphs are darker and less active. After the second instar the males form in rapid succession a dark false pupa followed by a pupa, which develops inside a white, cottony cocoon. After a complete metamorphosis, a winged male emerges from this pupa. The female second instar nymphs, on the other hand, settle on the leaf and begin to secrete wax, moulting to a third instar and then the adult female without a complete metamorphosis.

Different species of mealy bug are very difficult to distinguish from one another; furthermore the taxonomy of these species is not entirely clear. Most research has been conducted on the citrus mealy bug, *Planococcus citri*, one species that can be readily identified. In the case of *Pseudococcus* spp., however, a good deal of uncertainty still exists.

9.3. Eggs

9.4. Young nymphs

9.5. *Pseudococcus longispinus*

9.6. Colony of *Pseudococcus affinis*

Population growth

The nitrogen content of a plant has a great influence on the reproduction of mealy bugs: the higher the nitrogen content, the faster the population growth.

The citrus mealy bug has to mate before eggs can be laid, but other species also reproduce asexually. It appears that there often exist different biotypes or sexual and asexual lineages of a single species that differ in their patterns of reproduction.

Damage

Although most species of mealy bug feed on the aerial parts of the plant, some species extract their nourishment from roots, whilst others are gall-formers. A few species can also transmit harmful viruses.
Mealy bugs inflict damage on the crop in various ways:
- nymphs and females extract the sap from the plant, stunting growth and causing deformed and/or yellowing of leaves, sometimes followed by defoliation. The overall effect is to reduce photosynthesis and therefore the yield. Where flowers and fruit are concerned, these often drop off;
- plant sap is low in proteins and rich in sugars. In order to gain an adequate intake of protein, mealy bugs must ingest large quantities of sap, getting rid of the excess sugars in the form of honeydew. Characteristically, dark sooty moulds (*Cladosporium* spp.) are often found growing on this honeydew, which, as well as the white, waxy secretion of the mealy bugs reduces the ornamental value of the affected plants. Fruit and flowers are also fouled, rendering them unfit for sale, and the reduced level of photosynthesis in the leaves also reduces flower and fruit production;
- in ornamental crops, the mere presence of mealy bugs is sufficient to render the product unfit for sale. A very small population can thus cause considerable economic damage.

Distribution and dispersal

In contrast to all the other Coccoidea, mealy bugs retain their legs in all instars, nymphal and adult, and are capable of movement. However, most stages scarcely move at all, apart from the newly hatched first instar nymphs, the crawlers, which actively search for a place to settle and feed, and the winged adult males. Mealy bugs therefore do not disperse far from one generation to the next without the effects of human activities: where these involve handling the crop itself they can spread the insects over much greater distances.

Strangely, mealy bugs do not always occur in local concentrations, but are sometimes also evenly distributed through the crop. This depends both on the plant species and method of cultivation.

Mealy bug eggs can probably also be spread by the recirculation system. In glasshouses with these systems, mealy bugs are often far more highly dispersed.

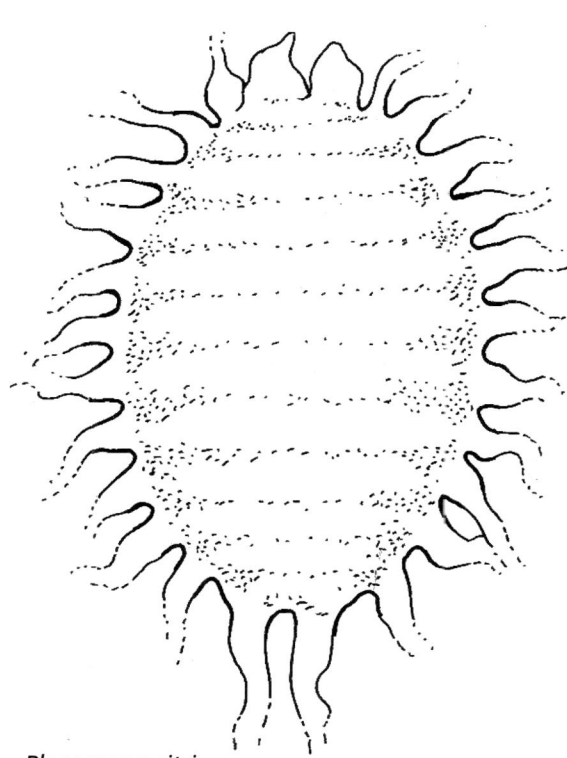

Planococcus citri

- Most important mealy bug in glasshouses
- Mainly found in growing tips and axils
- Occurs on all crops

Planococcus citri
Citrus mealy bug

The citrus mealy bug, *Planococcus citri*, has a world-wide distribution. This mealy bug has many and diverse host plants; in temperate regions of the world it is a problem in glasshouse horticulture, and in the tropics and sub-tropics a problem on outdoor crops. The insect causes damage especially in fruit trees and ornamental crops, particularly in pot plants such as ficus, palms, schefflera, croton and kalanchoë, but also in roses and gerbera. However, this mealy bug can also appear in cucumber, melon and aubergine. Citrus mealy bugs are able to adapt to very different situations. They are found on all parts of the plant, but especially on growing tips and in the axils.

Life-cycle and appearance
Adult females of *P. citri* are 2.5 - 4 mm long and 2 - 3 mm in breadth. Seen dorsally, they have an oval form; they are soft and covered with a fine waxy material. They move very little. They can be distinguished from other mealy bugs by their possession of 18 pairs of relatively short wax rods round the edge of the body, and two slightly longer 'tail filaments' that project at the posterior. The filaments are shorter near the head, and become longer towards the rear. They produce little wax, so that the light yellow to pink body is visible through its waxy covering. There is often a darker longitudinal stripe running over the body. An adult female *P. citri* is shown in plate 9.7.

The adult males are rather rare. They are smaller than the females, have two pairs of wings and two long tail filaments. They do not feed. Their sole task is to fertilize the females and as soon as a male emerges from its cocoon, it goes in search of a female.

A fertilized female lays her eggs in a kind of elongated cottony egg sac composed of white waxy threads produced by the female and which grows beneath and behind the female's body (see plate 9.8). The eggs themselves are a light yellow colour and oval to round. Each egg is separated from the others by a wax layer. Once the eggs are laid, the female shrivels up and dies.

In contrast to other mealy bug species, as far as is known, unfertilized females of the citrus mealy bug lay no eggs and therefore form no egg sac, but continue to feed and may in some cases live up to eight months. It may be, however, that there are biotypes which have a different pattern of reproduction.

The first instar nymph develops from the egg. As in other Coccoidea and in whiteflies, these first instar nymphs are known as 'crawlers'. They are highly active in their search for a new feeding place and are capable of moving a reasonable distance over the plant. The nymphs are yellow and have a waxy covering. They are 0.5 - 0.7 mm long and 0.2 - 0.3 mm in breadth. At this stage, the male and female nymphs are virtually identical.

Second instar nymphs are slightly larger, darker and are less active than the first instar, and at this stage the difference between males and females is visible. The male nymph attaches itself to the plant, whereas the females remain mobile throughout their entire development. After the second instar, a male nymph forms a dark brown 'prepupa' from which a pupa rapidly develops, inside a white cottony cocoon. During this stage complete metamorphosis takes place.

In the case of female individuals, there is no complete metamorphosis, but second instar nymphs begin to secrete the white waxy material and settle down to feed. They undergo little change in form, passing through a third instar after which they become sexually mature. In the third nymphal instar there are seven antennal segments, while adult females have eight antennal segments.

Population growth
The generation time of *P. citri* depends on temperature, the host plant and the relative humidity. Temperature influences the whole of development from egg to adult, whereas humidity mainly affects egg

9.7. *Planococcus citri*

9.8. *Planococcus citri* **female with egg sac**

development and hardly affects the nymphs at all. Development is more rapid, and the time for development from egg to adult is thus shorter at higher temperatures and at higher relative humidity (table 9.1).
In glasshouses, this development time is 1 to 2 months. Below approximately 8°C all development ceases.
Under normal circumstances, the progeny of *P. citri* consist of 50% males and 50% females. However, because adult males have a shorter lifespan, they are far less frequently encountered than females.
The number of eggs laid by a female depends mainly on the temperature and the host plant, varying from 100 to 600, laid during a 1 to 2 week period. Below 13°C egg-laying ceases.

Overwintering
P. citri spends the winter in the ground in the nymphal stage. There is no diapause. When temperatures are more favourable and there are host plants around, the mealy bugs emerge one by one from their hiding places.

Table 9.1. Influence of temperature on the egg to adult development time of *Planococcus citri* (Tingle, 1985)

temperature (°C)	generation time (days)
18	81
22	46
26	32
30	29

- **Very widespread**
- **Can occur on tomato, but also on ornamental crops**
- **Found on the stalks of tomato plants**

Pseudococcus affinis
Obscure mealy bug

Different species of the genus *Pseudococcus* can be found in glasshouse crops. They are all of very similar appearance to the citrus mealy bug, but their body is covered with a thicker layer of wax, the filaments around the body are longer and the two tail filaments are always markedly longer than the others. Most of them have a more restricted range of host plants than the citrus mealy bug.
The most important species is the obscure mealy bug, *Pseudococcus affinis*. There is still some confusion about the correct name of the obscure mealy bug. Originally, the species was named *Pseudococcus obscurus*, but was later split into two species, *P. affinis* and *P. maritimus*. Currently, the species is also known as *P. viburni*.
P. affinis is sometimes encountered on tomato and mainly at the foot of the stalk. With an overhead wire system of cultivation, the mealy bugs are often located on the horizontal part of the stalk (see plate 9.9). This species can also cause damage in ornamental crops such as passiflora. *P. affinis* is difficult to control.

Life-cycle and appearance
This mealy bug is very similar in appearance to *Planococcus citri*, but is slightly darker and slightly larger. The two tail filaments are longer than in *P. citri*, but not as long as in *P. longispinus*.

9.9. *Pseudococcus affinis* on tomato

The mealy bug passes the winter in the nympha stage either in the ground or hidden within the fabric of the glasshouse. With a rise in temperature and in the presence of host plants the mealy bugs emerge. The males are scarcely 1 mm long and much smaller than the females, which measure 3 - 5 mm long. The males possess wings and are mobile

but they have no mouthparts and do not feed.
Reproduction may be sexual, but is usually asexual. The eggs, 200 - 350 in number depending on the temperature, are laid in an egg sac made of white waxy threads.
As usual, the female dies after laying her eggs. Newly hatched crawlers are about half a millimetre long. The female nymphs already possess short wax threads that are absent in the males. Unlike most mealy bug species, females pass through four nymphal instars, while the males have three.

Population growth
The obscure mealy bug develops optimally at temperatures of 20 to 25°C. It is more tolerant of cold and therefore has a wider distribution than other species

Overwintering
Pseudococcus affinis can survive cold periods in the ground in any instar, but usually in the late third or early fourth nymphal instar. *P. affinis* is, however, sensitive to frost and there is no diapause. The pattern of overwintering thus corresponds with *Planococcus citri*.

9.10. Female adult *Pseudococcus affinis*

- Easily recognized by the long tail filaments
- Thrives in a warm, humid environment
- Can be a problem in ornamental crops

Pseudococcus longispinus
Long-tailed mealy bug

Pseudococcus longispinus occurs naturally in tropical and sub-tropical regions, but these days it is widespread through out the world. The range of host plants of *P. longispinus* is less extensive than that of the citrus mealy bug but it nevertheless encompasses many species of (ornamental) crops (*e.g.* croton). The species often inhabits concealed places such as auxiliary buds and loves a warm, humid environment.

Life-cycle and appearance
The long-tailed mealy bug is easily recognized by the long, posterior tail filaments, which are at least the length of the body itself (see plate 9.11). The female is 3 - 4 mm long and more elongated than *P. affinis*. The length of the other filaments is roughly half the width of the body. First instar nymphs (the crawlers) are small, pink and mobile, while later instars are more like the adult female. Males are small, delicate insects with long tail filaments.
Reproduction in *P. longispinus*, as in *P. affinis*, may be sexual or asexual, the latter apparently being the more common situation. Unlike most other mealy bugs, *P. longispinus* produces no egg sac. Instead, the female bears live young, depositing already hatched first instars which are at first kept beneath the body in a network of fine waxy threads. Over 2 to 3 weeks, a female produces roughly 200 nymphs. The rest of the life-cycle corresponds with that of *P. citri*.
In the summer, the life-cycle takes about 6 weeks to complete and in the winter about 12 weeks.

9.11. *Pseudococcus longispinus*

Knowing and recognizing **Mealy bugs and scale insects**

Soft scale insects

The most important species of soft scale in glasshouse crops are:
- *Coccus hesperidum*
- *Saissetia coffeae*
- *Parthenolecanium corni*

9.12. *Coccus hesperidum*

9.13. *Saissetia coffeae*

9.14. *Parthenolecanium corni*

Soft scale insects belong to the family Coccidae. In glasshouses, they are a less significant pest than mealy bugs, but still cause problems more frequently than armoured scales. They often damage perennial, woody crops, but some annuals can also be affected.
The most commonly occurring species in glasshouses will be discussed in detail below: *Coccus hesperidum* (the brown soft scale), *Saissetia coffeae* (the hemispherical scale) and *Parthenolecanium corni* (the brown scale). Other species that occur only occasionally in glasshouses are *Coccus viridis*, *Pulvinaria regalis* and *Ceroplastes* spp.

Population growth
There are generally only one or two generations per year outdoors, but in a glasshouse there can be between three and six, with generations overlapping each other.
Most soft scale insects are averse to extremes of temperature. They thrive best under relatively warm, humid and shady conditions. Reproduction is often parthenogenetic, with females giving rise to solely female progeny.

Damage
The kind of damage caused by soft scales is similar to that caused by mealy bugs. Infestations can be easily overlooked because the immobile soft scales are often situated in concealed places, are coloured grey or brown and are dispersed over the whole plant. They can settle on virtually any part of the plant, including the roots, but most species are to be found close against the veins of the leaf on upper and lower surfaces, or on leaf stalks, branches and (woody) stems. They feed mostly on leaves and twigs, causing discolouration of foliage, stunting and in extreme cases eventual defoliation. However, it is mainly the great quantity of honeydew that causes the damage because of the sooty mould that grows on it. Ants are also attracted by the honeydew. In fact, the honeydew with its sooty mould is usually the first sign of the presence of this type of pest. Among ornamental crops, ferns are particularly sensitive to soft scales, as are oleanders and orchids.

Life-cycle and appearance
Soft scale insect

The life-cycle of males differs from that of females. Adult females emerge after sometimes two, but more usually three nymphal instars, whereas in the males, after two nymphal instars, a prepupa followed by a pupa develop within the scale of the second instar. The adult male that emerges from this scale is small and winged and resembles a male mealy bug. From the second instar onward, female and male nymphs are distinct, the male usually more of an elongated oval shape, the female rounder. Compared with armoured scale insects, the scale of the soft scales is on the whole larger, higher and rounder, and darker in colour. In contrast to the armoured scales, the scale of a soft scale cannot be removed from the insect, even after death. All instars except the egg-laying female possess legs and are capable of movement.

The female generally lays a vast number of eggs under the hard scale (up to 3,000) over a number of weeks. From these eggs emerge the first instar 'crawlers' that disperse over the whole plant, find a feeding place and settle down. There is a very high level of mortality at this stage. Sometimes the females bear live young. In either case, the females die after producing the eggs or young nymphs.

9.15. Eggs

9.16. *Saissetia coffeae*

9.17. Developmental stages of *Parthenolecanium corni*

9.18. Opened female soft scale full of eggs

- Recognizable by its flat scale
- Common in ornamental crops
- Produces large quantities of honeydew

Coccus hesperidum
Brown soft scale

The brown soft scale *Coccus hesperidum* is very common in tropical and sub-tropical regions and is also a common pest in glasshouses throughout the world. *C. hesperidum* has many host plants including many sub-tropical fruiting and ornamental crops such as schefflera, ficus, hibiscus, oleander, palms, ferns and orchids. This species is also very common on indoor houseplants.

Life-cycle and appearance
The scale of the female is flatter than that of other soft scales, usually an elongated oval shape, 3 - 4 mm long, yellow-green or light grey in colour, with dark brown speckles or spots (see plate 9.20).
The colour of the scale darkens as the female ages. The size, shape and characteristics of the female also vary with different host plants. On citrus fruit for example, the female is flat and irregularly shaped with radiating lines, whereas on other plants it can be elongated and oval in shape without lines. It is possible that there are different lineages or biotypes. In all cases, however, the middle of the scale is darker than the margins.
Males live for only 1 to 2 days and are rarely seen.
Females are able to reproduce parthenogenetically, without mating, and may produce 80 - 250 young scales. It may appear that they bear live young, but in fact the first instar nymph emerges from the egg under the mother scale, the empty white egg-shells remaining under the female as shrivelled packets. Over a period of roughly 2 months, crawlers periodically emerge from beneath the female scale, eventually spreading over the whole plant.
The pink crawlers are active for 2 - 3 days and disperse locally before settling. Once fixed in position, their colour changes to a transparent, light brown and they begin to feed. The second instar nymph produces enormous quantities of honeydew. Plate 9.19 shows a nymph of the brown soft scale.
As with all other soft scales, and unlike the armoured scales, this species retains its legs and antennae through the last moult.

Population growth
Outdoors, under summer conditions, the life-cycle takes at least 2 months. Under glasshouse temperatures of 18 to 25°C, six or seven generations per year are possible.

Damage
C. hesperidum mainly sits on the leaf veins, either on the upper or lower leaf surface depending on the host plant. They produce copious quantities of honeydew, more than any other species of soft scale found in glasshouses.

9.19. Nymph of *Coccus hesperidum*

9.20. Adult female of *Coccus hesperidum*

- Distinguished by its high, round form and smooth surface
- Mainly found on the underside of leaves, close to the midrib
- Can be a problem in ornamental crops

Saissetia coffeae
Hemispherical scale

Saissetia coffeae (the hemispherical scale) occurs widely in tropical regions and in glasshouses throughout the world. It is particularly troublesome in several economically important crops, such as coffee and citrus. Like the brown soft scale, the hemispherical scale also regularly turns up on indoor houseplants. Many ornamental crops can be affected, including various different species of ferns, orchids, ficus, oleander, carnation and stephanotis.

Life-cycle and appearance
An adult female of *S. coffeae* has a hemispherical form with a smooth surface. The colour is usually dark to sometimes lighter brown or even almost yellow, with many evenly distributed light yellow flecks (see plate 9.21). The female of this species is more spherical and darker than the flatter, brown soft scale. The high, rounded form and smooth surface distinguish the species from most other soft scales. The size depends on the host plant and can vary from around 4.5 mm on cycas to only 2 mm on asparagus.

Reproduction is almost entirely parthenogenetic and at a sufficiently high temperature can continue throughout the year. A female lays some 500 - 2,500 eggs which are deposited under the scale, after which she dies.

Males hardly ever appear; when they do they live only 1 to 2 days.

The young nymphs are white to yellowish, more oval in form and with a raised H-pattern on the dorsal surface which is much fainter in the adult females.

All instars live on the underside of leaves, often close to the midrib. Only when an infestation is particularly severe do individual scales appear on stems, branches and twigs.

9.21. *Saissetia coffeae*

- Very common outdoors
- Variable appearance, sometimes rather inconspicuous
- Affects mainly ornamental crops in glasshouses

Parthenolecanium corni
Brown scale

Parthenolecanium corni (the brown scale, or on the American continent, the European fruit lecanium) is a very common scale in orchards, fruit nurseries and small fruit crops in the temperate regions of the world. In glasshouses it is mainly ornamental crops that are liable to infestation.

Life-cycle and appearance
The size, form, colour and other characteristics of *P. corni* can vary surprisingly with different host plants, the location, the time of year and the age of the insect. On leaves, this soft scale is often yellow-green and even rather transparent, while on twigs it is a spotted yellow-brown. On twigs, the second and third instar nymphs and young adult females that have not yet laid eggs are quite inconspicuous and easily overlooked. Full-grown females have a thick, brown scale measuring roughly 6 mm long and 4 mm wide.
The eggs are white to light yellow and lie beneath the body of the female. Male nymphs are elongated and roughly 2.5 mm long. After pupation, the small winged adult males emerge, but they are infrequent and do not appear every year. Colonies can be found on stems as well as on leaves and branches.

Population growth
A single female lays from 100 to sometimes 5,000 eggs. Both sexual and asexual reproduction is known. Sexual reproduction follows mating with a winged male, while parthenogenesis in this species leads to both female and male offspring.
P. corni overwinter in the second instar stage on twigs, continuing their development to become adult in the early spring. Eggs are laid from early May and first instar nymphs may be seen from mid June. There is usually a single generation per year, but more are possible, particularly in warm locations. Temperatures above 33°C are lethal for this soft scale.

9.22. Opened female brown scale filled with eggs

9.23. Different developmental stages of *Parthenolecanium corni*

9.24. *Parthenolecanium corni*

Armoured scale insects

The most important armoured scale insects in glasshouses are:
- *Aspidiotus nerii*
- *Diaspis boisduvalii*
- *Pinnaspis* spp.

9.25. *Aspidiotus nerii*

9.26. *Diaspis boisduvalii*

9.27. *Pinnaspis* spp.

Armoured scales belong to the family Diaspididae. This large family includes one third of the approximately 1,400 species of the superfamily Coccoidea. Armoured scales are mainly found on the underside of leaves and on the stem where they are clustered tightly together in white fluffy groups that can sometimes form thick crusts. They frequently colonize perennial and mainly woody crops. Where they have a woolly scale covering they can sometimes be confused with mealy bugs.
Aspidiotus nerii, the oleander scale, has a wide range of host plants, including orchids, acacia, oleander, hedera, rhododendron and palms.
Diaspis boisduvalii, the Boisduval scale, occurs particularly on the leaves of palms and orchids.
The different *Pinnaspis* species most commonly encountered in glasshouses are difficult to distinguish. There are probably three different species that occur on various crop plants, notably on ferns and aspidistra. There are quite a few other species that can occur under glass.

Damage
Armoured scales, unlike mealy bugs and soft scales, produce no honeydew. They feed by sucking the contents of epidermal cells, into which they inject toxic substances that cause yellow, red or brown patches to appear on leaves and fruit. This can eventually kill the leaf. Armoured scales are less common than mealy bugs and soft scales but can be very difficult pests to deal with.

Life-cycle and appearance
Armoured scale insect

As with mealy bugs and soft scales, in armoured scales the two sexes show different life-cycle traits. Females only have nymphal instars, the second of which resembles a small adult and develops directly into an adult female. Males have three nymphal instars, a prepupal stage and a pupa, from which the (usually) winged male develops.

Nymphs and adult females are small, with a scale whose shape and colour is distinct for different species. There are round, oval, elongated or oyster-shaped scales that may or may not be raised. The scale is not connected to the body and can be easily removed. Without its shield, however, the insect soon dehydrates. The colour of the scales varies from species to species and according to genus but is most usually white to grey-brown.

Only the first instar crawlers and the adult males are equipped with legs and have the ability to move. The crawlers disperse over the whole plant before settling down and fixing themselves firmly in the feeding position. During this motile phase the crawlers do not feed, but as soon as feeding begins the formation of the scale also begins. The cuticle of first and second moults remains on the body, giving the armoured scale its characteristic appearance. Sometimes the difference between males and females is already apparent in the nymphs. The scale of the male is smaller than that of the female nymph and often a more elongated oval shape. The colour is often white and the scale is often less hard than that of the female.

The adult female is by far the most important stage of the life-cycle. These females lack antennae, legs and wings. The greater part of the scale is formed during the adult stage and the size usually increases. The head, thorax and abdomen are fused into a single unit which is always flattened and measures approximately 1 - 3 mm. This adult female lays some 50 - 100 eggs, usually under the scale. Populations exist where no males have ever been seen and reproduction is exclusively parthenogenetic.

Males are very small (seldom longer than 1 mm), have a distinct head, thorax and abdomen and usually have one pair of wings. Their lifespan is from several hours to 2 days at the most.

Aspidiotus nerii, the oleander scale, is a greyish, dirty white or yellowy colour, round in shape, with a scale whose apex is off-centred. This scale resembles a miniature fried egg with a diameter of 2 mm. The newly hatched crawlers are roughly 0.3 mm long. They quickly attach themselves to the plant and form a scale. After 1 to 2 months, the female becomes adult and begins to lay eggs. An adult female may live for several months. Colonies mainly form on the underside of leaves.

Female nymphs and adults of *Diaspis boisduvalii*, the Boisduval scale, are flat with a yellow centre, while the male nymphs are smaller, woollier, and with a more elongated form. The male nymphs, female nymphs and adults often sit apart in separate colonies on the plant. The males can sometimes be confused with mealy bugs.

The different species of *Pinnaspis* that appear in glasshouses are difficult to tell apart. There are probably three different species. The females are a faded brown colour and oyster-shaped; the male nymphs have a white back and are similar to those of *Diaspis boisduvalii*.

9.28. Nymphs of *Aspidiotus nerii*

9.29. Adult females with eggs of *Aspidiotus nerii*

9.30. Males and females of *Pinnaspis* sp.

Knowing and recognizing Mealy bugs and scale insects

Natural enemies of mealy bugs and scale insects

The most important natural enemies of mealy bugs and scale insects in glasshouses are:
Cryptolaemus montrouzieri
Leptomastix dactylopii

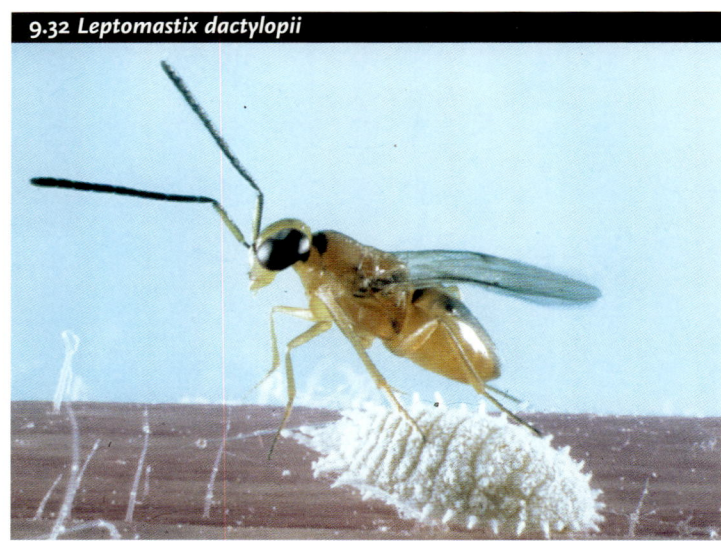

9.31. *Cryptolaemus montrouzieri*

9.32 *Leptomastix dactylopii*

To date, the most successful and therefore the most commonly employed natural enemy of mealy bugs has been the ladybird *Cryptolaemus montrouzieri*. Apart from this predator there are several parasitic wasps on the market that can be introduced to control particular species. Of these, only the wasp *Leptomastix dactylopii*, used against the citrus mealy bug, will be discussed here.

- Ladybird
- Larvae and adults eat all instars of their prey
- Polyphagous, but with a marked preference for mealy bugs

Cryptolaemus montrouzieri

Cryptolaemus montrouzieri is a beetle (order Coleoptera) belonging to the family of ladybirds, the Coccinellidae.
This ladybird of Australian origin was first employed against the citrus mealy bug in California, since when it has been introduced to combat various mealy bugs in many different parts of the world.
The family Coccinellidae has been described in chapter 7. In this chapter, an overview sets out data that enable *C. montrouzieri* to be compared with other frequently used or merely commonly occurring ladybirds. Soft scales are also eaten by *C. montrouzieri*.

Population growth

Temperature is a major factor influencing the generation time of *C. montrouzieri*: the higher the temperature, the shorter the time of development (see table 9.2). The times of development for the different stages are shown separately in table 9.3.
The point below which development is not possible is approximately 14°C. The optimal conditions for population growth are a temperature of 22 - 25°C and a relative humidity of 70 - 80%.
The sex ratio in *C. montrouzieri* is approximately 1:1. Males are sexually mature after 5 days. Females mate very soon after emerging from the pupal stage and begin to lay eggs about 5 days later, the eggs being deposited separately within the egg sac of the mealy bug. The eggs of unmated females are infertile. The total number of eggs laid by a single female depends strongly on her diet; any deficiency in the food supply leads to diminished egg production. The data shown in table 9.4 were recorded under optimal conditions.

Feeding behaviour

All moving stages of *C. montrouzieri* prey on mealy bugs, seizing their prey and consuming them entirely. Adult ladybirds and young larvae prefer the eggs and young larvae while older larvae appear to take all stages of mealy bugs equally. Adult ladybirds are good fliers and can disperse over a large area in search of prey.
Although *C. montrouzieri* prefers mealy bugs, this species is known to be a polyphagous predator. Sometimes other insects related to mealy bugs are eaten, such as soft scales or aphids. However this beetle is most effective in controlling large populations of mealy bugs. Once the supply of mealy bugs becomes inadequate, the ladybirds fly off in search of new prey. Sometimes the larvae resort to cannibalism under these conditions.
At a temperature of 21°C, during the whole period of its development to adult beetle, a larva will eat more than 250 second and third instar mealy bug nymphs. When mealy bugs are eaten, they are consumed entirely, and nothing remains. The larvae of the late third and fourth instars are the most voracious, consuming up to 30 mealy bugs per day.
Beetles and larvae are most active in sunny conditions. Above 33°C their searching behaviour comes to a standstill, while below 16°C they slow down until activity ceases at temperatures below 9°C. *C. montrouzieri* does not appear to kill more mealy bugs than it consumes. However, it does not do well on tomato, presumably due to the glandular hairs.

Table 9.2. The egg to egg development time of *Cryptolaemus montrouzieri* at different temperatures

temperature (°C)	generation time (days)
18	72
21	54
27	33
30	25

Table 9.3. Development times of the different instars of *Cryptolaemus montrouzieri* at 21 °C and 27°C and a relative humidity of 60% (Fisher, 1963)

instar	development time (days)	
	21°C	27°C
egg	8-9	5-6
larva 1	4-8	3-4
larva 2	4-6	2-3
larva 3	4-6	3-4
larva 4	5-6	4-6
pupa	14-20	7-10
egg-adult	43-47	28-29

Table 9.4. Egg production and lifespan of females of *Cryptolaemus montrouzieri* at ± 25°C (Fisher, 1963)

total number of eggs laid per female	440
number of eggs laid per female per day	8.6
number of days after which 50% eggs are laid	18
lifespan (days)	51

Life-cycle and appearance
Cryptolaemus montrouzieri

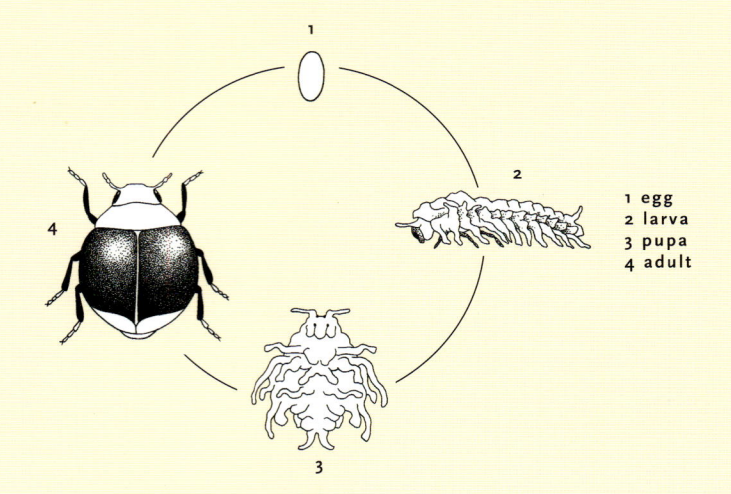

1 egg
2 larva
3 pupa
4 adult

The life-cycle of *C. montrouzieri* comprises the following stages: egg, usually four larval instars, a pupal instar and finally the adult beetle. The eggs, which are a yellow-white colour, smooth, oval and approximately 1 mm long, are laid in mealy bug colonies.

Larvae can be up to 13 mm long. The body is covered with waxy appendages which make the young larvae in particular resemble mealy bugs. They have, however, a more substantial covering of hairs and are much more mobile, and as they develop they become much longer. This resemblance enables them to hide among mealy bugs and thus avoid their own parasites and predators. Larvae cease feeding shortly before the last larval moult. They pupate in various places, on any nearby convenient stalks or on the underside of a leaf, near the midrib, or on glasshouse material.

The adult insect is a dark brown beetle roughly 4 mm in length, whose head and anterior thorax, the tips of the elytra and abdomen are a reddish brown. The shape of this adult beetle is like that of other ladybirds.

Males and females can be distinguished by the difference in colour of the first pair of legs. In the female, the middle segment of these legs is dark grey to black, whereas in the male this middle segment is yellow.

9.33. Larva of *Cryptolaemus montrouzieri*

9.34. *Cryptolaemus montrouzieri*

- Endoparasite of the citrus mealy bug
- Female lays eggs in third nymphal instar and adult females of the citrus mealy bug
- Efficient searching behaviour, and therefore suitable for controlling minor infestations

Leptomastix dactylopii

Leptomastix dactylopii is a parasitic wasp of the order Hymenoptera, of the superfamily Chalcidoidea (chalcid wasps), which is probably the largest superfamily of all insects. *L. dactylopii* is a member of the family Encyrtidae, a monophagous parasitic wasp for which *Planococcus citri* is the sole suitable host. *L. dactylopii* is an endoparasite, laying its eggs inside the host. There have been reports of other hosts, but these reports often involve eggs that are laid in other species which do not, or hardly ever, subsequently hatch.

L. dactylopii originally comes from South America, probably Brazil. The insect was introduced into the United States (California) in 1934 and has subsequently spread further to the Mediterranean and other regions of the world.

Table 9.5. Egg to adult development time for *Leptomastix dactylopii* at different temperatures (Tingle, 1985)

temperature (°C)	development time (days)
17.5	45.3
20	33.3
24	18.5
26	15.3
30	12.5
35	12.3

Population growth

The generation time of *L. dactylopii* is mainly dependent on temperature. Males generally develop faster than females, but the difference is slight. The development times at different temperatures are given in table 9.5. Under favourable conditions, a female parasitic wasp will produce 60 - 100 offspring in about 10 to 14 days. Most eggs are laid during the first weeks.

The sex ratio is generally 1:1, but the size of the host has an influence on the sex of the parasite: more males develop from smaller hosts, while more females develop from larger hosts. Unfertilized females can only produce male offspring. The average lifespan of an adult female under optimal conditions (26°C) is 27.5 days and 24 days for a male.

Distribution and dispersal

L. dactylopii goes mainly for the third nymphal instar and the adults of the citrus mealy bug. The females often try to lay eggs in second instar nymphs, but usually bore straight through these nymphs with their ovipositor, with the result that both nymph and wasp eggs die.

When laying an egg in a prospective host, the female wasp first spends several seconds investigating it before inserting her ovipositor between the waxy projections of the mealy bug. In general, one egg is laid in each host. When occasionally more are laid, only one develops.

The adult parasitic wasps are very good fliers. Their search behaviour is extremely efficient and they can thus be used to control low densities of mealy bugs.

Life-cycle and appearance

Leptomastix dactylopii

On the whole, *Encyrtidae* are very small wasps. Compared with other species of the family, *L. dactylopii* is rather large with a length of 3 mm. The life-cycle consists of an egg, four larval instars, a pupal instar and the adult wasp. The adults are a yellow-brown colour with very long, elbowed antennae. Males are slightly smaller than females and their antennae hairier. When at rest, the wings are held obliquely upright rather than folded over the body.

The female wasp lays eggs in the third nymphal instar and adult mealy bugs. After hatching from its egg, the wasp larva consumes the mealy bug from within, moults several times and pupates. The pupa (within the 'mummified' corpse of the mealy bug) appears at first white, because the mealy bug is still covered with white wax, but gradually becomes visibly brown with red eyes, eventually turning dark brown to black. The 'mummy' is hard on the outside and striped yellow-brown as the dark pupa of the wasp becomes visible in places where the wax covering is thinner. The adult wasp finally emerges from this mummy through a round hole.

9.35. Parasitizing *Leptomastix dactylopii*

Biological control of mealy bugs and scale insects

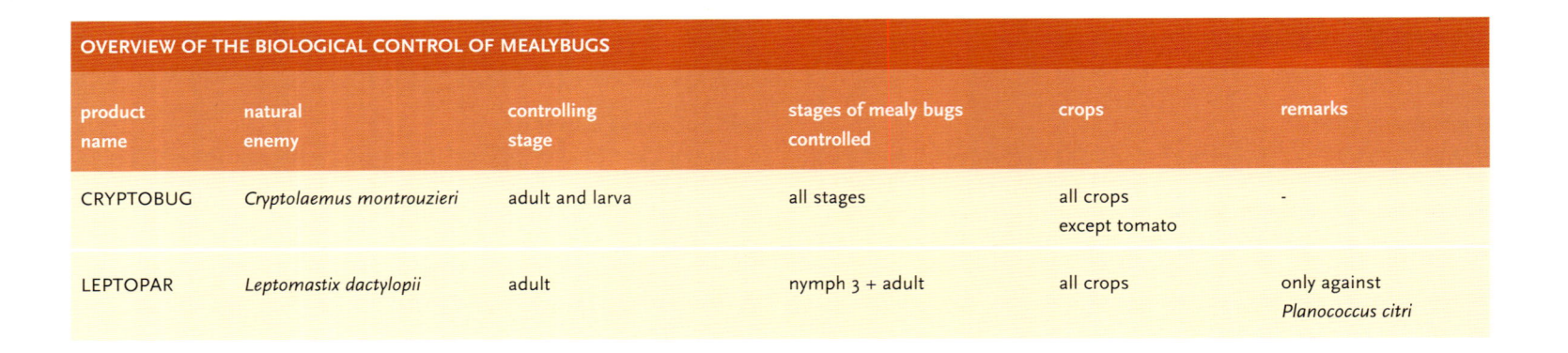

OVERVIEW OF THE BIOLOGICAL CONTROL OF MEALYBUGS					
product name	natural enemy	controlling stage	stages of mealy bugs controlled	crops	remarks
CRYPTOBUG	*Cryptolaemus montrouzieri*	adult and larva	all stages	all crops except tomato	-
LEPTOPAR	*Leptomastix dactylopii*	adult	nymph 3 + adult	all crops	only against *Planococcus citri*

It is possible to control mealy bugs with the help of the predatory ladybird *Cryptolaemus* montrouzieri and the parasitic wasp *Leptomastix dactylopii*. The strategy is comparable to that employed against aphids. That is, the ladybird is suitable for infestations involving colonies of mealy bugs, while the parasitic wasp is better able to find and parasitize more widely dispersed individuals. The adult and larval ladybirds will deal with all stages of mealy bugs, while the wasps show a distinct preference for certain stages. The beetle is able to deal with far more species of mealy bugs, whereas the wasp parasitizes solely and specifically the citrus mealy bug. The ladybird *C. montrouzieri* is not effective under all conditions: it cannot be successfully applied in tomato. It is supplied in the adult form under the international trade name CRYPTOBUG.

The parasitic wasp *Leptomastix dactylopii* is also supplied in the adult form, under the international trade name LEPTOPAR.

Because *L. dactylopii* only parasitizes the citrus mealy bug, it is important first to identify the species of mealy bug concerned.

Research is also being carried out on parasitic wasps that might potentially serve as biological control agents of different species of *Pseudococcus*. The problem is that different *Pseudococcus* spp. occur in glasshouses whereas most parasitic wasps are species-specific in terms of the mealy bugs they parasitize. Thus they are only effective against a single species of mealy bug. In addition, the effectiveness of the parasitic wasps depends on the host plant. Both the hairiness of the leaves and the development of the mealy bugs on the plant can affect the behaviour and survival of the wasp.

The biological control of mealy bugs and scale insects is successful in office plants, tropical swimming pools and tropical glasshouses. Several companies that specialize in this kind of planting sell a range of biological agents for the control of these pests. In commercial glasshouses, however, mealy bugs and scale insects are not tolerated, and for this reason scale insects are hardly ever controlled biologically. Furthermore, the biological control of mealy bugs is still not without its problems.

10. Beetles and their natural enemies

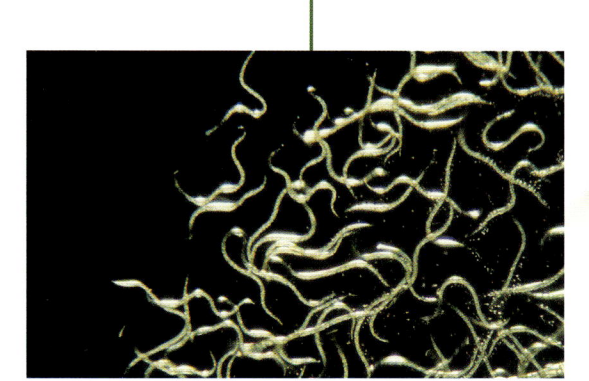

There are several species of beetle that can cause damage to horticultural crops, particularly ornamental crops. For example, *Otiorhynchus sulcatus*, the black vine weevil, is a particular problem in ornamental crops. Most damage is caused by the larvae, although adults can also damage plants. This species can be controlled with the help of the entomopathogenic nematode *Heterorhabditis bacteriophora*. Wireworms, the larvae of click beetles (Elateridae), can also be a pest in various crops, as can the Colorado beetle, *Leptinotarsa decemlineata*, and the palm seedborer *Coccotrypes dactyliperda*. Both sometimes occur in glasshouses and cause damage. However, beetles hardly ever occur in tomato because they have difficulty negotiating the plants glandular hairs.
All the species mentioned above are fully dealt with in this chapter.

Beetles

10.1. *Leptinotarsa decemlineata*

10.2. *Otiorhynchus sulcatus*

10.3. *Agriotes lineatus*

The most important beetles in glasshouses are:

Leptinotarsa decemlineata
Otiorhynchus sulcatus
Elateridae
Coccotrypes dactyliperda

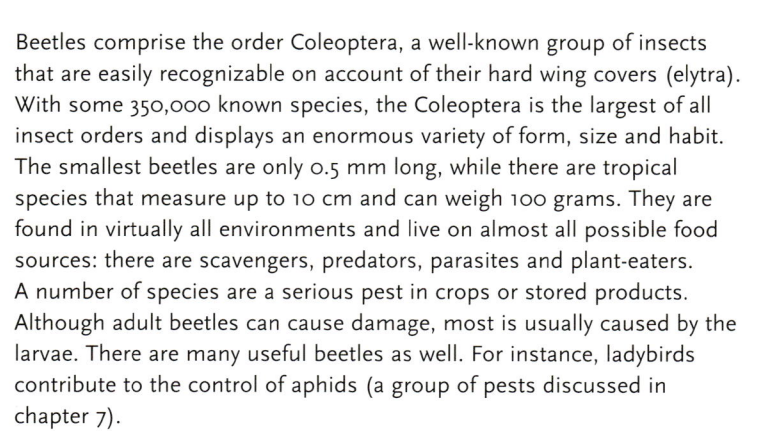

Beetles comprise the order Coleoptera, a well-known group of insects that are easily recognizable on account of their hard wing covers (elytra). With some 350,000 known species, the Coleoptera is the largest of all insect orders and displays an enormous variety of form, size and habit. The smallest beetles are only 0.5 mm long, while there are tropical species that measure up to 10 cm and can weigh 100 grams. They are found in virtually all environments and live on almost all possible food sources: there are scavengers, predators, parasites and plant-eaters. A number of species are a serious pest in crops or stored products. Although adult beetles can cause damage, most is usually caused by the larvae. There are many useful beetles as well. For instance, ladybirds contribute to the control of aphids (a group of pests discussed in chapter 7).

Beetles can often be confused with bugs (Hemiptera). One conspicuous difference is that the wing covers of a beetle lie against each other, whereas in bugs the semi-hardened wings always overlap each other. The mouthparts also differ: in beetles they are designed for biting, whereas in bugs they are modified for piercing and sucking.

Beetles undergo a complete metamorphosis, that is, the young larvae (popularly known as 'grubs') do not resemble the adult at all, and must pass through a pupal instar before reaching the adult stage.

Larvae have biting mouthparts and mostly feed on the same food as the adults. Sometimes, coleopteran larvae lack legs and thus resemble caterpillars (the larvae of the Lepidoptera). The pupae do not make a cocoon and already display many similarities with the adult form, particularly as many adult parts develop externally where they are clearly visible.

The adult beetles usually possess a pair of membranous wings beneath the wing covers. They are often very capable fliers.

- Quarantine organism
- Seen worldwide as a highly dangerous pest
- Can sometimes cause problems in sweet pepper and aubergine

Leptinotarsa decemlineata
Colorado beetle

The Colorado beetle, *Leptinotarsa decemlineata*, belongs to the family Chrysomelidae, the leaf beetles. This is a large family with more than 25,000 species, almost all of which are leaf-eaters. Most leaf beetles are small and attractive, with shiny metallic colouration. The Colorado beetle is a particularly serious pest in potato crops. As the name suggests, the insect originally comes from the eastern slopes of the Rocky Mountains in the west of the United States, but it was probably first collected in Iowa, not in Colorado. In the mid nineteenth century, the beetle began to transfer from its natural host plants (wild species of the order Solanaceae) to the recently introduced and cultivated potato. The beetle spread through the United States and Canada and was first observed in Europe (in Germany) in 1877. Today, the Colorado beetle is distributed through the United States, South Canada, Mexico, parts of central America and the Caribbean region, most of Europe and parts of Asia and Africa.

The bacterium *Bacillus thuringiensis tenebrionis* can be used effectively against several species of beetle, including
L. decemlineata. The larvae ingest the bacteria along with leaf material and are rapidly killed. Another strain of this bacterium, *Bacillus thuringiensis kurstaki*, is discussed in chapter 8.

Population growth
Larvae hatch from the eggs after approximately 7 days and immediately begin to feed. Larval development takes about 2 - 3 weeks, and pupation another 2 - 3 weeks. Adult beetles overwinter, digging into the ground to a depth of 25 - 40 cm in order to hibernate. In the glasshouse there may be more generations.

Damage
Colorado beetles have a preference for potatoes, but also feed on other solanaceous species, such as aubergine, sweet pepper and tomato. Both adults and larvae damage the plants by eating the leaves. They are such prolific feeders that plants can be rapidly defoliated, leaving only bare stalks and skeletonized foliage. In glasshouses, Colorado beetles cause problems only occasionally (particularly in aubergine) and the economic damage is only slight.

Life-cycle and appearance
Leptinotarsa decemlineata

The adult Colorado beetle is approximately 1 cm long, with an oval-shaped body. The colour is pale yellow to brown-yellow with irregular black spots on the head and thorax, and 5 black longitudinal stripes on the wing covers. The legs are predominantly yellow.

The beetles come to the ground surface in spring or early summer (depending on the climate) and if the temperature is high enough (above 10°C) they can fly. Males and females mate, following which a female is able to lay several hundred eggs. The eggs are a yellow to light orange elongated oval, and approximately 1.2 mm long. They are found in groups of about 20 on the undersides of leaves.

There are four larval instars. The young larva is red-brown, while older larvae are orange to red. Larvae have a conspicuous head with biting mouthparts, 3 pairs of legs and a hump-backed body with two rows of black dots on either side. When full-grown they fall to the ground and bury themselves in order to pupate. The life-cycle is completed when the adult beetle develops from this pupa.

10.4. Larva

10.5. Adult

- Very common weevil
- Can be a pest in a great many ornamental crops
- The larvae are the most damaging stage

Otiorhynchus sulcatus
Black vine weevil

The black vine weevil, *Otiorhynchus sulcatus*, is a pest in a great many ornamental crops, particularly in yew, rhododendron, cyclamen and azalea, as well as in strawberries and other fruit crops in both temperate and sub-tropical regions. The black vine weevil belongs to the order Coleoptera (beetles), the superfamily Curculionoidea (weevils and bark beetles) and the family Curculionidae, the weevils. This large family comprises more than 40,000 species, all of which possess a head that is elongated into a snout, or rostrum. At the end of this rostrum are the jaws, and inserted half-way along are the club-shaped, elbowed antennae. Most species of this family cause damage. The beetles of the genus *Otiorhynchus* are all short-snouted weevils.

The larvae of this beetle are difficult to control chemically, not only because of their secretive habit, but also because they are resistant to many pesticides. Apart from *O. sulcatus* there are several other species, such as *O. ovatus* (the strawberry root weevil), *O. singularis* (the clay coloured weevil) and *O. rugosostriatus* (the rough strawberry root weevil). All have virtually the same life habit and are all capable of causing great economic damage.

Population growth
The development time of the black vine weevil varies enormously and depends, among other factors, on climatic conditions. Depending on temperature, development from egg to hatched larva can last from 10 to 30 days. The larval period, under outdoor conditions, can last 9 months or exceptionally even 22 months, whereas in the glasshouse a larva can pupate after 2 months. The pupal stage lasts approximately 20 days.
A population of black vine weevils consists entirely of females and reproduction is by parthenogenesis. The lifespan of an adult female is 5 to 12 months, but some adults have been known to live for 3 years or more and still lay eggs. In any case, by far the most eggs are laid in the first year, with a female being capable of laying from 100 to several thousand. Egg-laying is influenced by both temperature and day-length and ceases below 11°C. Although 50 to 500 eggs hatch, the number of larvae that survive depends on the ecological conditions.
After the last eggs are laid, the beetle lives no more than 2 - 3 months. Table 10.1 gives data concerning the development of *O. sulcatus*. Egg-laying begins approximately 10 weeks after the females emerge from the pupae.
In the glasshouse, the timing of the life-cycle is very different from that outdoors. Development is much faster with each stage lasting a much shorter time. Because beetles can enter the glasshouse from outside, different life-cycles are in progress at the same time and all stages are present in the glasshouse throughout the year. During the winter the larvae continue their feeding as long as the temperature does not fall too low.

Damage
Both adult beetles and larvae damage plants. The beetles are only active at night, taking round bites out of the leaves, starting from the edge. This 'notching' reduces the value of ornamental plants (see plate 10.6).

10.6. Round notch from a leaf caused by adult *Otiorhynchus sulcatus* feeding

In shrubs and young trees, damage is sometimes caused by the adult beetles feeding on buds and the cambial tissue of young bark.
The greatest damage, however, is caused by the larvae. Young larvae at first feed on organic soil particles but rapidly turn to feeding exclusively on root hairs. The larger they grow, the larger the roots they eat. Even the base of the plant stem can be attacked. Thick roots are de-barked and the larvae often devour the whole neck of the root or stem base. This leads to the withering and eventual death of the affected plant. By ringing the base of the stem, a single larva is sufficient to kill a plant.

Table 10.1. Average development data (with range in parentheses) of *Otiorhynchus sulcatus* in the laboratory at 21°C and 60-70% relative humidity on yew (Casteels et al., 1994)	
po-period* (days)	27 (12-50)
oviposition period (days)	378 (5-551)
total lifespan (days)	404 (33-576)
total number of eggs per beetle	1,603 (63-3.216)
average number of eggs per beetle per day	4.6 (1,7-12,8)
number of productive days	208 (5-391)

* po-period = pre-oviposition period, *i.e.* period from becoming adult to first egg-laying

Life-cycle and appearance
Otiorhynchus sulcatus

1 egg
2 larva
3 pupa
4 adult

The life-cycle of the black vine weevil consists of an egg, 6 or 7 larval instars, a pupal instar and the adult beetle.

In temperate climates, the first adult weevils appear around May. They are approximately 7 - 10 mm long, brownish black and have dull yellow spots on their back. The antennae are elbowed. The wing covers are grooved and fused with the body. For this reason vine weevils cannot fly and are compelled to walk, which they do very efficiently. During a single night a weevil can easily cover tens of metres. They are strictly nocturnal insects, and even a full moon causes an immediate reduction in their activity. During the day they hide, and can often be found between the inside of a plant pot and its contents, under lumps of earth, in vegetation and under planks etc, although they are seldom seen in the open.

In temperate climates the small (0.7 mm), roughly spherical white eggs are laid from the beginning of July until around the end of October. These soon turn brown. The larvae that hatch have a white, translucent to (sometimes) pinkish body and a reddish brown head. These larvae live in the soil, are legless, roughly 1 mm long at hatching but grow to approximately 12 mm. A larva is often curled into the typical C-shape which it assumes whenever it is disturbed. The body is covered with stiff white to light brown, bent hairs. Overwintering occurs in the larval stage, usually in the mid (third to fifth) instars. Once the temperature rises, the larvae become active once more.

The full-grown larvae pupate in the spring, in the soil. Pupae are similarly white to cream coloured and 7 - 10 mm long. Outdoors there is a single generation each year, although the larval stage can sometimes be unexpectedly prolonged so that pupation does not occur until the following season.

Development is different in the glasshouse, since without the winter rest period the life-cycle is much shorter.

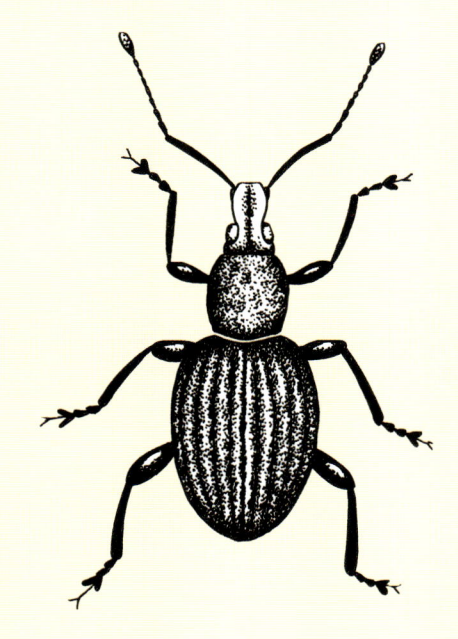

10.7. Larva

- Common insects
- Only cause damage to young, tender plants
- Biological control is not yet possible

Elateridae
Wireworms/click beetles

Wireworms are the larvae of click beetles (Elateridae) and are widespread in Europe, Asia and North America. The most common species in the United States and Canada belong to the genus *Limonius*. In Europe, it is members of the genus *Agriotes* that most commonly occur in crops, particularly *A. lineatus*, *A. obscurus* and *A. sputator*. Wireworms can be pests of many horticultural crops, such as cabbage, beans, lettuce, onion, strawberries, tomato, carnations and chrysanthemums, in all cases by devouring the roots.

Population growth
In the spring, female adults lay groups of 40 - 100 eggs in the upper layer of the soil. These eggs hatch in 4 to 6 weeks. Development from hatching to a full-grown larva can last as long as 2 to 5 years. In the summer of the last larval year, pupation takes place about 15 cm deep in the soil.

Damage
Wireworms devour plants below ground level and even bore holes into stems, bulbs and tubers. The adult beetles themselves do little damage. Curiously, in view of the extent of the damage they can cause in a crop, wireworms eat relatively little plant tissue. They make small holes in main root systems that can lead to the withering and eventual death of seedlings and young plants. The wireworm eats its way up into the stem, perhaps getting as far as 40 cm in large plants. During the first year of their lives, wireworms eat predominantly organic soil material, so that glasshouse crops are virtually untouched. In subsequent years, however, they eat the roots of almost any plants. The severity of the damage they inflict depends on the vigour, size and density of the plants at the time they are attacked.

Life-cycle and appearance
Elateridae

Click beetles are so named because of the ability of up-turned adults to right themselves by launching themselves in the air (with a loud accompanying "click"), and turning over before landing. This leap is made possible by the action of a spine on the ventral side of the first thoracic segment, which fits into an opening on the second thoracic segment. If the body is bent, this spine is suddenly released against the ground, hurling the beetle upward.

The beetles are elongated, black or brown to yellow in colour and roughly 7 - 12 mm long.

In temperate climates click beetles mate in May and June and the females lay their eggs either separately or in small groups just below the ground surface where there is a covering of grass or weeds. After approximately 4 to 6 weeks small (1 - 1.5 mm) white larvae hatch from these eggs. These larvae, known as wireworms, grow slowly, remaining in the ground for 2 - 5 years before they become adults. When full-grown they are 2 - 4 cm long and clear yellow to yellowish brown in colour. They are rather stiff, elongated and cylindrical in shape and have three pairs of short legs close behind the head. The epidermal cuticle is remarkably hard and smooth. Further, they have large, powerful biting mouthparts used to eat living or dead plant material, and thus potentially damage crops. Wireworms are most active in spring when many crops are most susceptible. In temperate climates larvae migrate vertically in the soil during the year, being found closest to the surface from March to May and from July to September. At other times of the year they bury down to a depth of more than 60 cm. In July and August full-grown larvae make small holes in the soil at a depth of 10 to 15 cm in which they pupate. Adults can hatch from these pupae within a month, although they usually overwinter below ground unless disturbed, in which case they move to the surface and seek a new location.

10.8. Wireworm

10.9. Click beetle

- Introduced with plant material
- Can be a problem especially in palm
- Biological control is not yet possible

Coccotrypes dactyliperda
Palm seedborer

The palm seedborer, *Coccotrypes dactyliperda*, belongs to the family Scolytidae - the bark beetles. The species originally comes from the Middle East but can now be found throughout most sub-tropical and tropical parts of the world. It is mainly ornamental crops that are affected by this insect, specifically the palms, such as various *Areca, Chamaedorea, Chamaerops, Cocos, Howeia* (*Kentia*) and *Phoenix* spp.

Population growth
The generation time (from egg to egg) can take 5 months.
A population consists of 80 - 95% females. Mated females produce both males and females, whereas unmated females produce only males. The pre-oviposition period (the period from becoming adult to the first egg-laying) is 0.5 to 3 days.
The average number of eggs laid by a mated female at 28°C and a relative humidity of 70% (30.4 eggs) is considerably higher than the number laid by an unmated female (6.6 eggs). Under the same conditions, however, unmated females live longer. Some 70 to 80 individuals can develop in a single date stone, completely removing the contents. The adult beetles then go in search of new seeds.

Damage
C. dactyliperda is specifically a pest in palms. These beetles are found both as larvae and adults in and on the seeds. Because it is not always evident that imported seeds are infected and because the life-cycle is so slow, it may be several months before any infestation is noticed. When plants develop from such infested seeds, they grow very poorly and often die.
C. dactyliperda can become established and survive in the glasshouse. The species is very difficult to control because the larvae develop inside the seeds, and the adult beetle is mainly found there also.

Life-cycle and appearance
Coccotrypes dactyliperda

The palm seedborer is an elongated beetle, more or less oval, glossy red-brown in colour and covered with long, fine downy hairs. The length of the thorax is slightly greater than its breadth and is rounded at the anterior end. The beetle cannot fly. The female is 2 mm long, the male slightly smaller (1.5 mm). The beetles are often found inside palm seeds, feeding on the seed itself.
Eggs are deposited beside the palm seeds. The larvae are able to bore into unripe and ripe seeds and undergo development there. Larvae have a white body and a brown head.

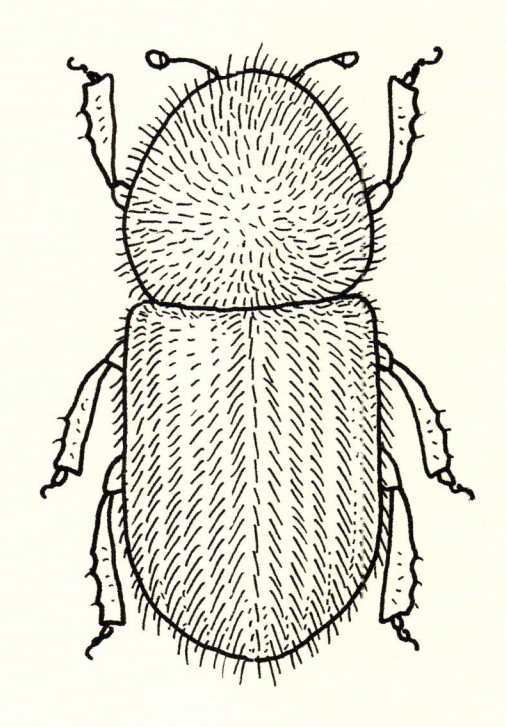

Reference: Lepesme, 1947

Natural enemies of the black vine weevil

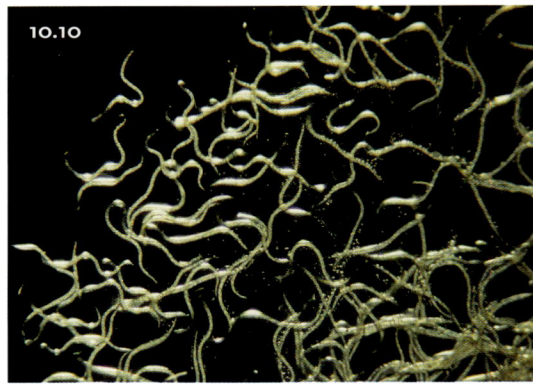

Entomopathogenic nematodes

Nematodes belong to the class Nematoda, also known as roundworms. Within this class there are predators, parasites and saprophytes (feeders on decaying organic material). Among the parasites are species that can infect plants, humans, other mammals and insects. In horticulture, parasitic nematodes are often associated with problems that are difficult to control, but it is less well known that they can also combat insect pests.

The best known nematodes used in the biological control of insects belong to the order Rhabditida, of which the families Steinernematidae and Heterorhabditidae are important.

Entomopathogenic nematodes occur throughout the world and are polyphagous, *i.e.* different *Heterorhabditis* and *Steinernema* species can parasitize a wide range of insects. However, there are great differences in the efficacy of the various species and strains. No single strain is able to combat all insect pests.

In this chapter, *Heterorhabditis bacteriophora* is discussed, a nematode used to help control the black vine weevil. In chapter 11, *Steinernema feltiae* is discussed, a nematode introduced to combat sciarid flies (fungus gnats).

- Commercially introduced
- Invades the larvae of *Otiorhynchus sulcatus*
- Lives in symbiosis with a bacterium that kills the insect

Heterorhabditis bacteriophora

Although different species belonging to the families Steinernematidae and Heterorhabditidae can invade the black vine weevil, *Heterorhabditis bacteriophora* is best suited to the biological control of this insect.

Population growth

The growth of a population of nematodes within the host depends on two factors: the number of infectious (third stage) nematodes that penetrate the host insect, and the size of the host. Both factors determine how many of these third stage nematodes develop further and eventually leave the host. Several thousand nematodes can eventually leave a large-sized weevil larva.

In the soil, the population growth of these nematodes depends mainly on the type and condition of the soil, temperature, and humidity. Natural enemies, such as moulds and predators, can also reduce numbers. Several days after introducing nematodes, the number of those surviving will have fallen to only a few percent of those originally introduced, and over time this will drop even further.

The population of nematodes in the soil will only grow when there are sufficient hosts present. Without hosts, hardly any nematodes will remain in the soil.

Life-cycle and appearance
Heterorhabditis

The nematodes of the genus *Heterorhabditis* are small, un-segmented roundworms.

Infectious nematodes in the third larval stage of development penetrate the host larva either through one of the body openings or by breaking through the epidermis (see 1, life-cycle). In the nematodes gut there are bacteria (*Photorhabdus* sp.) that are released into the host. The parasitic nematode functions in principle as a vector for these bacteria, introducing them into the insect and by-passing the insect's immune system. In exchange, the bacteria cause the insect's death through the release of toxins into the blood and thus provide food for the nematodes.

The bacteria spread through the insect and multiply rapidly. The beetle larva is killed within 48 hours and turns a brown colour (plate 10.12).

The bacteria break down host tissue, converting cellular material into organic products that are easily ingested and absorbed by the nematodes. The fourth stage develops within the insect and finally gives rise to the hermaphrodite adult (see 2, life-cycle). These can grow to a length of 6 mm and produce both male and female sex cells, the eggs being fertilized internally. Depending on food supply, these hermaphrodites lay up to 1,500 eggs, although many of these will not survive. From the eggs that remain viable hatch larvae that rapidly develop to the fourth stage and subsequently to the second generation adults. This generation of adults consists of both males and females that are capable of sexual reproduction (see 3, life-cycle). After mating, the males die. In the presence of sufficient food a female lays her eggs inside the host. If there is insufficient food however, the eggs and first two larval stages develop inside the female nematode. There are generally four larval stages, the first being completed within the egg before the second stage larvae emerge (see 4, life-cycle).

As soon as the larvae reach the third stage they leave the old host and go in search of new ones. Only these infectious larvae can survive outside the body of the host, since they alone have no need of food (see 1, life-cycle). These larvae are around 1 - 1.5 mm long and are characterized by the old skin of the second stage that remains loosely attached after the moult.

These infectious (L3) larvae actively search for a host, penetrate the insect and discard the old L2 cuticle. Theoretically, a single nematode larva is enough to kill an insect.

Insects infected by *Heterorhabditis* change colour from pink to red and may even turn brown. The symbiotic bacteria of this nematode break down the insect cuticle only very slowly, so that the insect remains fairly intact.

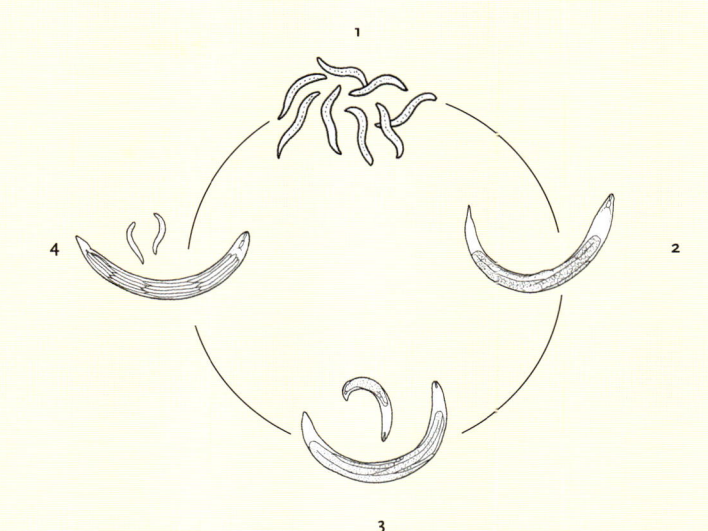

1 free-living infectious nematodes (third larval stage)
2 hermaphroditic parasitic nematode
3 female and male of the second generation
4 development of third stage in the body of a female

(reference: Gaugler & Kaya, 1990)

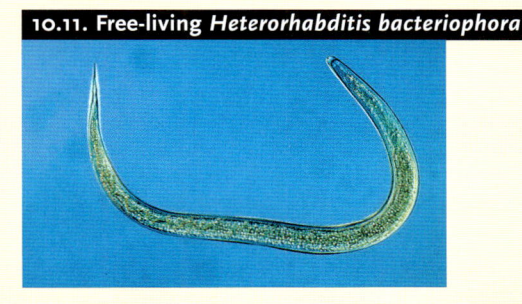

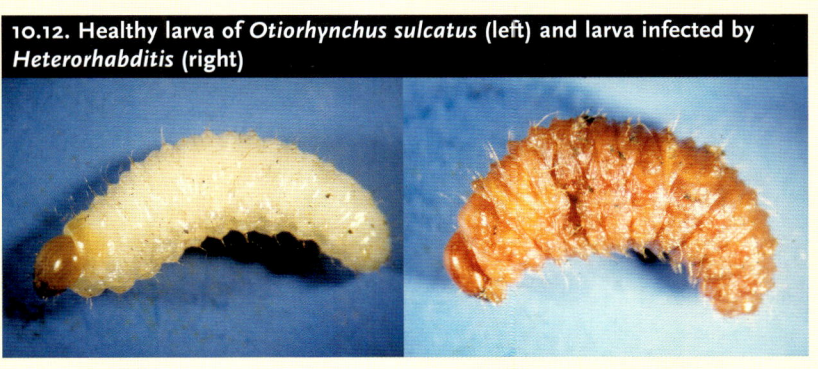

10.11. Free-living *Heterorhabditis bacteriophora*

10.12. Healthy larva of *Otiorhynchus sulcatus* (left) and larva infected by *Heterorhabditis* (right)

Biological control of beetles

OVERVIEW OF THE BIOLOGICAL CONTROL OF BEETLES

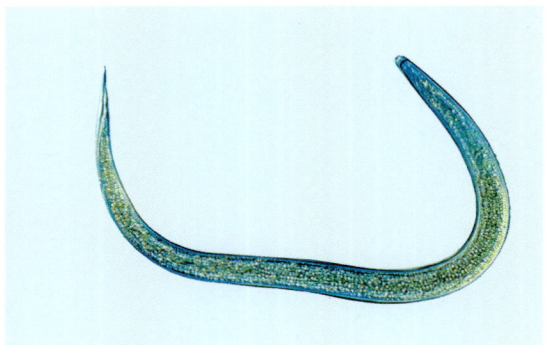

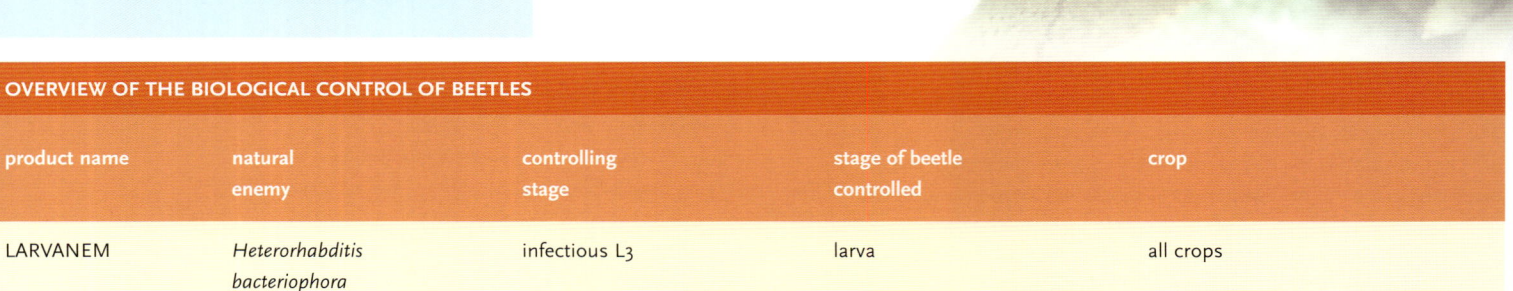

product name	natural enemy	controlling stage	stage of beetle controlled	crop
LARVANEM	*Heterorhabditis bacteriophora*	infectious L3	larva	all crops

The biological control of the black vine weevil by means of *Heterorhabditis bacteriophora* is employed in the cultivation of various nursery trees (including yew, thuja and rhododendron), in strawberry crops, in pot plants and in roses under glass. The nematodes are supplied as infectious third stage (L3) larvae. They can be kept for a short time at a low temperature and are best introduced by means of a sprayer. A soil temperature of approximately 14°C or more is required for the nematodes to work effectively. Their activity will be negatively affected by lower temperatures, the presence of natural enemies, and by UV light. Application in full sunlight should therefore be avoided as far as possible.

Although in principle new nematodes can develop from infected insects and then go on to kill more insects, this will only occur if there are sufficient hosts.

Infectious nematodes are capable of actively going in search of a host, and can travel several tens of centimetres during their lifetime, depending on the soil structure and humidity. Migration of nematodes in sticky clay for example, will be much slower than in loose sand or loam. In rock wool or other artificial substrates the possibility of flushing away nematodes should be borne in mind, particularly if using ebb and flow irrigation systems.

The instructions accompanying the product indicate how the nematodes should best be introduced.

The international trade-name of the product is LARVANEM.

Wireworms cannot be controlled with nematodes alone.

It is possible to control the larvae of the palm seedborer effectively with nematodes, but the accessibility of larvae that are inside the seeds can be a problem.

The larvae of the Colorado beetle can be controlled with the bacterium *Bacillus thuringiensis tenebrionis*

11. Flies and their natural enemies

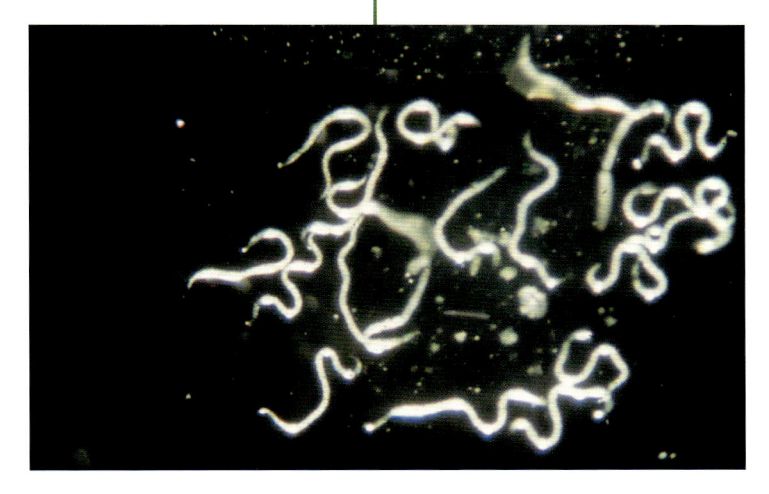

There are several flies that are pests in glasshouse horticulture. A particularly important group is the family Agromyzidae (the leaf miners) which, together with their natural enemies have been discussed in detail in chapter 6. Other species can also cause damage or nuisance, particularly small flies of the families Sciaridae (sciarid flies), Ephydridae (shore flies), Drosophilidae (fruit flies), Psychodidae (moth flies or sand flies) and other lesser known families. The larvae of these insects generally live on rotting plant material and are mostly found in the soil. They feed mainly on organic remains and the fungi that grow on them, although some species can also eat living plant material. The adults mostly prefer damp habitats. Although there are specialist feeders within this group, their traits are often comparable. In this chapter, more specific information is given on several species that cause damage in various cultivated crops. In the overview, the various families of the order Diptera whose members can occur in glasshouses are compared. Finally, the entomopathogenic nematode *Steinernema feltiae* and the predatory mites *Hypoaspis miles* and *H. aculeifer* are discussed. These organisms are introduced specifically to control sciarid flies, but they also combat other dipteran larvae and soil-dwelling pests.

Knowing and recognizing **Flies**

Flies

11.1. Sciarid fly

11.2. Shore fly

The following flies are important in glasshouse crops:
Sciaridae (sciarid flies)
Ephydridae (shore flies)

Flies belong to the order Diptera, a very large order including some 100,000 known species. Their common characteristic is that they have only one pair of wings (Diptera = two wings). The original hind wings are modified to small pin-shaped organs called halteres, which act like gyroscopes and provide stability in flight. There are also some wingless species.

The order is divided into three sub-orders: the Nematocera, including gnats, midges and mosquitoes, and the Orthorapha and Cyclorrhapha, whose members are collectively known as flies. Species from all three sub-orders may be encountered in glasshouses, but it is species of the sub-orders Nematocera and Cyclorrhapha that are of most importance.

Feeding
Almost all adult flies and gnats feed on fluids. As exceptions to this rule, a few hoverflies are capable of crushing pollen grains, and there are some flies that do not feed at all as adults. The fluids that are ingested by the rest, depending on the species, may be juices exuded from rotting matter, blood, plant sap or nectar.

In species that cause damage in cultivated crops it is mainly the larvae that do the damage. The larvae have biting mouthparts and mostly live on rotting material or fungi, although some species can consume tender living plant material. Some species make galls in plants providing their larvae with both food and shelter.

Sometimes both larvae and adults can cause damage indirectly by transmitting harmful organisms.

Life-cycle and appearance
Flies

There is enormous diversity within the order Diptera, and adults can vary from minute to large, and from stout to extremely slender. The head of a fly or gnat is usually large, much of it taken up by the conspicuous compound eyes that in some cases are so large that they meet in the middle.

The structure of the antennae is highly important for the identification of dipterans. The Nematocera are mostly slender insects with long, multi-segmented, narrow antennae that are usually longer than the head and thorax. The antennal segments are rather uniform. The Orthorapha are mostly well-built flies whose antennae are shorter than the head and thorax. Their antennae comprise three segments, but the third is sharply ringed, suggesting several more segments. The antennae are shorter and stouter than those of the Nematocera and often have a terminal bristle or spur. In the Cyclorrhapha the antennae are less conspicuous. The three main segments are usually directed downward, with the third often bearing a long bristle.

Most species of Diptera lay small, elongated eggs from which, in favourable conditions, pale, legless larvae (maggots) soon hatch.

The most important differences between the three sub-orders concern the larvae. In the Nematocera, there are usually four larval instars, whereas in the Cyclorrhapha there are only three, and in the Orthorapha five to eight. There is enormous diversity in the form and life habits of the larvae; some are terrestrial, others aquatic, some live in plants while others parasitize animals. Several of them are harmful to cultivated crops. The larvae have no true legs, although many species have stumps that enable them to crawl. In gnats and midges (Nematocera) the head is striking and possesses biting jaws. Orthoraphan larvae also have biting mouthparts, although the head is much smaller and can be withdrawn into the thorax. The larvae of Cyclorrhapha are small, anteriorly tapered maggots that have no obvious head. They have small jaws at the pointed anterior end with which they scrape at their food, releasing digestive juices and subsequently sucking up the fluids.

The pupae of the Nematocera and Orthorapha remain mobile, whereas in the Cyclorrhapha the pupa is immobile and surrounded by the hardened cuticle of the last larval instar.

The total development time of flies can be very short, sometimes less than a week, but this of course depends on the species and the prevailing conditions.

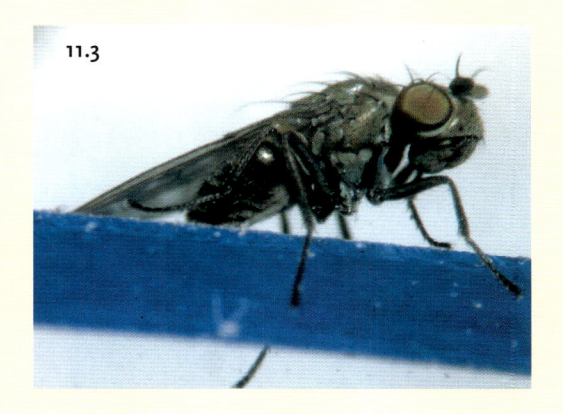

11.3

- Very common, but not always a pest
- Particularly harmful in cuttings and seedlings
- Appear particularly in humid environments

Sciaridae
Sciarid flies

Sciarid flies constitute the family Sciaridae, a family of the sub-order Nematocera (gnats and midges). The species that cause most damage belong to the genera *Bradysia, Lycoriella* and *Sciara*. Species identification is very difficult. The species *Bradysia paupera* can appear in vast numbers in glasshouses.

Although they generally do little harm to large healthy plants, sciarid flies are particularly problematical in cuttings and other young growing plant material, especially in poinsettia, ferns, lisianthus and other ornamental crops. They can be enormously damaging in the mushroom cultivation. Sciarid flies are found in damp, humid environments, and appear very commonly in glasshouses throughout the world. Previously, when most cultivation was carried out in the ground, growers had more trouble from these gnats than they do today. The larvae nestle within the root system and, in the absence of decaying matter, they devour the roots of the plant. In cases where there is a large population on vulnerable or weak plant material, they can cause considerable damage.

Population growth

The development of sciarid flies from egg to adult takes approximately 6 weeks at 15°C, decreasing to 3 weeks at 20 – 24°C. The life span of an adult is about one week.

Mating takes place shortly after adults emerge from the pupa. The males approach the waiting females with fluttering wings and the ensuing mating can last from several seconds to around one minute. The male proceeds to mate with more females, which by releasing a pheromone, are in turn approached by more males. Eggs are sometimes laid on the same day as mating, but usually on the following day. The number of eggs laid can vary from 50 - 300, and is strongly dependent on the size and species of the female.

Damage
Sciarid flies can cause damage both directly and indirectly. Indirect damage is caused by the larvae spreading mites, nematodes, viruses and fungal spores, and also by the adult gnat transferring various fungi. Direct damage is the immediate result of larvae chewing on the roots. Young larvae feed mainly on rotting plant material and on algae and moulds present in the soil. Some species, however, can also feed on root hairs, rootlets, and tender root, stem and leaf tissue. Later instars may even devour plant stems. When the root system is eroded in this way, the crop has difficulty taking up water and minerals, and dies off. The points of attack by feeding larvae also provide invasion routes for various harmful fungi. Because the larvae move very little, plant death is mostly local, particularly with young cuttings, seedlings or pot plants. Young plants that are kept humid and well watered are particularly at risk. Sciarid flies can be a problem in cucumber cultures in cases of thick root syndrome. An environment rich in dead organic material is ideal for the gnat's development, and a substrate with thick cucumber roots provides precisely that. A sciarid fly can amplify the effect of thick roots by weakening the plant further.
Sciarid flies show no particular preference for plant species, as long as they can exploit a humid, hummus-rich environment. Damage often appears when plants grow poorly or when conditions are too damp. Large, healthy plants are not generally harmed.

Distribution and dispersal
Sciarid flies are only found in the immediate vicinity of damp or humid places. In summer, they also occur outside in damp places, and can enter glasshouses via ventilation openings to reach the crop.
Maintaining a lower level of water in the substrate will ensure that most larvae are unable to survive.

Life-cycle and appearance
Sciaridae

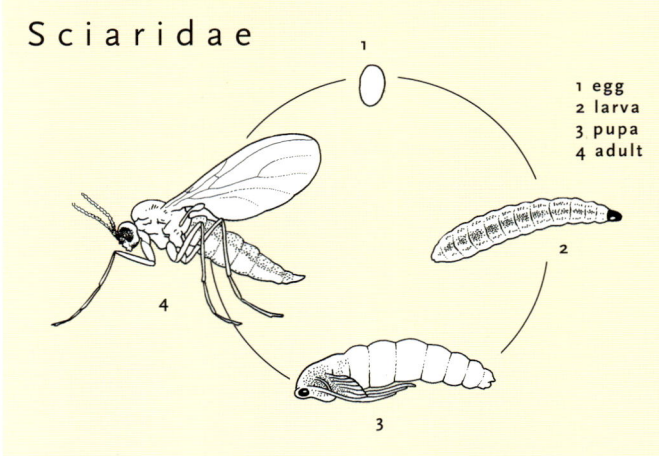

1 egg
2 larva
3 pupa
4 adult

The life-cycle of sciarid flies consists of an egg, four larval instars, a pupal instar, and the adult gnat.
The adult insects are 1 to 5 mm long, grey-black gnats with long, beaded antennae consisting of 14 segments. They have relatively long legs, and the wings show very clear venation. The gnat has a small head equipped with sucking mouthparts, although they hardly feed at all during their short lifetime. The males are usually smaller than the females. They are not fast fliers, and prefer dark places. They occur the whole year round.
A female mates within a few hours of emerging from the pupa and lays some 50 to 300 eggs. These are minute (0.1 to 0.25 mm) and yellowish white in colour, but may differ between species. The eggs are laid on the ground surface close to plant roots, the females being attracted to damp places with much organic material, such as dying plants or rotting plant fragments.
The larvae can grow to between 5 and 12 mm in length and 0.5 to 1.5 mm in diameter. They are legless and have a conspicuous black head.
There are four larval instars that are morphologically identical, each a larger version of the previous instar, although there are visible colour changes. At first they are entirely transparent with a brown head capsule, which subsequently darkens until it is black. The gut contents become brown (or green according to the nature of the digesting food). In the fourth (last) larval instar, the body is milky white and almost opaque. The body is cylindrical, smooth and comprises 12 segments. The head is equipped with biting-chewing mouthparts.
Pupation takes place in a small hole in the ground. Pupae are 2 to 5 mm long and 0.3 to 1.5 mm in diameter. Initially they are white, but later become yellow to brown. The posterior end of the pupa remains mobile.
Sciarid flies appear outside only in the summer months. In the glasshouse, however, they may appear throughout the year, but population increase is evident mainly in the spring and summer.
The different species of sciarid fly are all very similar in appearance.

11.4. Larva

11.5. Adult

- A nuisance to workers
- Often confused with sciarid flies
- *Scatella stagnalis* is a very common species under glass

Ephydridae
Shore flies

Shore flies are small to minute insects that occur very commonly, often in the vicinity of water. They constitute the family Ephydridae, of the sub-order Cyclorrhapha. They are very common on the coast and on the shores of lakes and pools. They feed mainly on rotting material, although some species are predatory. Among the terrestrial and aquatic larvae there are leaf miners, predators and detritus eaters. *Scatella stagnalis* is a species that occurs commonly in glasshouses, and is discussed in more detail below.

Scatella stagnalis

Scatella stagnalis appears commonly in glasshouses where there are green algae. Although they do not damage plants directly, they do transmit plant diseases. Frass deposited on leaves or flowers in ornamental crops and herbs also causes cosmetic damage, and sometimes this can occur on such a scale that growth, particularly of young cuttings, can be retarded. The flies can occur in such vast numbers that they become highly irritating for nursery workers in glasshouse.
Although shore flies can occasionally be confused with sciarid flies, they are more stoutly built and have shorter, un-beaded antennae.

Population growth
Tables 11.1 and 11.2 give data relating to the population growth of *S. stagnalis*.

Table 11.1. The population growth of *Scatella stagnalis* at a fluctuating temperature of 23-34°C (Vänninen & Koskula, 1996)

development time (days)	
egg	1.15
larva	5.2
pupa	3.8
total	10.1
% eggs hatched	86.7
% eggs developing to adult	50-55

Table 11.2. The population growth of *Scatella stagnalis* at 25°C in the laboratory (Vänninen & Koskula, 1996)

lifespan ♀ (days)	14.5
lifespan ♂ (days)	22.8
po-period * (days)	3.2
total number eggs/♀	315
eggs/♀/day	21.6
% ♀ in population	53.4

*po-period = pre-oviposition period, *i.e.* period from becoming adult to first egg-laying

Life-cycle and appearance
Scatella stagnalis

The life-cycle of *Scatella stagnalis* consists of an egg, three larval instars, a pupal instar and the adult fly.
The eggs are laid separately and are slightly bean-shaped, 0.1 mm long and approximately 0.05 mm in diameter. A first instar larva is on average 1.9 x 0.4 mm in size and translucent white. Second instar larvae are brownish and less translucent, measuring approximately 2.8 x 0.6 mm. The average size of third instar larvae is 5 to 5.4 mm long and 1 mm in diameter. The body is cylindrical, and the colour an opaque brown. The larvae are found in the top layer of damp soil, where they also pupate.
The pupa forms within the cuticle of the previous, third larval instar. This cuticle hardens to form the pupal case. The pupa is approximately 2.3 x 0.9 mm, much shorter than the full-grown larva. The head has a pair of conspicuous eyes. Small antennae are also visible, as are the three pairs of legs and a pair of wings adhering closely to the body.
The adult shore fly is black, about 4 - 5 mm long, and stout with short antennae and short legs. The wings have grey-brown patches.

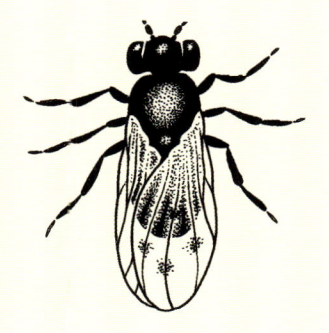

11.6. Adult

Overview of different flies

	SCIARIDAE (sciarid flies)	**EPHYDRIDAE** (shore flies)	**DROSOPHILIDAE** (fruit flies)	**CECIDOMYIIDAE** (gall midges)
important species	*Bradysia paupera*	*Scatella stagnalis* (*Scatophila* spp.)	*Drosophila melanogaster*	*Feltiella acarisuga* *Aphidoletes aphidimyza*
sub-order	Nematocera	Cyclorrhapha	Cyclorrhapha	Nematocera
larval habit	feed mainly on rotting plant material but also on algae and fungi in or on the ground; occur preferably in damp substrate containing much dead and dying organic material	larvae may be leaf miners, predators or scavengers, both terrestrial and aquatic; *Scatella stagnalis* lives on green algae	most larvae live on rotting plant material, but also several predatory species, or parasites	mainly on plants where they often cause plant galls and are harmful; sometimes predators; *F. acarisuga* preys on spider mites, *A. aphidimyza* on aphids
larval appearance	maximum 5-12 mm long, at first transparent so that gut contents are visible, later milky white and opaque; brown to black head capsule	maximum 5-8 mm, white transparent to brown opaque; no obvious head	maximum 3-5 mm, whitish with black mouthparts	variable size and colour, body organs and gut contents often visible through the epidermis
adult habit	gnats suddenly appear when disturbed, run around on the ground or fly, dancing erratically; flight behaviour is more floating than shore flies; the gnats do not usually fly high or far; they prefer dark places	the flies feed mainly on rotting material; some species are predators; they display a 'jumping' flight behaviour: when disturbed they briefly fly up and immediately descend again to the resting position	all species of this family are attracted by fermenting material; in gerbera, sugar rot is transmitted by fruit flies	variable
adult appearance	1-5 mm, slender, grey-black in colour with long antennae resembling a string of beads, with long legs, the wings are wholly transparent with a "V" in the posterior wing venation	4-5 mm, black colour, with short antennae and short legs and with grey-brown flecked wings	2-3 mm, yellow and brown with bright red eyes, short antennae and clear, transparent wings	3-4 mm, yellowish to orange-brown; long thin legs, fairly long beaded antennae with clear segments; wings with fringe of hairs, severely reduced venation, anterior veins not strong, maximum of four veins reach the wing margin

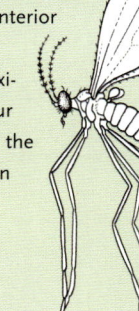

	SYRPHIDAE (hoverflies)	**AGROMYZIDAE** (leaf miners)	**PSYCHODIDAE** (moth flies)	**MUSCIDAE** (house flies)
important species	*Episyrphus balteatus*	*Liriomyza* spp. *Chromatomyia syngenesiae*	various	*Coencsia* spp.
sub-order	Cyclorrhapha	Cyclorrhapha	Nematocera	Cyclorrhapha
larval habit	range of habits; sometimes living on compost of vegetable material, sometimes predatory on insects; *E. balteatus* is a predator of aphids	larvae mostly leaf-miners; some live in seeds and stems, most species are monophagous, but most well-known harmful species are polyphagous	live in rotting material or in water and can reach very high numbers in sewage beds	highly varied habits; larvae of *Coenosia* spp. live in soil and prey on dipteran larvae, e.g. sciarid fly larvae, sucking out body fluids
larval appearance	fusiform, anterior smaller than posterior end, white to light orange with black or orange-brown stripes	variable, white to yellow in colour	maximum 0.6 cm long, rather flattened with a kind of ventral sucker	approx. 1 cm, yellowish white, translucent
adult habit	striking flight behaviour: alternate periods of motionless hovering with sudden darting flight	females make holes in leaves with their ovipositors, both for feeding and egg-laying; males make use of these feeding spots to obtain food	variable; mostly nocturnally active, resting during the day; can be found in shady places in the daytime	highly varied habits, *Coenosia* spp. prey on sciarid flys, adult whitefly, leaf miners, cicadas and other small Diptera species; their long fore-legs enable them to spring in the air from stand-still and seize flying prey, which they consume at their leisure
adult appearance	variable in length with large eyes, conspicuous, often yellow and black markings, sometimes mimicking wasps or bees; short antennae, venation of wings characteristic; several veins run parallel to the hind wing margin and form a false margin; sometimes almost horizontal posture	small to minute, at most a few millimetres, mostly yellowish or blackish colour, look like miniature house flies; the antennae are short and the wing venation is variable	very small, strongly haired, moth-like flies that appear greyish or dark because of hairs; the antennae bear hairs and appear beaded; wings with many hairs and many veins, folded over body like a tent when the fly is at rest; rather long legs also with hairs	2.5-4 mm, short antennae; *Coenosia* spp. possess very long fore-legs

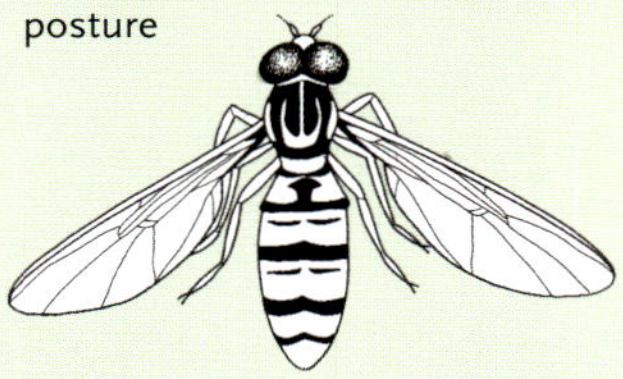

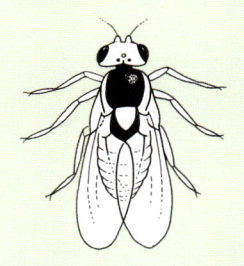

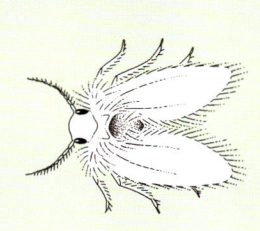

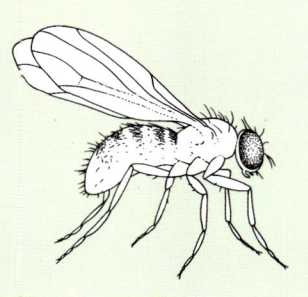

Knowing and recognizing Flies

Natural enemies of flies

The following species are important natural enemies of the flies dealt with in this chapter:
Steinernema feltiae
Hypoaspis spp.

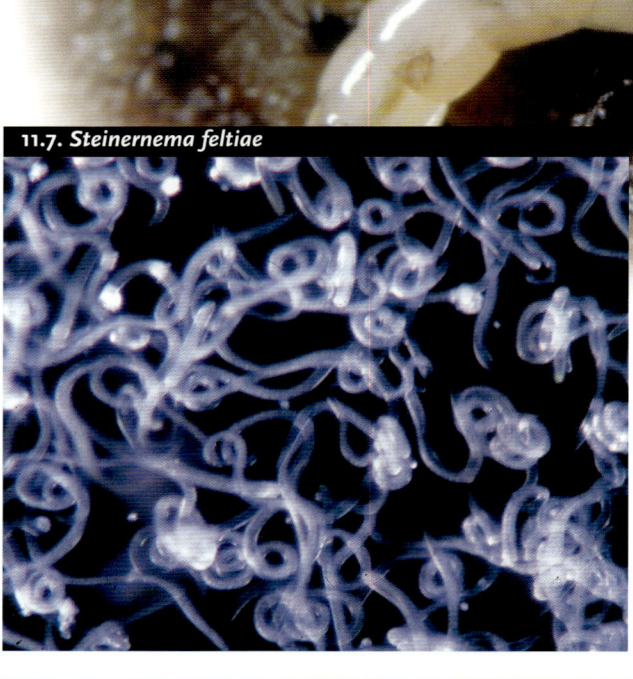

11.7. *Steinernema feltiae*

11.8. *Hypoaspis* spp.

The nematode *Steinernema feltiae* and the predatory mites *Hypoaspis* spp. are mainly introduced against the larvae of sciarid flies, but they can also combat other soil pests. They will be discussed in more detail in this chapter.

Leaf miners can be controlled with several of the parasitic wasps discussed in chapter 6.

- Entomopathogenic nematode
- Infects the larvae of fungus gnats
- Lives in symbiosis an insecticidal bacterium

Steinernema feltiae

Steinernema feltiae belongs to the family Steinernematidae. Like the family Heterorhabditidae, this family contains entomopathogenic nematodes. Although historically there has been some confusion over the name of this nematode, it is now accepted to be *Steinernema feltiae*. It has also been known as *Steinernema bibionis* and *Neoaplectana bibionis*. It is a common species, capable of parasitising various different insects. Although the species has been sold for use against sciarid flies since 1984, it can also infect larvae of the shore fly and the banana moth, *Opogona sacchari* if these are present.

More information on entomopathogenic nematodes can be found in chapter 10, where the biology of *Heterorhabditis bacteriophora* is described.

11.9. Free-living larvae of *Steinernema feltiae*

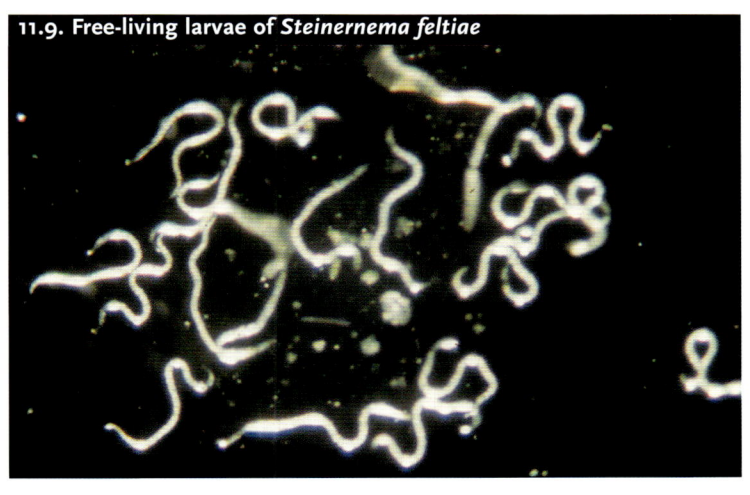

11.10. Sciarid fly larva infected by *Steinernema feltiae*

Life-cycle and appearance
Steinernema feltiae

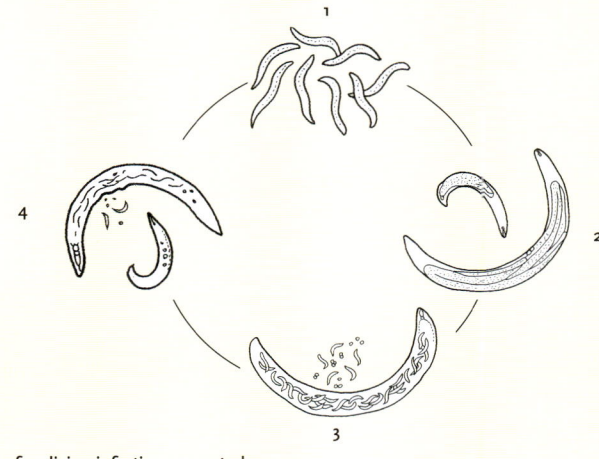

1 free-living infectious nematode
2 male and female of the first generation
3 fertilized female
4 male and female of the second generation, their progeny develop into infectious larvae
(reference: Gaugler & Kaya, 1990)

The life-cycle and appearance of *S. feltiae* are more or less identical to those of *H. bacteriophora*.

Unlike *H. bacteriophora*, however, *S. feltiae* has no hermaphrodite generation, but produces males and females. Unlike *Heterorhabditis* spp., *Steinernema* spp. are only able to enter the insect via the body openings and not by the epidermis. Insects parasitized by *S. feltiae* turn to a yellow or light brown colour. The symbiotic bacteria released from the gut of this nematode (*Xenorhabdus bovienii*) break down the cuticle, as a result of which the infected insect rapidly disappears and is often difficult to locate. Sciarid fly larvae infected by *S. feltiae* are an opaque white to light yellow in colour. They may be parasitized in any larval instar

- Predatory mites
- May be introduced, but also occur naturally
- Feed on soil-dwelling organisms

Hypoaspis spp.

Hypoaspis miles and *Hypoaspis aculeifer* are predatory mites that occur naturally in Europe and North America. *Hypoaspis* spp. belong to the family Laelapidae of the sub-order Mesostigmata. The family Phytoseiidae, containing the predatory mites discussed in chapter 2, also belongs to the sub-order Mesostigmata. There are some 30 species in the family Laelapidae that occur naturally in Europe.

They are typical soil mites and are seldom found in plants. They live on soil-dwelling organisms, such as springtails, the larvae of flies, gnats, and beetles, nematodes, various other species of mite, and thrips pupae. Both species of predatory mite are able to establish themselves in various root environments and growing media, including potting compost and rock wool. They are especially fond of dark, damp places and are not highly tolerant of dry conditions. Both species can be observed outdoors during the summer, with very few individuals remaining active outside this period. Only females overwinter, although there is no diapause.

With sufficient food the predatory mites remain fairly inactive, but in high densities or with insufficient food they become restless and disperse. They can move rapidly, and the depth at which they are found in the soil depends on its compactness, humidity and type. In a compact soil the mites are less able to disperse widely, whereas they penetrate a loose soil more easily. They are hardly ever found in plants.

Two species, *Hypoaspis aculeifer* and *Hypoaspis miles*, have been used since 1995 for the control of various soil-dwelling organisms and thrips in glasshouses. They are discussed in more detail below.

Hypoaspis miles

Hypoaspis miles (also known as *Geolaelaps miles*) is a typical soil mite, but may also be found in the rodent nests or grain stores.

Population growth

At low temperatures, population growth is slow. Although adults can still feed and lay eggs at 12°C, there is no population growth at 10°C. At 32°C the level of egg mortality becomes very high and any females that develop from such eggs are barely capable of laying eggs.

The mites have a long lifespan and can survive for some time without food. Adults can survive 3 - 4 weeks without food, and much longer if they have fed adequately beforehand. However, in the absence of food, egg-laying ceases. Egg-laying is also dependent on the nutritional quality of the food eaten. Thus, at 20°C with straw mites as prey, approximately 10 eggs are laid during a lifetime, whereas on sciarid fly larvae about 32 eggs are laid (see table 11.4). The number of eggs laid by a female is not influenced by mating. However, unfertilized females produce only males, while fertilized females produce both males and females. A population consists of 50 - 70% females.

Data relating to the development time of *H. miles* at different temperatures are given in table 11.3, with further data concerning development on different foods given in table 11.4.

Table 11.3. Development time (in days) of the different instars of *Hypoaspis miles* at different temperatures and with all stages of *Acarus siro* (flour mite) as prey (Wright & Chambers, 1994)

	temperature (°C)			
	15	20	24	28
egg	11.9	5.9	4.0	3.1
larva	3.1	1.9	1.0	1.0
nymphs	18.6	9.7	6.5	5.5
total egg-adult	33.6	17.5	11.5	9.6

Table 11.4. The development of *Hypoaspis miles* at 20°C with *Tyrophagus* spp. (bran mites) or sciarid larvae (fungus gnats) as prey (Enkegaard et al., 1996)

	prey	
	bran mites	sciarid larvae
development time (days)		
egg	3.6	2.9
larva	1.4	1.2
protonymph	7.5	5.9
deutonymph	4.5	4.6
total	17	14.6
po-period*	8.9	5.1
egg-laying period	68.5	53.2
% mortality	20	3.5
lifespan ♀ (♂) (days)	110 (219)	82 (168)
sex ratio (% ♀)	54	66

* po-period = pre-oviposition period, *i.e.* period from becoming adult to first egg-laying

Feeding behaviour
H. miles is a polyphagous mite, and the larvae, nymphs and adults are all predatory. There is an order of preference when alternative foods are presented simultaneously, with sciarid fly larvae taken preferentially, followed by thrips pupae, springtails, nematodes, leaf miner pupae, gall midges. Thus, if sciarid fly larvae are present, *H. miles* will devour these first, even if it comes across an nematode whilst feeding.
When sciarid fly larvae of different sizes are encountered, the smaller ones are taken first, perhaps unsurprisingly as a full-grown sciarid fly larva is ten times larger than *H. miles*.

Each day, *H. miles* can consume up to 8 small (first instar) sciarid fly larvae, or 0.2 -0.6 fourth instar larvae. Eggs of sciarid flies are scarcely ever eaten as they are probably not recognized as prey.
H. miles can consume 9 -10 *Tyrophagus* mites per day.

Life-cycle and appearance
Hypoaspis spp.

Hypoaspis species have the same life-cycle as the Phytoseiidae and thus pass through the following stages: egg, 6-legged larva, 8-legged protonymph and deutonymph, and adult.
H. miles is a rather large mite with a body that may be as long as 1 mm. The body and legs are brown, whilst the jaws are a darker brown. The females are larger than the males, and have a visible white stripe on the terminal body segment. The eggs, larvae and first nymphal instars are white, with nymphs becoming light brown as they age.
H. aculeifer resembles *H. miles* so closely that the two species cannot be distinguished with the naked eye. A practised observer can see the differences under a microscope at 20x magnification. *H. aculeifer* is slightly more slender than *H. miles*, and is slightly more glossy. It also has conspicuously more pointed, almost thorny hairs on the legs, whereas the legs of *H. miles* are covered with soft hairs. The dorsal shield also differs: the brown shield of *Hypoaspis miles* is pointed whereas the dorsal shield of *Hypoaspis aculeifer* is round. The length of the body of the female is about 0.6 mm. The eggs are white, oval and smooth and are about 0.35 mm in length. The first larval instar is white and on hatching has an average length of 0.37 mm. This instar is slow moving and does not feed. However, the larva grows rapidly to about 0.44 mm. The protonymph is white and 0.5 – 0.62 mm in length. Deutonymphs are 0.6 – 0.8 mm long with a pale yellow to light brown dorsal shield.

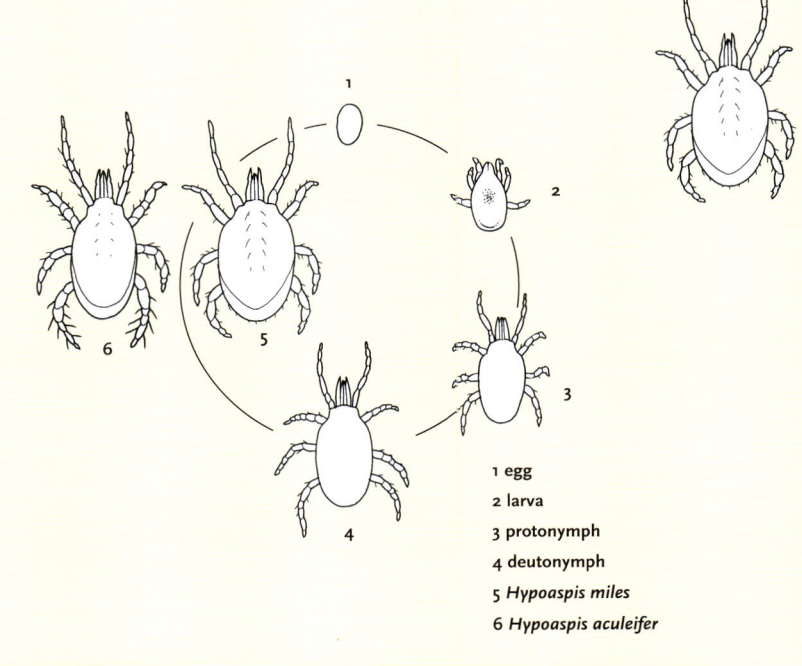

1 egg
2 larva
3 protonymph
4 deutonymph
5 *Hypoaspis miles*
6 *Hypoaspis aculeifer*

11.11. Hypoaspis miles

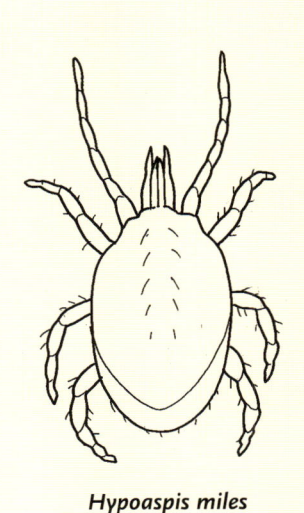

Hypoaspis miles 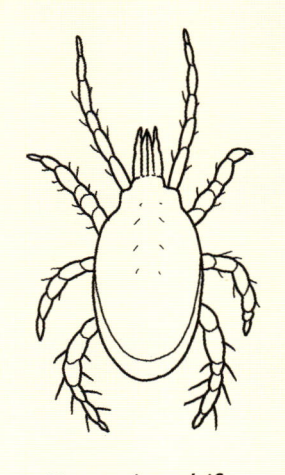   **Hypoaspis aculeifer**

Hypoaspis aculeifer

Hypoaspis aculeifer (also known as *Geolaelaps aculeifer*) is sometimes found in flower bulbs, where it lives on the bulb mite *Rhyzoglyphus robini*.

Population growth
The optimal temperature for the population growth of this predatory mite is 22°C. Data relating to development at different temperatures are given in table 11.5.
Eggs are laid separately, and reproduction can be either sexual or asexual. An unfertilized female will produce only males progeny, and will lay half the number of eggs as a mated female, who will produce both males and females.
The optimal temperature for *H. miles* is higher (25°C) than that for *H. aculeifer* (22°C). The actual differences in efficacy between the two species are not yet clear.

Feeding behaviour
Like *H. miles*, *H. aculeifer* is a polyphagous mite, feeding on the bulb mite *Rhyzoglyphus robini*, sciarid fly larvae and eggs, springtails, nematodes, storage mites, thrips pupae and other soil-dwelling organisms. If other prey is lacking they can even resort to cannibalism, with females eating their own eggs, and sometimes even males. Sometimes several mites may feed on a single prey item.

Table 11.5. The population growth of *Hypoaspis aculeifer* on springtails of the genus *Onichiurus* as food source at different temperatures (Chi, 1980)

	temperature (°C)		
	15	22	28
development time (days)			
egg	10.9	5.1	3.5
larva	3.8	1.6	1.1
protonymph	12.2	4.4	3.3
deutonymph	13.0	4.6	3.9
total egg-adult	39.9	15.7	11.8
% ♀♀	65	80	30
% survival	72	52	56
total number eggs/♀	82	87	51
egg/♀/day	0.42	1.9	2.3
lifespan (days)	194	45	23

11.12. *Hypoaspis aculeifer*

Knowing and recognizing Flies

Biological control of flies

OVERVIEW OF THE BIOLOGICAL CONTROL OF GNATS AND FLIES

product name	natural enemy	controlling stage	crop	remarks
ENTONEM	*Steinernema feltiae*	infectious larva	all crops	- against fungus gnat larvae - soil water level must be high
ENTOMITE ACULEIFER ENTOMITE MILES	*Hypoaspis aculeifer* *Hypoaspis miles*	all moving stages	all crops	against fungus gnat larvae, thrips pupae and other soil-dwelling organisms

The nematode *Steinernema feltiae* is mainly used by nurseries against the larvae of sciarid flies. The nematodes are supplied as infectious larvae in the third (L3) stage, and can be watered onto the crop. A high water content and a soil temperature of around 15°C are necessary for success. When introducing the nematodes, it is important to take into consideration the soil structure and to avoid strong (sun)light, since they are UV-sensitive. The international trade name of the product is ENTONEM.

The predatory mites *Hypoaspis aculeifer* and *H. miles* are sold in a shaker containing all stages. The international trade names are ENTOMITE ACULEIFER and ENTOMITE MILES. The mites are sold for the control of sciarid flies and thrips pupae, but other soil-dwelling organisms may also be predated.

Knowing and recognizing Other bugs

12. Other bugs

Bugs belong to the order Hemiptera, many species of which damage plants. The order consists of insects with piercing mouthparts that enable them to suck the juices or fluids of plants or animals. They are a highly diverse group of organisms, differing widely in size, form and lifestyle.
The order Hemiptera divides into two clearly distinct sub-orders, the Heteroptera and the Homoptera. This latter group includes the leafhoppers, treehoppers, froghoppers, scale insects, aphids and whiteflies. The division between the two sub-orders is based on the fore-wings, which in the Heteroptera are partially hardened, with a tough, leathery basal part and a membranous distal part ("hetero" comes from the Greek word meaning "different"), and in Homoptera are either wholly membranous or entirely hardened ("homo" meaning "same").
The heteropteran bugs will be discussed first, looking in greater depth at three regularly occurring species of capsid bug that cause plant damage: the common green capsid, *Lygocoris pabulinus*, the common nettle capsid, *Liocoris tripustulatus*, and the (European) tarnished plant bug, *Lygus rugulipennis*. Also *Nezara viridula* the southern green stink bug, an important pest species belonging to the family Pentatomidae is discussed. The leafhoppers will then be discussed, in particular the grape leafhopper (*Empoasca vitis*) and the common frog hopper (or spittle bug), *Philaenus spumarius*. Finally a parasite of leafhopper eggs, *Anagrus atomus*, will also be discussed.

Knowing and recognizing **Other bugs**

Heteropteran bugs

12.1. *Lygocoris pabulinus*

12.2. *Liocoris tripustulatis*

The most damaging bugs in glasshouses are:
Lygocoris pabulinus
Liocoris tripustulatis
Lygus rugulipennis
Nezara viridula

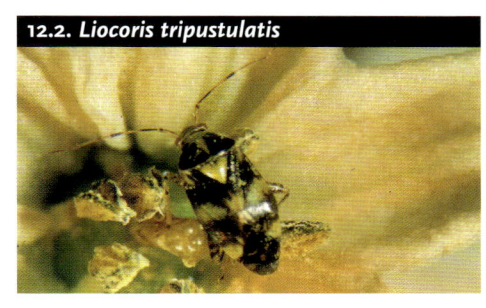

12.3. *Lygus rugulipennis*

12.4. *Nezara viridula*

Heteropteran bugs form a highly diverse group of insects, with both harmful and useful species found in glasshouses. The heteropteran bugs already discussed in previous chapters, such as *Orius* spp. and *Macrolophus caliginosus*, are very useful biological control agents. This chapter, however, deals with the most important of the harmful bugs. Phytophagous bugs damage plants by puncturing leaves and growing points with their piercing-sucking mouthparts, resulting in deformations and holes. The most obvious damage is the deformation of growing points. They can appear in vast numbers, especially in the summer, and can thus affect flowers. They sometimes also cause tissue damage and deformed growth as a result of toxic substances that they secrete in the saliva while extracting plant juices. Where such toxins are involved, the population threshold for damage is much lower.

Three species of bug discussed below belong to the family Miridae, the plant bugs or capsid bugs. This is a large family consisting of small and medium-sized bugs with rather soft bodies. Although most of these can damage plants to a greater or lesser extent, the family also contains species such as *Macrolophus caliginosus* which predate pest mites and insects.

Also *Nezara viridula* a member of the family Pentatomidae or stink bugs, is discussed. This family also contains both harmfull species and species which can predate pests, like *Podisus maculivertris*.

Biological control
Various different natural enemies of harmful bugs are known, but so far none have been produced on any great scale. Eggs and small nymphs may be preyed on by predatory bugs present in the crop, but the value of this contribution is unclear.

Life-cycle and appearance
Heteropteran bugs

1 egg
2 young nymph
3 older nymph of common nettle capsid
4 adult common nettle capsid

Heteropteran bugs are insects which often have flattened bodies, which fold their wings flat over the body when at rest. The wings overlap each other exposing the scutellum, a prominent triangular region of the thorax between the wing bases. This distinguishes them from beetles, which have hard, leathery forewings that meet, but do not overlap, in the dorsal midline. Some bugs give off a pungent odour.

The life-cycle of a bug consists of an egg, usually 5 nymphal instars and the adult insect. The eggs of bugs often have a special cap, or lid, to help the young nymph to emerge. The eggs are often deposited in plant tissue such that the lid is visible as a kind of plug. The young nymphs usually look rather different from the adult insects, but change gradually as one instar moults to the next. Often, a moult is followed by a change of colour, particularly at the last moult. The wings usually only become clearly visible in the fourth instar. Nymphs normally feed on the same food as the adults.

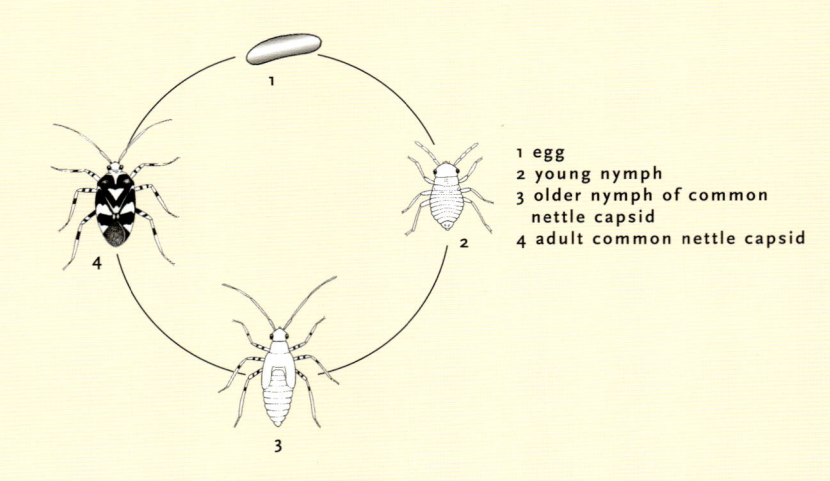

12.5. Nymph *Lygocoris pabulinus*

12.6. Nymph *Liocoris tripustulatis*

12.7. Nymph *Lygus rugulipennis*

12.8. Adult *Lygocoris pabulinus*

- Occurs very commonly
- Can appear in the glasshouse in mid spring
- Can be a problem, especially in sweet pepper

Lygocoris pabulinus
Common green capsid

The common green capsid (*Lygocoris pabulinus*, previously known as *Lygus pabulinus*) can occur as a pest in many crops. The species is both widespread and common in Europe. Damage occurs in woody crops such as apple, pear, cherry and various currants, but also in sweet pepper, aubergine and other crops. The growth of malformed fruit caused by this bug is the main cause of serious economic damage.

Life-cycle and appearance

The common green capsid is a shy and highly mobile insect. The nymphs are pale green or bright green with orange-red tips to the antennae. They somewhat resemble aphids, but are far more mobile and no not possess siphunculi. They have red eyes and completely green legs.

The adults are flat, rather elongated insects, some 5 - 7 mm in length. The fore-wings are partially leathery and when at rest lie folded over each other. The hind-wings are membranous. Adults are bright green with relatively long antennae (see plate 12.9). The piercing mouthparts are conspicuously large. These bugs are often only noticed after the damage has been seen.

In late summer the females lay their eggs in their winter hosts, which are usually perennial woody crops such as apple trees. The eggs are 1 - 1.5 mm long, banana-shaped, cream-coloured and glossy, rounded at one end and with a small cap. They are deposited in the stem of the plant such that the top (the cap end) lies more or less on a level with the surface of the bark and appears like a small plug. These eggs overwinter. As soon as the nymphs emerge from the winter eggs in the spring, they move to the top of a shoot or an inflorescence, where they suck plant juices from the growing point or the youngest leaves. From the third instar onward (there are five in total) the nymphs penetrate deeper and show a preference for the young fruit forming at the shoot tips.

From mid May, the older nymphs and adults migrate to herbaceous plants - the summer hosts. The exact timing of the bugs' departure from the woody crops can vary from one year to the next. Under certain conditions, part of the population remains on the winter host plants for the whole year. Sweet peppers and other glasshouse crops are summer host plants for *L. pabulinus*.

In temperate climates the first eggs are laid in June or July on the summer host plants. Outdoors, there is generally one summer generation per year, but in the glasshouse there are usually more because the insects reproduce well in humid, warm conditions and develop rapidly. Outside, the adults of the second generation appear around mid August, and seek the woody winter host plants again where the females will lay their eggs.

This species can be distinguished from *L. rugulipennis* (the tarnished plant bug) by its evenly bright green colour, in contrast to the less even light green to dark brown colouring of the tarnished plant bug. The scutellum of the latter species, however, is always a yellow-green colour.

12.10. Damage caused by the common green capsid

Leaf damage (sweet pepper)

Fruit damage (sweet pepper)

12.9. *Lygocoris pabulinus*

Damage

Although in temperate climates the first bugs may be observed in the glasshouse from mid May onward, the first damage is only apparent several weeks later. The puncture damage is concentrated on the tender parts of the plant, such as the growing points, young leaves and developing fruit. In pierced leaves, small holes develop that can later expand to become quite large (see plate 12.10).

This leaf damage is seldom economically significant in fruiting vegetables. Reddish-brown specks appear at the base of the leaves and in serious cases the leaves are twisted and deformed. The pattern of damage is similar to that caused by caterpillars, except that calluses appear along the margins and there is no visible frass.

What is economically significant, however, is the damage to fruit as a result of malformation or small round holes. The malformations arise through the secretion of toxic substances in the saliva that inhibit local growth. As a result, punctured fruit develops unevenly and at harvest shows corky outgrowths on the skin where it was punctured (see plate 12.10). In these corky areas the point of entry of the bug's stylets is always visible as a small pimple.

Distribution and dispersal

Nymphs readily fall from the leaves, particularly when they are full grown, and disperse over the ground from plant to plant. Adults can also disperse by flying.

- Very common outdoors
- Can cause problems in the glasshouse in summer
- Can sometimes reduce yield in sweet pepper

Liocoris tripustulatis
Common nettle capsid

The common nettle capsid, *Liocoris tripustulatus*, is a common insect that is widespread over the greater part of Europe and further east, including Syria and Turkey. This bug specifically attacks the aerial parts of nettles. The bug also sometimes occurs in cucumber, sweet pepper and aubergine. In sweet pepper in particular it can sometimes lead to an obvious loss of produce.

Life-cycle and appearance

Nymphs of the common nettle capsid are green with red-brown spots and have long striped legs (see plate 12.11).

Young adults are a light yellow-brown. After the adults have overwintered they are a dark chocolate brown colour with orange spots. Adult nettle bugs are slightly larger than common green capsids and have long, yellow-brown striped legs. On their back a yellowish heart-shaped scutellum is clearly visible (see plate 12.12).

Both male and female adults overwinter in sheltered places such as leaf litter or the dried remains of stinging nettles. Once days begin to lengthen and the temperature starts to rise, they emerge from their winter diapause. In temperate climates in about June, and after mating,

12.11. Nymph of *Liocoris tripustulatis*

the females lay their eggs in the stalks of nettles. The males have by this time largely disappeared. When the following generation approaches adulthood (around mid July) there are still several females remaining from the winter generation. The nymphs of the summer generation are all adult by the end of August. June, July and August are thus the months of most activity for common nettle bugs. From September, shorter day-length triggers diapause.

Common nettle bugs occur in glasshouses mainly in the summer. Conditions under glass are more favourable than outdoors and allow more generations to be produced. Although July and August are the months when the greatest numbers occur, they can be found under glass even in winter. These are mainly individuals carried over from the last crop. Under short days they develop very slowly, although the females can still lay eggs. How many depends on the food supply, the temperature and the day-length. Food and temperature are not limiting factors in a glasshouse, but day-length is, and therefore nettle bugs start to lay more eggs as summer approaches. Because the bugs can fly, they can easily spread throughout the glasshouse.

Damage

Both the nymphs and adults of the common nettle bug cause damage. The nettle bug is usually readily found in the top of the plant, but it may be weeks before damage becomes visible in the form of bunched, squashed growth caused by the bug having punctured the growing point. This puncturing also seems to stimulate vegetative growth whilst inhibiting flowering. The fruit that eventually forms is often highly deformed. Once the nettle bugs have been dealt with it can still be weeks before bug-free plants can grow and produce normally once more. The damage can be considerable.

In sweet pepper and cucumber, direct damage can be inflicted on older fruit through the bugs puncturing the fruit wall causing round, dark brown or black places to develop on ripe fruit (see plate 12.13). These patches quickly go rotten, and so the fruit must be very carefully sorted. In ornamental crops the greatest damage arises from shrivelling of the flower buds at an early stage of development. The bug punctures the flower stalk whilst the buds are barely visible, and the flower is doomed from the start. In cucumber a clear wound is visible in the stalk where the bug has punctured the tissue. The main shoot or the side shoot grows crooked as a result, becoming brittle and liable to breakage.

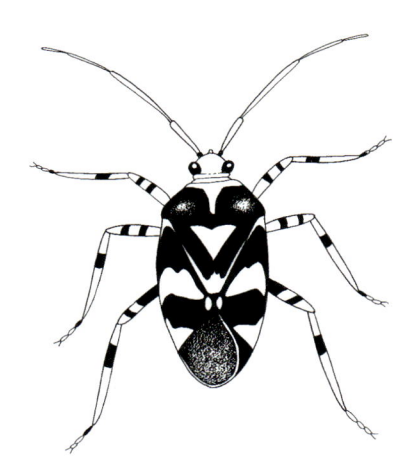

12.12. *Liocoris tripustulatis*

12.13. Damage in sweet pepper caused by *Liocoris tripustulatis*

- Occurs very commonly
- Can appear in the glasshouse from mid spring
- Can be a problem particularly in young fruiting vegetable crops

Lygus rugulipennis (European) tarnished plant bug

The European tarnished plant bug is widely distributed over the temperate regions of the world. The species will henceforth be referred to as the tarnished plant bug, although in the United States there is a related species, *Lygus lineolaris*, which has the same common name. The species is polyphagous, and causes problems particularly in cucumber, aubergine and sweet pepper, also attacking various other vegetable and ornamental crops such as gerbera and chrysanthemum.
Sometimes two other *Lygus* species are encountered, *L. pratensis* and *L. maritimus*. These appear particularly in sweet pepper and are often thought to be *L. rugulipennis*, which they closely resemble.

Life-cycle and appearance
The colour of the tarnished plant bug can vary from light green to dark brown depending on the crop in which they occur. As a rule the females are lighter in colour than the males, which can sometimes be almost black. In cucumber the adult bug is a dull colour somewhere between brown and grass-green. An adult specimen of the tarnished plant bug is slightly larger and more slender than the common nettle bug. Plate 12.14 shows an adult tarnished plant bug. In spring they emerge from their overwintering places, the females mate and lay eggs for a new generation.
The eggs, up to 100 per female, are laid in the stalks and flower buds of the host plant. The eggs are cream-coloured, bottle-shaped and 1.0 x 0.25 mm in size.
The nymphs are smaller and green and run very rapidly over the plant. There are five nymphal instars. As the nymphs become older the wing buds become more visible. Fifth instar nymphs are 4 to 5 mm in size, with five clear black dorsal dots (see plate 12.15), which distinguish them from the nymphs of the common nettle bug and the common green capsid, which have no dots.
The nymphs become adult in summer, after which a subsequent generation develops. The nymphs of this generation become adult in the autumn, after which they seek out hiding places in which to overwinter. Outside, there are thus two generations per year, although in the glasshouse there may be more.
In the autumn the adult insects hide away and can overwinter in the glasshouse. In the open air, overwintered adults are highly active in the spring and capable of flying large distances to reach new host plants (and glasshouses). Adults overwinter between dead leaves.

Damage
When nymphs or adults feed they secrete with their saliva a toxic substance that kills the cells surrounding the point of entry of the mouthparts. Usually, the first signs of damage are small brown spots on young leaves. Because the area surrounding each spot where the bug feeds then dies, the affected plant can no longer properly develop.
It would seem that the nymphs cause more damage to the crop than the adults. Where the nymphs accumulate, more holes appear in the leaves and the stalk is clearly damaged by being punctured and having the sap withdrawn. When the punctured leaves develop further the holes become enlarged and the leaf is pulled out of shape. When flowers are punctured by the bug there may be no further development of fruit. Puncturing the head of the plant leads to arrested growth. In the stem, part of the bark may be destroyed, and gum secreted by the plant. As a result the normal longitudinal growth of shoots and runners is curtailed and heads remain squashed. In cucumber and aubergine it may also be seen that the axils of affected plants are empty, as the fruits have been aborted at a very early stage. Sometimes a plant will bear no more fruit along a stem, the entire sequence of axils remaining barren. The holes in the leaves appear later than the damage to the stem and the fruit. The common nettle bug

12.14. Adult of *Lygus rugulipennis*

12.15. Nymph of *Lygus rugulipennis*

12.16. Damage to cucumber caused by *Lygus rugulipennis*

also damages the stalks, but the places of damage are larger and fewer in number than for the tarnished plant bug. The latter does not often inflict much damage to a full-grown crop, but in a young crop the damage to the fruit and the plant can be considerable. Plate 12.16 shows damage caused by the tarnished plant bug to cucumbers and cucumber leaves.

- **Stink bug**
- **Very common in tropical and sub-tropical areas**
- **Can cause damage in, amongs others, aubergine and sweet pepper**

Nezara viridula
Southern green stink bug

Stink bugs belong to the family Pentatomidae, the true bugs with a pentagonal prothoracic shield or pronotum, of which *Nezara viridula* is one of the most important pest species. Although it is believed to have originated in Ethiopia, it is now distributed throughout almost all tropical and sub-tropical areas, and continues to spread. *N. viridula* causes problems in many important food and fiber crops, including key glasshouse crops such as aubergine and sweet pepper.
Stink bugs derive their name from the strong odor that they emit from scent glands when disturbed.

Life-cycle and appearance
Eggs are laid in clusters of 30 – 130 on the undersurface of leaves and fruits, in the upper portions of crops and weeds. The eggs are glued firmly to each other and to the substrate. Newly laid eggs are about 1 mm in height, are pale yellow and barrel shaped with flat tops, and have a conspicuous girdle of spines. Over time they change colour, eventually becoming clear orange (see plate 12.17).
Nymphs hatch from the egg by opening the disc shaped cap. Eggs of the same cohort emerge simultaneously, with the emergence of the first nymphs triggering the emergence of the others. Nymphs are reddish in colour with red eyes and transparent legs and antennae. First instars do

12.17. Eggs of *Nezara viridula*; newly laid (left), old (right)

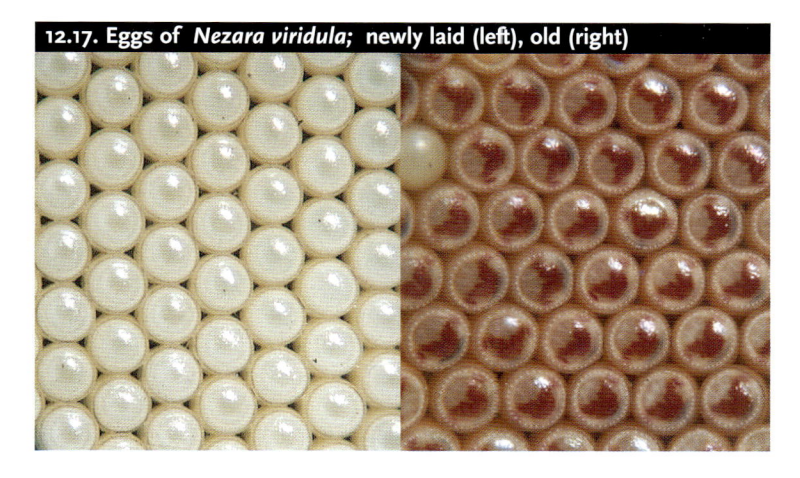

not feed and, unless they are disturbed, aggregate around the empty eggs. The possible advantages of aggregating in this manner are increased protection from drought and high temperatures (to which they are very sensitive at this stage), and increasec protection from natural enemies by pooling their chemical defences.
Second and third instars do feed, but the aggregation behavior continues

through to the fourth nymphal stage. Fourth stage nymphs disperse, and can be found in the top of crops in the morning. The second instar has a black head and legs, a black thorax with yellow spots on each side, and black antennae with red between the segments. The abdomen is red. The third and fourth instars differ from the second in size and colour, which becomes greenish all over. The fifth instar is characterized by wing buds and a yellowish green abdomen with red spots along the median line. Adults of *N. viridula* are shield-shaped bugs with an overall dull green colour (becoming brownish in winter), about 13 mm in length, with 3 - 5 pale dots on the front of the pronotum and dark red or black eyes. Adjacent antennal segments alternate between dark and light, and although the wings completely cover the abdomen, small black spots can be seen on either side of the body.

Population growth
The total development time of *N. viridula* from egg to adult is dependent on climatic conditions and can vary from 3 to 10 weeks. The southern green stink bug is known to have three or four generations per year, with a fifth generation sometimes occurring in tropical areas. The species overwinters as an adult in the bark of trees, in leaf litter, or in other locations offering protection from the weather. During diapause, the adults are brownish in colour. Harsh winters can cause high mortality, although the adults are able to feed during mild spells. As temperatures begin to rise in spring, *N. viridula* leaves its winter shelter and starts searching actively for food. The adults mate almost immediately and usually at night, after which the females start searching for oviposition sites. Females who do not overwinter can lay eggs two to four weeks after adult emergence. Under optimal conditions, the incubation time for the eggs is five days, increasing to two to three weeks at lower temperatures. Females lay hundreds of eggs in batches of between 30 and 130, with the average batch size being 60 eggs. Several days may elapse between the production of batches, and the fecundity of individual females can vary enormously. The optimal temperature for development of *N. viridula* is 30°C.

Damage
N. viridula feeds on all plant parts, preferring growing shoots and developing fruit. Feeding damage to fruit resembles hard brownish or black spots, with attacked shoots usually withering away, or in extreme cases dying off. Young fruit growth is retarded, and the withered fruits may drop from the plant.
Fifth instars and adults cause the most damage, whereas younger instars are less significant.

Biological control
Biological control of the southern green stink bug relies on parasites, such as the tachinid fly *Trichopoda pennipes* which parasitizes adults and nymphs, and the parasitic wasp *Trissolcus basalis*, which parasitizes eggs (see plate 12.20).

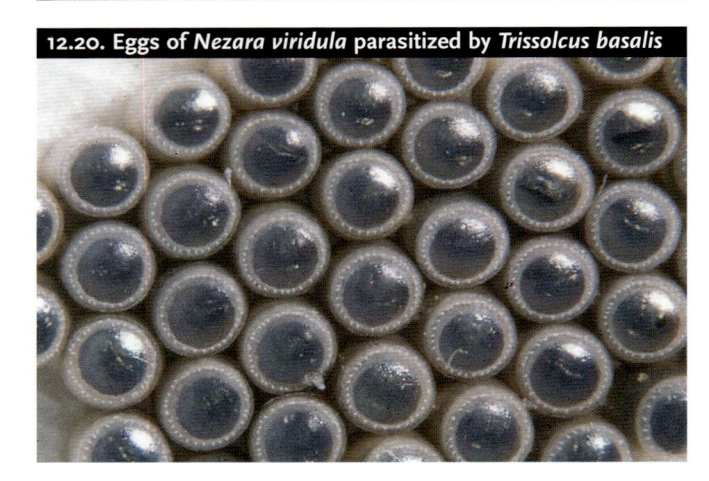

12.18. Emerging eggs

12.19. Adult on tomato

12.20. Eggs of *Nezara viridula* parasitized by *Trissolcus basalis*

Differences between harmful bugs

Lygocoris pabulinus common green capsid	**Liocoris tripustulatis** nettle bug	**Lygus rugulipennis** tarnished plant bug	**Nezara viridula** southern green stink bug
adult appearance			
5 - 7 mm in size, bright green; with relatively long antennae and large, conspicuous mouthparts; glossy	4 - 5 mm in size, at first light green-brown; after over-wintering dark chocolate brown with orange spots; glossy	6 - 9 mm in size, colour varies from light green to dark brown depending on crop; matt	± 13 mm in size dull green colour, becoming brownish in winter, with 3-5 pale dots on the front of the pronotum and with dark red or black eyes
nymph appearance			
pale or bright green with orange-red tips to antennae and red eyes	green with red-brown spots and long striped legs, glossy	green with five clear black dots dorsally in the fifth nymphal instar	nyphs vary in colour, with younger instars being reddish and black with yellow spots and older instars becoming greenish over all
important host plants			
sweet pepper and other fruiting vegetables	cucumber, sweet pepper and aubergine	cucumber, aubergine, sweet pepper, chrysanthemum, gerbera	aubergine and sweet pepper
damage			
leaves with holes, twisted or deformed; deformed fruit with corky growths on skin	crowded and squashed growth in heads with severely deformed fruit; damage to the stem; shrivelled flower buds in ornamental crops	holes in leaves, arrested growth, damage to stems, aborted stem fruit in cucumber and aubergine	dark spots on fruits, young fruit growth is retarded and the withered fruits may drop from the plant, attacked shoots usually wither away

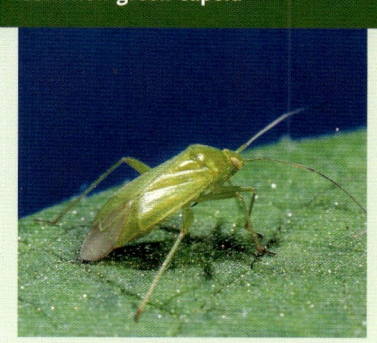

Knowing and recognizing **Other bugs**

Leafhoppers

The most important leafhoppers in glasshouses are:
Empoasca vitis
Philaenus spumarius

12.21. *Empoasca vitis*

12.22. *Philaenus spumarius*

The common name 'leafhoppers' is sometimes used below to refer specifically to the family Cicadellidae, but also more generally to the wider group of small, cicada-related insects comprising leafhoppers, treehoppers and froghoppers (or spittle bugs). Leafhoppers, like aphids, whiteflies and scale insects, belong to the sub-order Homoptera of the order Hemiptera. Within this sub-order, aphids, whiteflies and scale insects constitute the group Sternorrhyncha, while the cicadas and the leafhoppers form the group Auchenorrhyncha. The former group have long, threadlike antennae, whereas Auchenorrhyncha have short antennae, plus relatively large eyes situated at the sides of the head. Among the leafhoppers are several species that cause damage. In glasshouses, however, the only species of importance are *Empoasca vitis*, the grape leafhopper, of the family of Cicadellidae (the leafhoppers), and *Philaenus spumarius*, the spittle bug, and a few other froghopper species (family Cercopidae). In addition *Eupterix* sp. (family Cicadellidae) is very occasionally found in sweet pepper, cucumber and chrysanthemum. The family Cicadellidae is vast, with some 8,500 species worldwide. Some species occur almost everywhere, others in restricted areas or climatic zones. The family of froghoppers (Cercopidae) is best known for the frothy substance in which the nymphs live, the 'spittle' or 'cuckoo spit', for which they get their popular name of spittle bugs. This froth, which protects the nymphs against desiccation and to some extent against predators, is produced by the nymphs themselves by blowing air into a fluid exuded from the anus. Leafhoppers and spittle bugs appear in glasshouses on various crops but seldom constitute a real problem.

Damage
Leafhoppers suck the sap of leaves and sometimes also of fruit. The contents of the cells once sucked out are replaced by air, giving them a greyish white appearance. These greyish white spots, about 1 mm in diameter and resembling the feeding spots of leaf miners, can sometimes be present in huge numbers, sometimes randomly distributed and sometimes in stripes.
Certain species can also cause damage by transmitting viruses, or defects resulting from the action of their toxic salivary juices. In other parts of the world these species cause major problems.
In various crops the spittle bug is responsible for cosmetic damage, the 'spittle' reducing the ornamental value and sometimes causing crooked growth of shoots and lumpy leaves.
It is possible that this kind of damage is often seen without it being realized that leafhoppers are the culprits. This may be because these insects are relatively unfamiliar, and because they are fairly inconspicuous. The adult insect jumps or takes flight immediately if disturbed.

Life-cycle
Leafhoppers

1 egg
2 nymph 5
3 adult

Leafhoppers pass through the following stages: an egg, five nymphal instars and the adult. Adult leafhoppers are usually rather slender, and at most only a few millimetres long. At rest, the wings cover the insects like a roof. Their colour depends on the species but is often light green or white with darker spots. The abdomen tapers to a point. The adults immediately take flight or jump when disturbed, often leaping a metre or more.

The light coloured nymphs live mainly on the undersides of leaves. As they get older they develop wing buds that gradually develop into complete, adult wings.

The eggs are laid in the tissue of the leaf veins or leaf stalks and are invisible to the naked eye.

Depending on species and conditions, leafhoppers can produce one or more generations per year. Although overwintering is possible in any instar, it is usually the eggs or adult individuals that overwinter. Warm, dry weather promotes population growth.

All instars are capable of moving very rapidly over the leaf surface.

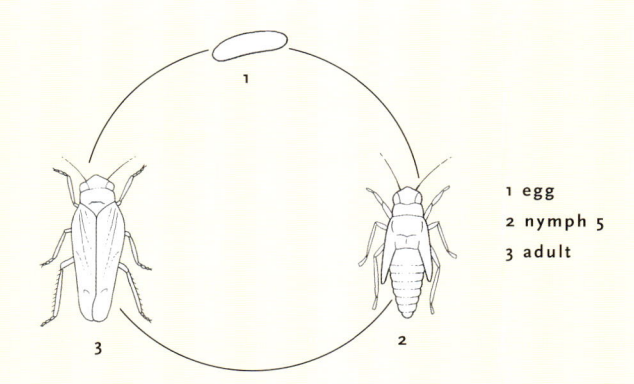

12.23. Young adult of *Philaenus spumarius*

12.24. Old adult of *Philaenus spumarius*

12.25. Nymph of *Empoasca vitis*

12.26. Adult of *Empoasca vitis*

- May appear in the glasshouse from spring onwards
- Can be a problem, particularly in sweet pepper
- Only a real problem when few insecticides are used

Empoasca vitis
Grape leafhopper

Empoasca vitis can be a pest particularly in sweet pepper, but ornamental crops can also be affected. It has become a pest in sweet pepper ever since natural enemies were employed against many other harmful insects. Previously, this insect was controlled incidentally by chemical insecticides used against other pests. Because this species has a wide range of host plants, it is quite likely that it can also cause damage in other crops. In some regions (notably the United States) this leafhopper transmits viruses.

Life-cycle and appearance
Adults of *Empoasca vitis* are light green in colour (plate 12.27).
In temperate climates, the adults overwinter under the protection of evergreen plants, from which they emerge in the spring. Eggs are laid in June on the underside of leaves. They hatch after about 2 weeks to produce nymphs that are highly active, running sideways over the leaf surface when disturbed.
However, nymphs cannot jump as the adults do. After approximately 5 weeks they develop into adults. Eggs of the second generation are then laid in August. In some seasons there may even be a (partial) third generation.
Adults and nymphs sit on the undersides of leaves, particularly of the young growing shoots.

Damage
Both nymphs and adults feed on plant sap. In sweet peppers the fruit is punctured giving rise to stripes ranging from several millimetres to several centimetres in length. Affected leaves are mottled.
It has occasionally been found that fuchsia fails to survive an attack by *Empoasca vitis*. At first the leaf margins "burn", after which the whole leaf turns brown and eventually the plant dies. The cause of death is the toxic salivary fluid introduced into the plant cells.

12.27. *Empoasca vitis*

12.28. Nymph of *Empoasca vitis*

12.29. Sweet pepper fruit damaged by *Empoasca vitis*

- Can cause damage in crops from late spring onwards
- Highly polyphagous, but is seldom a real problem in the glasshouse
- Distinctive 'cuckoo spit' with which the nymph surrounds itself

Philaenus spumarius
Common froghopper, spittle bug

Philaenus spumarius is found on many different species of trees, shrubs and herbaceous plants, particularly on strawberry and other fruit crops, such as apple, cherry and raspberry. In glasshouses, this froghopper is found in small fruit crops, ornamental crops and sometimes in vegetable crops such as cucumber.

Life-cycle and appearance
Philaenus spumarius attracts attention by the frothy 'spittle' with which the nymph surrounds itself for its protection (see plate 12.30).
Adults are 5 - 7 mm long. They are highly variable in colour and may be yellow, greenish or brown, even almost black. The head is blunt and the eyes are conspicuous (see plate 12.31).
In temperate climates, eggs measuring 1 mm in length are deposited on stalks in packets of up to 30. In September the eggs overwinter, with nymphs emerging in the spring. The nymphs sit stationary and surround themselves with 'spittle' which they produce themselves. They feed on the sap of the host plant and pass through 5 instars to reach the adult stage in June. The duration of the nymphal period may be 4 - 13 weeks, depending on the temperature.

Damage
The spittle bug causes small white spots on the leaves of cucumber. In the main, they merely blemish the plants but sometimes the shoots grow crookedly, and bumpy leaves are produced. Feeding on the plant usually results in very little if any harm. The quantity of 'spittle' can be a nuisance, for instance when harvesting strawberries. The 'spittle' also reduces the value of ornamental crops.

12.30. *Philaenus spumarius* 'spittle' in strawberry

12.31. *Philaenus spumarius*

Just emerged adult in 'spittle'   Older adult

Natural enemies of leafhoppers

Since both larval and adult leafhoppers are highly mobile and fast runners they are unlikely to be easily caught by predators. Some parasites, such as the egg parasite *Anagrus atomus*, are known to be capable of parasitizing *Empoasca vitis*. This species can occur naturally. Several more polyphagous predators, such as the predatory bug *Macrolophus caliginosus* (see chapter 4) and the lacewing *Chrysoperla carnea* (see chapter 7), can also contribute to the biological control of leafhoppers.

- Egg parasite
- Can occur naturally
- Adult females parasitize the eggs of Cicadellidae

Anagrus atomus

Anagrus atomus is an egg parasite that belongs to the family Mymaridae of the superfamily Chalcidoidea. This superfamily of the order Hymenoptera also contains many of the parasitic wasps discussed in this book. The Mymaridae are all egg parasites. Some species are smaller than 0.25 mm; in fact the smallest known insects belong to the Mymaridae.
A. atomus occurs naturally and may enter the glasshouse throughout the summer.

Life-cycle and appearance
An adult is brown, roughly 0.6 mm long, with two very slender wings and a similar wingspan of 0.6 mm (see plate 12.32).
The parasitic wasp runs actively over the leaf surface and immediately takes flight if disturbed. Adult females mostly produce female offspring by parthenogenesis, although males may also be found. If conditions are suitable, the wasp can reproduce throughout the whole year.
The eggs are mainly laid separately in the eggs of leafhoppers situated in the leaf veins. Full-grown larvae are about 0.7 mm long. The last larval instar and pupa are conspicuously red, making it easy to identify the parasitized leafhopper eggs in the leaf veins. These parasitic wasps may enter the glasshouse naturally, but never in sufficient numbers for effective control of *Empoasca vitis*.

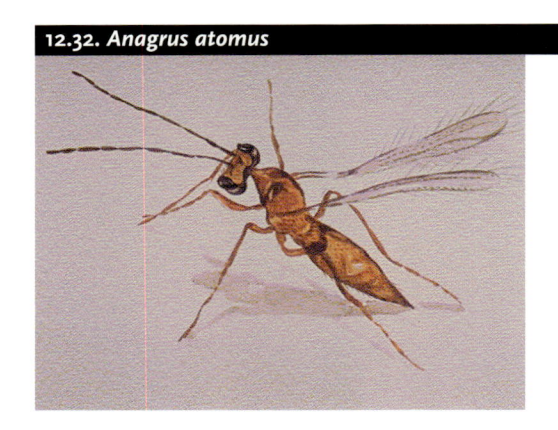

12.32. Anagrus atomus

Population growth
The complete life cycle takes roughly 16 - 21 days at temperatures between 18 and 24°C.
On emerging from the pupa a female lays a maximum of 9 eggs. Adults live for only a few days.

Knowing and recognizing Other harmful organisms

13. Other harmful organisms

This chapter deals with the following:
springtails
ants
earwigs
crickets
dust lice
millipedes (Diplopoda)
symphylans
slugs and snails (Gastropoda)
woodlice (Isopoda)

13.1. Springtail

13.2. Ant

13.3. Cricket

13.4. Garden symphylan

13.5. Snail

13.6. Woodlouse

This chapter deals with several groups of organisms that have not so far been covered, but which can sometimes damage glasshouse crops. Not all these organisms are insects or even arthropods, and they by no means always cause damage on an economic scale. However, when present in very large numbers, or where plants are weak, their activity can lead to reduced production and/or ornamental value. Firstly, various groups of insects will be discussed, namely springtails (order Collembola), ants (order Hymenoptera, family Formicidae), earwigs (order Dermaptera), crickets (order Orthoptera, family Gryllidae) and dust lice (order Psocoptera). Several other organisms will be dealt with that belong to other arthropod classes: the millipedes (class Myriapoda, sub-class Diplopoda), the garden symphylan (*Scutigerella immaculata*, class Myriapoda, sub-class Symphyla, family Scutigerellidae) and woodlice (class Crustacea, order Isopoda). Slugs and snails, which are not arthropods but molluscs (Phylum Mollusca, class Gastropoda), will also be discussed.

- Seldom constitute a serious problem
- Cause damage particularly to seedlings and young or vulnerable plants
- Can sometimes occur in huge numbers

Collembola
Springtails

Springtails comprise the order Collembola, some 6,000 different species of which have been described worldwide. They are small, wingless animals that differ radically from the rest of the Insecta. Some zoologists even doubt whether they should be included among the insects at all. However, it is generally assumed that they are a primitive side-branch in the evolutionary radiation of the insects. They like a damp environment and are often encountered in glasshouses where they live on organic material such as algae, pollen grains, mosses and moulds.

Springtails seldom become a serious problem, although many different species are found in enormous numbers in all glasshouse soils that are not regularly sterilized. Nevertheless, certain species can sometimes cause damage. When these species appear in vast numbers the soft parts of roots can be damaged, with consequent stunting or death of seedlings and young plants. Predatory mites of the genus *Hypoaspis* can contribute to the control of springtails, but are incapable of providing complete control.

Damage

The damage caused by springtails consists of small round holes on developing seedlings or young roots. In the main, this damage only arises with germinating seeds and seedlings, *i.e.* in plants at their most vulnerable stage. Where plants grow slowly, or are grown under cold conditions, or if the level of attack on rootlets is very high, the damage can be fatal. The holes are ports of entry for all kinds of pathogens, and sometimes this is the cause of the plant's death rather than the immediate damage itself. Once they have grown to a certain size, most plant species are no longer vulnerable to damage by springtails.

13.7. Different species of springtail

Life-cycle and appearance
Collembola

Springtails are the most numerous insects in the soil. They are small, highly variable in colour, and have a rounded body. They vary in size from 0.3 mm to more than 10 mm, but most species are 1 - 2 mm long. Some species are shown in plate 13.7.

The posterior of the abdomen bears the so-called "springfork", by means of which the insect is able to make huge leaps when disturbed. Under normal circumstances the animal moves by walking or running in short bursts. The antennae are constantly in motion, feeling the way and searching for food. Springtails are sometimes mistaken for thrips, particularly when under conditions of higher humidity they appear higher up in the plant or in the flowers.

Fertilization is indirect. The male gametes are enclosed in a spermatophore which the male deposits on the ground for the females to take up. Some species reproduce asexually. In adult springtails, shorter phases of feeding and longer phases of reproduction alternate with each other. These phases are separated by a moult, and in some species it is alleged that the two phases also differ in appearance.

The eggs are oval, pale and mostly rather smooth. They are deposited either singly or in groups, either in the ground or in leaf litter.

Springtails are wingless insects and show no marked morphological changes during their development. That is, with the exception of perhaps the first instar, a young nymph does not differ in appearance from the adult insect. After about 5 - 10 moults, the nymphs become adult, and these adults are sexually mature before they are fully grown. The adults thus continue to moult, and do so throughout their life. Some species may moult as many as 50 times, although springtails do not usually grow any more after the 15th moult.

Springtails live in the ground, in leaf litter and vegetation. They cannot tolerate drought. In spring and summer they keep within the top 15 cm of the soil, while in the winter they migrate deeper to avoid frosts. Depending on soil temperature, the development from egg to adult insect takes about six weeks.

All species of springtails have biting mouthparts. They can be divided into different groups based on their feeding behaviour, and fungal spores and hyphae, bacteria, living, dead or rotting plant material, pollen grains, algae and even dead and living soil-dwelling animals (including some harmful organisms which they sometimes prey on) may all provide a source of food. It is only the group that can also feed on living plant material that causes damage.

- Attracted by the honeydew of aphids
- Hardly ever cause direct damage
- May interfere with the biological control of aphids

Formicidae
Ants

Ants belong to the order Hymenoptera (to which sawflies, bees and wasps also belong). They form the family Formicidae, a very large family with some 15,000 species described worldwide. The species of ant most commonly encountered in glasshouses is *Lasius niger* (the black ant or garden ant). *Hypoponera punctatissima* is also sometimes encountered and is considered a nuisance mainly because the queens of this species can sting humans. Several exotic species occur in tropical (ornamental) crops.

Damage
The honeydew secreted by aphids, with its rich sugar content, is an important source of food for ants. To obtain this honeydew, a worker ant uses her antennae to tap on the abdomen of an aphid, stimulating the aphid to exude a drop of honeydew. This behaviour is known as "milking". Because the drop is then removed by the ant, the reproductive capacity of the aphids is enhanced, as the honeydew, when it accumulates, is invaded by moulds and other micro-organisms that contaminate the food and reduce the aphids' reproductive capacity. Ants can protect aphid colonies and withstand their natural enemies. In fact, introduced parasitic wasps and gall midges may be actively repulsed. Furthermore, ants carry objects that they find in their path back to their nest, including the larvae and pupae of the gall midge. This sabotage of biological aphid control is probably the greatest problem caused by a heavy infestation of ants.

In a sweet pepper crop, particularly at the beginning of the season, a light form of direct damage may sometimes be observed. If there is little blossom, the ants find their way to the few flowers that are open in search of nectar, where they may cause damage to the ovaries. This becomes visible later as callusing, after the setting and growth of the fruit. When the fruit swells these callused spots become large enough to constitute damage, although it is of little economic significance. If they are present in large numbers, however, ants can also damage young fruit by gnawing the margins of the crown. Even damage caused in this manner is not serious.

Flying ants are often a nuisance as they can sting humans.

Life-cycle and appearance
Formicidae

Ants live mostly in large colonies and construct their nests in various places, including rock wool mats and under substrate film. They are distinguished from other insects by the peculiar 'petiole' or 'waist' that separates the abdomen and thorax, which is doubly constricted so that they appear to be constructed of more than three parts. The very obvious 'elbow' in the antennae is also distinctive.

A colony consists of a queen and workers. During some periods of the year, males may also appear. Only males and young queens are winged, with the queen loosing her wings after mating. After being fertilized, a queen may enter an existing nest, in which case the nest will carry on with more than one queen, or she may wait for the following spring in order to start a new colony. The first batch of eggs is fed with her own saliva. As soon as the first ants (all workers) become adult, they take over the task of building and maintaining the nest so that the queen can devote herself exclusively to laying eggs. The workers construct a series of chambers and a network of passages around the queen's chamber, collect food, raise the young and keep the nest clean. These workers are wingless females that have not fully developed and are thus usually incapable of laying eggs (see plate 13.8). They are also smaller than the queen. The larvae are raised on honey and the larvae of other insects, and are often moved from chamber to chamber during their growth.

Males are much less common and only appear at certain times. They do not assist with the work of the colony, their only function being to mate with the new queens. They develop from unfertilized eggs.

An ant colony can be maintained for a number of years, with a new queen replacing the old one from time to time.

13.8. Worker ant

- Only come out at night
- May cause light damage or exacerbate existing damage
- Also feed on insect pests such as aphids

Dermaptera
Earwigs

Earwigs form the order Dermaptera, a very small order with about 1,800 known species. In general, earwigs do not withstand cold well, and for this reason only a few species live in temperate climatic zones. The familiar common earwig, *Forficula auricularia*, is very common. Earwigs can sometimes inflict damage to plants or, like ants, exacerbate existing damage. However, earwigs sometimes eat aphids and other harmful insects, and their presence is sometimes something to be encouraged, particularly in apple crops where they combat the woolly apple aphid.

Damage
Earwigs can inflict damage by chewing holes in leaves, or by gnawing into buds or stems where they install themselves. They often enter via existing wounds and in this way merely exacerbate the damage.

13.9. *Forficula auricularia*

Life-cycle and appearance
Dermaptera

Earwigs are elongated, usually brown insects, often (but not invariably) with short fore-wings that touch in the dorsal midline over the thorax and almost reach back to the abdomen (see plate 13.9).

The hind-wings, which are also not always present, are larger and when opened almost form a semi-circle. They are very thin and after each flight they are carefully and elaborately folded, like a fan, beneath the fore-wings. The order derived its name from the delicate nature of the wings; Dermaptera means 'skin-winged' (from the Greek: derma = skin and ptera= wing). At the posterior end of the body earwigs have a characteristic pair of pincers. These are strongly curved in the male, and almost straight in the female. These pincers are used to catch prey, for attack or defence and sometimes to assist in the folding of the hind-wings. The broad head bears long, slender antennae that may consist of up to 50 segments. Earwigs have large eyes, relatively short legs and biting mouthparts. They are mostly omnivorous and are capable of living on either dead or living plant material, carrion and living insects. They avoid the light and are only active in the evenings and at night. During the day they rest in dark and confined crevices, in soil or in plants.

Depending on the species and on conditions, a female may lay 20 - 80 eggs. These are egg-shaped, smooth, translucent white and are deposited in a cavity in the ground. The female keeps the eggs together, sometimes moving them to more suitable places, and cares for them meticulously by regularly licking them to keep them free of fungi and other harmful micro-organisms.

The young nymphs hatch early in the spring but only leave their hatching place after moulting to the second instar. They are often fed and cared for by the mother, even when they come above ground, until they are capable of looking after themselves. Apart from the gradual development of wings, the growing nymphs change little in appearance; they are lighter than the adult and have shorter antennae, and their pincers only gradually assume the adult form. In other respects they closely resemble the adult. There are usually 4 nymphal instars, and outdoors they become adults in late summer. The adults then mate before looking for an overwintering place where the females can lay and care for their eggs.

Earwigs love damp and warm environments.

- Conspicuous by their song, but usually remain well hidden
- Can reproduce very rapidly
- Can inflict damage if present in vast numbers

Gryllidae
Crickets

The true crickets, as distinct from the green bush crickets, constitute the family Gryllidae within the order Orthoptera, which also includes the grasshoppers. Some 3,500 species of cricket have been described, most of them found in tropical and sub-tropical regions. The house cricket, *Acheta domestica*, a species found world-wide, often turns up in glasshouses.

Damage
Many species of cricket, including the house cricket, are omnivorous and can feed on either plant or animal material. They can damage crops when they occur in very large numbers.

13.10. House cricket (*Acheta domestica*)

Life-cycle and appearance
Gryllidae

Crickets are better known for their song than for their appearance. Only the males are able to sing, which they do by rubbing their wings against each other, a behaviour known as "stridulation". The wings are held a little above the body and a toothed ridge on the underside of the right fore-wing, the 'file', is pulled over the posterior edge of the left fore-wing, the 'scraper'. The song of a cricket lasts a considerable time and is rather musical, with a higher tone than that of a grasshopper.

True crickets are somewhat broader and more flattened than the other members of the Orthoptera, with relatively shorter fore-wings (where present) and rolled up hind-wings that project posteriorly from the abdomen, rather like tails. They range in size from small to rather large insects, from 4.5 mm to 5 cm. They possess long, slender antennae and a rounded head. Not all species can fly; in some the hind-wings are lacking. Plate 13.10 shows a cricket.

The eggs are usually laid separately in the ground or in crevices. A single female may lay thousands of eggs, the number depending mainly on the feeding of the female during her earlier development.

There may be 4 - 15 nymphal instars (although between 5 and 10 is more usual), differing from each other only in size.

The manner of feeding varies enormously and includes vegetarian species, carnivores and omnivores. Most crickets are active at dusk, but those species that occur most commonly outdoors in temperate regions are mostly active during the day.

- Commonly occurring, small insects
- Sometimes mistaken for aphids or plant lice
- Do not cause economically significant damage

Psocoptera
Dust lice

Dust lice constitute the order Psocoptera, a small order with approximately 3,000 known species of which about 100 occur naturally in Central Europe.

The species that live indoors are mainly wingless or with short wings and live in dried materials such as books and behind wall paper, hence their common names of dust lice or book lice. They feed on the small spores of fungi on paper.

Most species, however, live outside where they feed on algae, pollen or fungal spores. They live on plants or on dry materials. Most of these species are equipped with fully developed wings. They are quite common in glasshouses but cause no damage.

Life-cycle and appearance
Psocoptera

Dust lice are small, rapidly moving insects, rarely longer than 6 mm, with a soft body and long, whip-like antennae. The head of a dust louse is broad and mobile, often with large compound eyes situated on either side. In winged species, there are two pairs of membranous wings, the fore-wings being larger than the hind-wings. When at rest, the wings are folded roof-like above the body, giving dust lice their characteristic appearance. Most species fly very little and crawl away when disturbed. The legs are well adapted for fast running.

Most dust lice lay small pearl-like eggs that may be protected by webs or by a crust of small food particles or frass. The eggs are usually laid separately in the ground or in crevices. A single female may lay thousands of eggs, the number depending mainly on the feeding of the female during her earlier development.

There may be 4 - 15 nymphal instars (although between 5 and 10 is more usual), differing from each other only in size.

In the successive nymphal instars the size increases, the nymph develops wings (in the case of winged species) and the number of antennal segments increases.

13.11. Dust louse

- Common, but cause no damage
- Live mainly in and on the soil
- A few species can consume living plant material

Diplopoda
Millipedes

Millipedes form the sub-class Diplopoda, one of the sub-classes of the Myriapoda. Most of the 10,000 or so known species live on dead plant or animal remains, but a few species can consume living plant material. Millipedes are generally considered to be useful because they digest dead organic material. They are very common, and can readily be found in dark, damp places.

Millipedes are sometimes confused with centipedes. The latter, however, are predatory animals that are incapable of living on plant material. They form the sub-class Chilopoda of the class Myriapoda. Millipedes and centipedes can be distinguished from each other by the fact that centipedes have one pair of legs per segment whereas millipedes have two pairs. Two millipede species that appear regularly from time to time in the glasshouse are *Ommatoiulus sabulosis* and *Proteroiulus fuscus*, but many other species may appear sporadically.

Damage

Millipedes feed mainly on rotting plant material and under some circumstances also on rotting carrion. As far as is known, only a few species can also consume living plant material. Harmful millipedes mainly affect the parts of the plant that are below ground, but sometimes they may damage the fruit. In either case they rarely give rise to any economic damage.

Life-cycle and appearance
Diplopoda

Millipedes have a cylindrical body that comprises many segments. Many millipedes have a conspicuous colour. They are distinguished from other myriapods by their club-shaped antennae, which usually consist of eight segments. In addition, they have a short thorax of four simple segments with a single pair of legs on the second to the fourth segment. The long abdomen consists of between 9 to over 100 double segments, with two pairs of legs on each. The head attaches to the thorax with its axis at ninety degrees to the longitudinal axis of the body, so that the mouth opening is situated ventrally rather than anteriorly. Plate 13.12 shows a millipede.

Mating occurs mostly in the spring and autumn. The female makes a nest in the ground in which to lay her eggs, depositing them either singly or in small groups. When she has finished, there may be from several dozen to several hundred eggs. This nest is often well guarded. Egg development lasts approximately 2 to 4 weeks before the young millipedes hatch. These usually have only 7 segments and 3 pairs of legs, but otherwise resemble the adult. More segments and legs are formed during the 10 – 15 subsequent moults. Development to the adult stage lasts about 18 months, while the total lifespan is often 3 - 5 years.

Millipedes live in damp, dark places under stones, and under or in rotting plant material. They move slowly, feeling their way with their antennae. Their many legs appear to move in a series of waves passing from the back forward. Some species roll up into a ball when disturbed, while species with a long body form a spiral. Members of some orders have stink glands, which can secrete a fluid capable of repelling enemies.

13.12. Millipede

- **Can be a serious pest in chrysanthemum**
- **Particularly damaging to young and vulnerable plants**
- **Very common**

Scutigerella immaculata
Garden symphylan

The garden symphylan (*Scutigerella immaculata*) belongs to the class Myriapoda, a class which, like the Insecta belongs to the phylum Arthropoda. This species belongs to the sub-class Symphyla (the symphylans) and to the family Scutigerellidae. Despite first appearances, *Scutigerella immaculata* is not one of the centipedes, another myriapod group constituting the predatory sub-class Chilopoda. In older literature the symphylans have sometimes been considered as a class separate to the Insecta and Myriapoda.

The garden symphylan is a polyphagous organism that inhabits passages and cavities in the soil. The species is capable of feeding on dead plant remains, yeasts, fungal spores and fungal hyphae, compost and even dead or wounded individuals of its own species. To reproduce, however, fresh plant material is necessary.

Garden symphylans are considered harmful creatures because they can eat the rootlets of young plants, particularly chrysanthemums. They can also be a problem in sweet pepper, tomato, cucumber, rose, carnation and occasionally in pot plants. Damage is more often evident in soil-based media than in artificial substrates.

Geographically, the garden symphylan is probably distributed throughout the world. It is found on a large scale in Europe and in North America and it is certainly known in other parts of the world. It also occurs in very different climatic conditions, which is undoubtedly due to the fact that it lives in the soil and can migrate vertically according to changes in humidity and temperature.

Population growth

The development time of the garden symphylan is closely related to temperature. The time taken for eggs to hatch can vary from 1 to 4 weeks. At 21°C, the hatched symphylan moults within another 2 days. The period between moults, particularly after the fifth moult, also depends very strongly on temperature. At 28°C an individual moults on average every 22 days, while at 10°C moults occur on average every 100 days.

Under optimal conditions a garden symphylan can live 4 - 7 years, but this too is highly dependent on temperature. Development stops at just above 2°C, although they can survive temperatures lower than this as long as they are able to acclimatize gradually. A slow transition to a high or low temperature can be tolerated much better than a rapid transition. A temperature of 37°C, however, is lethal even if the animal has adapted for some time to high temperatures.

Garden symphylans are almost immobile at 5°C, only becoming active above 12°C. A population grows fastest at a temperature between 21°C and 27°C.

The sex ratio of *S. immaculata* is 1:1. Between two successive moults, adult males deposit approximately 150 spermatophores complete with small stalks. The females consume many of these spermatophores, some of them being digested, while others are preserved in a kind of pouch in the mouth. Fertilization takes place during the deposition of eggs. With her mouthparts, a female takes an egg that has been expelled from the sex organ and sets it down on the substrate, following which she transmits sperm from the oral pouch by means of a kind of chewing action over the egg surface.

When there is no available living plant material there is little if any reproductive activity, although the garden symphylan can survive for months in such situations.

Damage

In general, an attack by garden symphylans means a reduction in yield because the young rootlets are eaten. Indirect damage can also occur through fungi and bacteria entering the damaged root system. Garden symphylans can live on a wide range of different plant species, but these are not all equally susceptible to damage. Many species can resist attack, or only sustain damage if attacked when the plants are very young or if the attack is exceptionally heavy. Once full grown, however, plants are much more capable of resisting an attack.

The extent of an attack depends not only on the plant species, but also on the cultivar.

Distribution and dispersal

Symphylans are photophobic, and move rapidly through the larger pores and channels in the soil, but do not make their own tunnels. They like to follow the passages of roots and can be found up to a depth of ten centimetres. The soil must be fairly water retentive, have a reasonably high average temperature, and a suitable texture with cracks and fissures that can be used for moving to lower ground layers. The garden symphylan is therefore mostly found in clay, peaty or composted soils and rarely in light, sandy soils. The greatest population densities are reached in a warm, humus-rich substrate with adequate moisture and food. They cannot survive in rock wool.

The differences in humidity and more particularly in the temperature of the soil through the course of the year are the factors that determine the symphylans' vertical migrations within the soil. In warm, humid conditions in the spring and autumn they are found in the top layers of the soil, moving to the lower levels in hot dry summer conditions and throughout the winter.

Life-cycle and appearance
Scutigerella immaculata

Following the egg stage, symphylans pass through many mobile stages. A symphylan continues to moult throughout its life, even after becoming sexually mature. The most significant developmental changes occur in the first 6 moults, with less obvious changes continuing thereafter. It is not known exactly how many moults an individual can undergo, although it can exceed 50.

When just laid, the eggs are pearly white, spherical and covered with tiny ridges. They darken with age until they are a light yellow-brown colour by the time they hatch. Eggs are usually deposited in groups of 9 to 12 in cavities in the ground.

Newly hatched garden symphylans differ from the adults in having only 6 pairs of legs and 6 distinct antennal segments. They are also relatively inactive. At this stage they can sometimes be confused with white springtails, although examination under a magnifying glass will always reveal a sprintail's fork under the abdomen and its three pairs of legs. Within 48 hours of hatching (at room temperature) the garden symphylan undergoes a first moult to an immature instar whose general appearance and behaviour now closely resembles that of the adult, although its activity is still relatively low. The number of antennae, pairs of legs and body segments increases with each moult.

After the sixth moult (seventh instar) morphogenesis is complete with 15 segments, 12 pairs of legs and usually 23 - 26 antennal segments, although the total number of antennal segments can vary from 18 to 60. This variance is the result of each successive moult being accompanied by the addition of new segments, and also of broken segments being gradually replaced.

Adults are a light cream colour and so transparent that the food in their gut is visible. Symphyhlans have no eyes and find their way by using their antennae. They have chewing mouthparts.

The garden symphylan increases in size with each moult up to a certain point, after which moults continue without further growth. This may be the reason why different authorities specify different lengths. Alternatively, given the enormously wide distribution of the species, different races may differ in size. Some authors have reported a length of 2 - 3 mm, while others cite 5 - 8 mm. Males are slightly smaller than females.

13.13. Garden symphylan

- Polyphagous
- Cause direct damage and contamination, and also spread fungal pathogens
- Mainly nocturnal

Gastropoda
Slugs and snails

Slugs and snails can cause enormous damage in both agricultural and horticultural crops. Although they are not generally selective in the plants they attack, some are particularly attractive, and in dense humid crops they can cause considerable damage. Slugs and snails are invertebrates belonging to the phylum Mollusca, (as distinct from the insects, which belong to the invertebrate phylum Arthropoda), and to the class Gastropoda. They have a flaccid, pulpy un-segmented body with neither internal nor external skeleton. Slugs have no shell, whereas snails possess a shell into which they can withdraw when threatened by enemies or desiccation. The brown slug, *Deroceras laeve*, is the most common species observed in glasshouses, while *Deroceras reticulatum* (the grey field slug) and *Arion hortensis* (the garden slug) and the snails *Helix aspersa* (the garden snail) and *Trichia striolata* (the strawberry snail) also occur quite commonly. (Sub)tropical species are also sometimes brought in with imported plant material, and can subsequently cause damage in the crop concerned.

Population growth
Damp conditions promote the activity of slugs and snails. They are most at home in a humid atmosphere among dense crops rich in foliage at a temperature of 18 - 30°C, although several species can manage perfectly well at lower temperatures. Temperature affects the rate of development as well as activity: at higher temperatures the growth of the population clearly accelerates.

With crops grown under glass where there is a continuously damp and humid atmosphere throughout the year and temperatures are usually quite high, there is a continual threat of damage. Outdoors, snails usually overwinter within their shell in groups. Adult slugs and snails can live from several months to several years.

Damage
Snails rarely constitute a major problem in glasshouses, but slugs can sometimes cause damage. Both groups have a tongue-like radula, a file-like organ with rows of hard ridges with which they rasp their food when feeding. Damage to plants caused by slugs and snails can always be identified by the trail of slime they leave behind. They begin feeding from the moment they hatch from the egg, on organic material and humus in their immediate environment. They then go in search of organic material on the ground or in the crop, where they are capable of attacking both the sub-surface and aerial parts of the plant. However, they have a preference for young shoots and leaves.

Many species are particularly active at night, although under favourable conditions they can also inflict damage during daylight hours.

In glasshouse horticulture, and most especially in the cultivation of ornamental flowers (*e.g.* gerbera or orchids) even slight damage has serious economic consequences. When damage is inflicted soon after sowing or planting this can lead to a total loss of the crop. Slug and snail damage can also promote infection by fungal pathogens, while contamination by faeces and slime is a nuisance that can reduce the ornamental value of plants. Efficient hygiene measures to make the environment immediately surrounding the glasshouse more hostile to slugs and snails can prevent a great deal of damage.

Distribution and dispersal
Depending on the species, slugs and snails are distributed over the ground surface (*e.g.* the grey field slug) or within the soil (*e.g.* the garden slug). In any particular crop they may be distributed in groups, or evenly distributed over the whole crop.

They are most active at night, although under cloudy conditions they will also feed for part of the day. Mostly, though, they spend daylight hours in damp, sheltered places, only emerging in the evening when the temperature falls and humidity rises.

Because the eggs are deformable, they can withstand being washed around in rain pipes and in this way can be spread through the glasshouse.

Life-cycle and appearance
Gastropoda

Slugs and snails form the largest group among the molluscs. They move by means of a muscular foot and have conspicuous eyes at the tips of long retractable tentacles, which function like periscopes.

Slugs and snails are hermaphroditic: each adult possesses both male and female sex organs and produces both eggs and sperm cells. Even though they are not dependent on mating with a partner in order to reproduce, mating does usually take place. After mating, both animals are ready and able to produce fertilized eggs that are laid in piles of 5 to 50 in a damp, protected place. Slugs and snails may lay eggs up to three times a year, laying each time a maximum of 500 eggs depending on the species. The time taken for eggs to hatch depends to a large extent on climatic conditions and can vary from less than 3 weeks in a warm spring to many months in the case of eggs laid outside in the autumn. The eggs have a tough shell, but are jelly-like and can be deformed (see plate 13.14). They are highly susceptible to desiccation and frost.

Young slugs and snails are like smaller versions of the adults, only usually lighter in colour. Plate 13.15 shows a brown slug. Snails are identified by the colour and pattern of their shells, whereas slugs are distinguished by the size, colour and form of their body.

13.14. Slug eggs

13.15. Brown slug

- Very common, but rarely cause damage
- Live mainly in rotting plant material
- Can be a nuisance when they multiply rapidly

Isopoda
Woodlice

Woodlice are not insects, and form the order Isopoda of the class Malacostraca (along with crabs and shrimps etc.), sub-phylum Crustacea, phylum Arthropoda. Those woodlice that appear in glasshouses belong to the sub-order Oniscoidea, the terrestrial woodlice, which is virtually the only group of crustaceans that have adapted themselves to life on land. The rest of the Crustacea are aquatic animals. Nevertheless, woodlice are highly sensitive to desiccation and can only survive in a humid environment. Although their presence can sometimes be a nuisance, with the exception of some soil-grown crops they cause hardly any damage. The commonest species are *Porcellio scaber*, *Trachelipus rathkii* and *Oniscus asellus*. There are also species that come from warmer regions and in temperate climates are only found in glasshouses, such as *Armadillo officinalis*.

Damage
Woodlice are omnivorous but mainly feed on rotting plant material. Very occasionally they may feed on young plants but they cause almost no economic damage. They sometimes multiply so rapidly that they become a nuisance.

Only in soil-grown sweet pepper, tomato and cucumber do they on occasion multiply to such huge numbers that young plants can be totally destroyed. In such situations, older plants can also suffer damage.

Life-cycle and appearance
Isopoda

Woodlice have a maximum length of 1.3 cm, and are grey to blue-ish or even purple in colour. They have a head with 2 pairs of antennae, only one of which is well developed and clearly visible. Some species lack eyes, and some species possess eyes consisting of a number of ocelli. Dorsally, the body is rounded, while ventrally it is flat. The thorax consists of seven distinct segments and an eighth (the first anterior segment) that is fused with the head and therefore not visible. Woodlice have seven pairs of legs which, in females, have plate-like appendages that overlap each other rather like roof tiles beneath the body. The space thus created between the ventral body surface and these appendages serves as a brood chamber in which the eggs and subsequently the young can be carried. The abdomen consists of six segments. Plate 13.16 shows a woodlouse.

Woodlice moult a number of times during their growth.

Some species reproduce parthenogenetically, with unmated females producing only female offspring, but in most species mating is necessary. Between 5 and 200 eggs are laid in the brood chamber, hatching into small, 1 mm long white coloured young woodlice which will remain in this chamber for some time. These subsequently grow larger and become progressively grey in colour. A woodlouse lives for about 3 years.

Woodlice are very common wherever there is sufficient moisture, and are characteristically found under pots, in moss and between the leaves of plants.

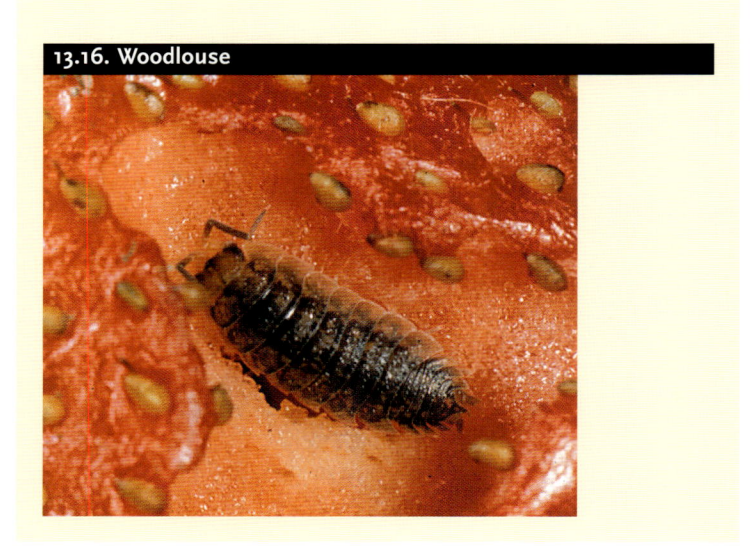

13.16. Woodlouse

14. Observing and monitoring

14.1

Efficient monitoring is of paramount importance in the management of a successful integrated pest control system. Only when the pest situation is known appropriate decisions can be made. This requires information on the location, number and type of pests present in the crop, and which natural enemies are being used. In addition it is often important to know which stage of the pests life-cycle is prevalent in the crop, and which stage is most effectively tackled by the predators and parasites available.

The ability to identify pests, predators and parasites within the crop accurately, as well as an understanding of the damage each pest causes, is highly advantageous.

Although it is important to look for plant damage as well as pests and natural enemies whilst working in the crop, it is equally important to record these observations accurately and to incorporate them in crop management planning. Monitoring of pests and their enemies can be carried out using sticky traps, pheromone traps and light traps. This can be undertaken by dedicated personnel within the company, or by an external 'scout' employed specifically for this purpose.

Sticky traps

Flying insects are often attracted to specific colours. Yellow and blue are particularly attractive, and these are the most common colours for sticky traps. The traps have an adhesive layer, like flypaper, to which the insect sticks. These traps can be used either to monitor and identify pest populations, to give early warning of pests migrating into the crop from outside, or to contribute to control by mass trapping.

Pheromones

Pheromone traps use the species-specific sex pheromone produced by female moths to attract male moths to a sticky trap. A small capsule of artificially synthesised pheromone is placed in a triangular delta trap (plate 14.1, page 275). In addition to identifying the species of moth present, such traps allow the size of the population to be estimated.

Light traps

Light traps use ultraviolet light to attract male and female moths to an electrified grid. The moths are killed as soon as they touch it. Although this method gives a good indication of the numbers of moths present in the crop, it can make accurate identification difficult.

Laboratory observations

It is sometimes impossible to get an accurate picture of how well biological control is progressing simply by observing the crop or by using the monitoring methods described above. For example, whenever an organism is too small to be seen with a hand lens, or where an organism must be dissected in order to identify its parasite (as with leaf miner larvae), samples can then be sent away for laboratory examination. In such cases, it is very important to follow the correct procedure for taking crop samples.

Pests that can be monitored using sticky traps		
pest	appearance on sticky trap	
whitefly	**yellow sticky trap** adult is white-yellow, discolouring to yellow-brown after several days on the sticky trap	
thrips	**yellow sticky trap** **blue sticky trap** adults are elongated, brown or brown-black	
leaf-miner	**yellow sticky trap** adults with yellow dot dorsally, or completely black	
aphid	**yellow sticky trap** the wings of adults are 2x the length of the insect itself; variable colour	
fungus gnat	**yellow sticky trap** adult conspicuous by long legs, abdomen tapers to a point	
leaf-hoppers	**yellow sticky trap** adults are green or brown, wings often extended, no antennae visible	

References

Abou-Awad, B.A., 1979. The tomato resset mite, *Aculops lycopersici* (Masee) (Acari: Eriophyidae) in Egypt. Anz. Schadlingskde. Pflanzenschutz Umweltschutz 52 : 153-156.

Adrichem, P. van, 1997. Behaarde wants is lastpost. Groenten + Fruit / Glasgroenten 33: 4-5.

Alauzet, C., D. Dargagnon & J.C. Malausa, 1994. Bionomics of a polyphagous predator: *Orius laevigatus* (Het.: Anthocoridae). Entomophaga 39 (1): 33-40.

Albert, R., H. Schneller & E. Renner, 1995. Biologische Schädlingsbekämpfung: Arbeitshilfen für Beratung und Betriebsführung. Ministerium Landlichen Raum Landwirtschaft Forsten - Stuttgart: 277 pp.

Albert, R., H. Schneller, P. Detzel & K. Schrameyer, 1993. Schädlinge und Nützlinge erkennen. Nicht alle sind Trauermücken.
Deutscher Gartenbau 47 (10): 629-633.

Albert, R., W. Paul & H. Schneller, 1993. Biologische Bekämpfung des Bananentriebbohrers. Gartnerbörse Gartenwelt 30: 1372-1374.

Alford, D.V., 1984. A color atlas of fruit pests; their recognition, biology and control. Wolfe Publishing - London: 320 pp.

Alofs, W., J. Derckx, J. Hoogstrate, C. Peters, M. Simonse, E. van der Ven, J. van Rijn, H. Schepers, 1989. Ziekten en plagen glasgroenten in beeld. Ministerie LNV / CAD Gewasbescherming - Wageningen: 59 pp.

Ankersmit, G.W., H. Dijkman, N.J. Keuning, H. Mertens, A. Sins & H.M. Tacoma, 1986. *Episyrphus balteatus* as a predator of the aphid *Sitobion avenae* on winter wheat. Entomologia Experimentalis et Applicata 42 (6): 271-277.

Anonymous, 1973. Glasshouse symphylid. MAFF Advisory leaflet 484. HMSO Press: 6 pp.

Anonymous, 1992. De vuilboomluis: een nieuwe plaag in de aardappelteelt. Agro Chemie Koerier 1: 2pp.

Anonymous, 1994. Bladluizen en hun natuurlijke vijanden. ZENECA Agro - Ridderkerk: 48 pp.

Anonymous, 1996, 1997, 1998. Annual Report Diagnostic Centre 1996, 1997, 1998. Plant Protection Service, Wageningen, the Netherlands.

Anonymous, 1998. Komkommer: behaarde wants (*Lygus rugulipennis*). Groenten + Fruit / Glasgroenten 7: 2.

Anonymous, 1998. Tomaat: tomatenwolluis (*Pseudococcus affinis*). Groenten + Fruit / Vollegrondsgroenten 39: 5.

Arakawa, R., 1982. Reproductive capacity and amount of host-feeding of *Encarsia formosa* Gahan (Hym., Aphelinidae). Zeitschrift für angewandte Entomologie 93: 175-182.

Arnett, R.H., 2000. American insects; a handbook of the insects of America North of Mexico. CRC Press - Boca Raton: 1003 pp.

Badii, M.H. & J.A. McMurtry, 1984. Life history of and life table parameters for *Phytoseiulus longipes* with comparative studies on *P. persimilis* and *Typhlodromus occidentalis* (Acari: Phytoseiide). Acarologia 25: 111-123.

Bakker, F.M. & M.W. Sabelis, 1989. How larvae of *Thrips tabaci* reduce the attack success of phytoseiid predators. Entomologia Experimentalis et Applicata 50: 47-51.

Balachowsky, A.S., 1972. *Helicoverpa armigera* Hb.: 1431-1445. In: Entomologie: Appliquée a l'agriculture. Part II Lépidopteres vol. 2. Zygaenoidea – Pyraloidea – Noctuoidea. France: Masson et Cie

Barrancos, G., 1998. Schade door rouwmuggen (Sciaridae). Stageverslag Rijn Ijssel College, Plantenziektenkundige Dienst, sectie Entomologie - Wageningen: 25 pp.

Bay, T., M. Hommes & H.P. Plate, 1993. Die Florfliege *Chrysoperla carnea* (Stephens): Uberblick über Systematik, Verbreitung, Biologie, Zucht und Anwendung. Mitteilungen aus der Biologischen Bundesanstalt fur Land und Forstwirtschaft Berlin Dahlem, heft 288 - Berlin: 175 pp.

Ben-Dov, Y. & C.J. Hodgson (eds.), 1997. World Crop Pests vol. 7 (A and B); Soft Scale Insects - their biology, natural enemies and control. Elsevier - Amsterdam.

Benuzzi, N. & G. Nicoli, 1988. Lotta Biologica e integrata nelle colture protette (strategie e tecniche disponibili). Centrale Ortofrutticola alla Produzione - Cesena: 166 pp.

Bertaux, F., 1994. Les thrips sur cultures ornementales. Association nationale pour la protection des plantes. Journée Information sur les thrips: 84-89.

Billen, W., 1987. Informationen zum Bananentriebbohrer (*Opogona sacchari* Bojer, 1856) (Lepidoptera: Tineidae). Gesunde Pflanzen, 39/11: 458-465.

Blackman, R.L., & V.F. Eastop, 2000. Aphids on the world's crops. An identification and information guide. John Wiley Ltd - Chichester: 466 pp.

Blind, M., 1998. Cicaden: (G) een probleem. Groenten + Fruit / Glasgroenten 23: 22-23.

Blind, M., 1998. Cicaden: (g) een probleem? Vakblad voor de Bloemisterij 23: 46-47.

Blind, M., H. Cevat & J. Derckx, (red), 1995. Ziektebeelden in de bloementeelt. Misset; DLV – Doetinchem: 182 pp.

Blumberg, B. & M. Kehat, 1982. Biological studies of the date stone beetle *Coccotrypes dactyliperda*. Phytoparasitica 10(2): 73-78.

Boers, J., 1997. *Drosophila* vliegt natrot van krop naar krop. Groenten + Fruit / Vollegrondsgroenten 8: 18-19.

Bonde, J., 1989. Biological studies including population growth parameters of the predatory mite *Amblyseius barkeri* (Hughes) (Acarina: Phytoseiidae) at 25°C in the laboratory. Entomophaga 34 (2): 275-287.

Boogaard, M., 1991. Op zoek naar nieuwe wapens tegen cyclamenmijt. Vakblad voor de Bloemisterij 48: 52, 53, 55.

Bouwman-van Velden, P., 1995. Mieren koesteren luizenkolonie. Groenten + Fruit / Glasgroenten 13: 8-9.

Bouwman-van Velden, P., 1998. Verborgen leven koolbladroler maakt bestrijding ingewikkeld. Groenten + Fruit / Glasgroenten 8: 20-22.

Boxtel, W. van, J. Woets & J.C. van Lenteren, 1978. Determination of host-plant quality of eggplant (*Solanum melongena* L.), cucumber (*Cucumis sativus* L.) tomato (*Lycopersicum esculentum* L.) and paprika (*Capsicum annuum* L.) for the greenhouse whitefly (*Trialeurodes vaporariorum* (Westwood) (Homoptera: Aleyrodidae). Mededelingen Faculteit Landbouwwetenschappen Rijksuniversiteit Gent 43/2: 397-408.

Brasch, K., J.C. van Lenteren, J. van Boisclair & H. Henter, 1994. Biological control of *Bemisia tabaci* with *Encarsia formosa*: a realistic option? Mededelingen Faculteit Landbouwwetenschappen Rijksuniversiteit Gent 59/2a: 325-332.

Bravenboer, L., 1959. De chemische en biologische bestrijding van de spintmijt *Tetranychus urticae* Koch. Proefschrift. PUDOC, Landbouwdocumentatie - Wageningen: 85 pp.

Brödsgaard, H.F., M.A. Sardar & .A. Enkegaard, 1996. Prey preference of *Hypoaspis miles* (Berlese) (Acarina: Hypoaspididae): non-interference with other beneficials in glasshouse crops. Bulletin IOBC/WPRS 19 (1): 23-26.

Brown, J.K., D.R. Frohlich & R.C. Rosell, 1995. The sweetpotato or silverleaf whiteflies, biotypes of *Bemisia tabaci* or a species complex. Annual Review of Entomology 40: 511-534.

Buczacki, S & K. Harris, 1998. Pests, diseases & disorders of garden plants. HarperCollins Publishers - Londen: 640 pp.

Butcher, J.W., R. Snider & R.J. Snider, 1971. Bioecology of edaphic Collembola and Acarina. Annual Review of Entomology 16: 249-288.

Calabro, M. I miridi predatori: loro biologica e possibili applicazioni in lotta biologica-integrata esperienze maturate in Sicilia. (Manuscript)

Carter, D., 1984. Pest Lepidoptera of Europe; with special reference to the British Isles. Series Entomologica vol. 31. Dr. W. Junk Publishers - Dordrecht: 431 pp.

Castagnoli, M. & S. Simoni, 1991. Influenza della temperatura sull'incremento delle popolazioni di *Amblyseius californicus* (McGregor) (Acari: Phytoseiidae). Redia 74 (2): 621-640.

Casteels, H., J.S. Miduturi, R. Moermans & R. de Clercq, 1994. Laboratory studies on the oviposition and adult-longevity of the black vine weevil *Otiorhynchus sulcatus* F. Mededelingen Faculteit Landbouwwetenschappen Universiteit Gent 59(2a: 189-195.

Cevat, H., 1998. Nieuwe trips bedreigt kasteelten. Vakblad voor de Bloemisterij 16: 49.

Cevat, H. & M. Roosjen, 1994. De belangrijkste plaaggeesten van 1993. Vakblad voor de Bloemisterij 14: 30-35.

Chambers, R. & E. Wright, 1993. The birth of a new mighty mite. Grower 9: 16,18.

Chi, H., 1980. Feind-Beute-Beziehungen zwischen *Onychiurus fimatus* Gisin (Collembola, Onychiuridae) und *Hypoaspis aculeifer* Can. (Acarina, Laelapidae) unter Einfluss von Temperatur und Insektiziden. Dissertation, Georg August Universität, Göttingen: 77pp.

Chinery, M. (Ned. Bew. R. de Jong), 1983. Elseviers insektengids voor West-Europa. Elsevier - Amsterdam: 411 pp.

Chinery, M., 1986. Collins guide to the insects of Britain and Western Europe. W. Collins Sons & Co. - Londen: 320 pp.

Christiansen, K., 1964. Bionomics of Collembola. Annual Review of Entomology 9: 147-178.

Clemens, S. & R. Keijzer, 1994. Plantparasitaire mijten: onderzoek naar begonia-, cyclamen- en stromijt. Stageverslag., Agrarische Hogeschool Delft: 45 pp.

Clercq, P. de, 1993. Biology, ecology, rearing and predation potential of the predatory bugs *Podisus maculiventris* and *Podisus sagitta* in the laboratory; Proefschrift Universiteit Gent: Faculteit landbouwkundige en toegepaste biologische wetenschappen: 144 pp.

Clercq, P.M.H. de, 1993. Biologie en kweek in kaart gebracht: Roofwants doodt rups Floridamot. Gewas 3: 16-19.

Coaker, T.H. & C.A. Cheah, 1993. Conditioning as a factor in parasitoid host plant preference. Biocontrol Science and Technology 3 (3): 277-283.

Cock, M.J.W., 1986. *Bemisia tabaci*: a literature survey on the cotton whitefly with an annotated bibliography. Ascot: CAB International Institute of Biological Control & Food and Agricultural Organisation of the United Nations FAO - Berks: 121 pp.

Conijn, C. & I. Lesna, 1997. Roofmijt houdt kasgrond vrij van bollenmijt. Vakblad voor de Bloemisterij 24: 62-63.

Cooper, S., 1993. The biology and application of *Anagrus atomus* (L.) Haliday. Bulletin IOBC/WPRS 16(8): 42-43.

Cornelius, S.J. & H.C.J. Godfray, 1984. Natural parasitism of the Chrysanthemum leaf-miner *Chromatomyia syngenesiae* H. (Dipt.: Agromyzidae). Entomophaga 29 (3): 341-345.

Corsten, R., 1991. Floridamot jaarlijks terugkerend in chrysant; Problemen verschillen sterk per bedrijf. Vakblad voor de Bloemisterij 33: 34-35.

Crépin, O., 1997. *Thrips tabaci*, le plus redoutable ravageur du poireau. PHM Revue Horticole: 380: 38-43.

Croft, P. & J.W. Copland, 1994. Larval morphology and development of the parasitoid *Dacnusa sibirica* (Hym.: Braconidae) in the leafminer host *Chromatomyia syngenesiae*. Entomophaga 39 (1): 85-94.

Croft, P. & M. Copland, 1993. Size and fecundity in *Dacnusa sibirica* Telenga. Bulletin IOBC/WPRS 16(8): 53-56.

Cross, J.V. & P. Bassett, 1982. Damage to tomato and aubergine by broad mite, *Polyphagotarsonemus latus* (Banks). Plant Pathology 31: 391-393.

Cuijpers, M., 1994. De juiste aanpak van wantsen. Groenten + Fruit / Glasgroenten 28: 12-13.

Czajkowska, B & D. Kropczynska, 1983. The influence of different host plants on the reproductive potential of *Tyrophagus putrescentiae* (Schrank) and *Tyrophagus neiswanderi* (Johnston and Bruce) (Acaridae). The Acari; Hst. 23. Eds. R. Schuster & P.W. Murphy. Chapman Hall - London: 313-317.

Danilov, L.G., 1990. Infestation and subsequent development of the nematode *Neoaplectana carpocapsae* strain "agriotos" in insects under free contact between host and parasite: 59-65. In: Helminths of insects. M.D. Sonin (ed.). E.J. Brill - Leiden.

Davidson, R.H. & W.F. Lyon, 1987. Insects pests of farm, garden and orchard, 8th edition. Wiley - New York: 640 pp.

Derckx, J., 1993. Schone paprika's in het MBT-blok. Groenten + Fruit / Glasgroenten 30: 20-21.

Diethelm, V., 1990. Nützlingseinsatz bei Gerbera: Bekämpfung von weissen fliegen mittels Schlupfwespen. Gartenbau Magazin 2: 52-57.

Dixon, A.F.G., 1985. Aphid ecology. Blackie – Glasgow: 157 pp.

Dorsman, R. & M. van de Vrie, 1987. Population dynamics of the greenhouse whitefly *Trialeurodes vaporariorum* on gerbera. Bulletin IOBC/WPRS 10 (2): 46-51.

Edelson, J.V & J.J. Magaro, 1988. Development of onion thrips, *Thrips tabaci* Lindeman, as a function of temperature. The Southwestern Entomologist, 13/3: 171-176.

Eggenkamp-Rotteveel Mansveld, M.H., J.C. v. Lenteren, J.M. Ellenbroek & J. Woets, 1982. The parasite-host relationship between *Encarsia formosa* (Hymenoptra: Aphelinidae) and *Trialeurodes vaporariorum* (Homoptera: Aleyrodidae). XII: Population dynamics of parasite and host in large commercial glasshouse and test of the parasite-introduction method used in the Netherlands. Zeitschrift für angewandte Entomologie 93: 113-130 (part1) & 258-279 (part 2).

Enkegaard, A., 1993. The bionomics of the cotton whitefly, *Bemisia tabaci* and its parasitoid, *Encarsia formosa* on poinsettia. Bulletin IOBC/WPRS 16(8): 66-72.

Enkegaard, A., M.A. Sardar & H.F. Brödsgaard, 1996. Life tables of *Hypoaspis miles* preying on mushroom sciarid larvae, *Lycoriella solani*, and mould mites, *Tyrophagus putrescentiae*. Bulletin IOBC/WPRS 19 (1): 35-38.

Evans, G.O., J.G. Sheals & D. Macfarlaine, 1961. Eriophyna: 120-125. In: The terrestrial acari of the British Isles I. Adbard and Son, Bartholomew Press - Dorking.

Farrar, C.A., T.M. Perring & N.C.Toscano, 1986. A midge predator of potato aphids on tomatoes. California Agriculture 40 (11-12): 9-10.

Fauvel, G., J.C. Malausa & B. Kaspar, 1987. Etude en laboratoire des principales caracteristiques biologiques de *Macrolophus caliginosus* (Heteroptera: Miridae). Entomophaga 32, 5: 529-543.

Fisher, T.W., 1963. Mass culture of *Cryptolaemus* and *Leptomastix* - natural enemies of citrus mealybug. Bulletin California Agricultural Experiment Station 797: 38 pp.

Frankenhuyzen, A. van, 1996. Schadelijke en nuttige insecten en mijten in aardbei en houtig kleinfruit. Nederlandse Fruittelers Organisatie - 's Gravenhage: 316 pp.

Fransen, J.J., 1994. Management of *Bemisia tabaci*: *Bemisia tabaci*, here to stay?. UK; Management *Bemisia tabaci*, abstracts.

Fransen, J.J. et al., 2000. Wol-, schild- en dopluis: signalering en bestrijding. Vakblad voor de Bloemisterij 16: 40-41.

Fransen, J. & A. van Nieuwenhoven, 1992. Problemen met kas- en tabakswittevlieg: door ze te kennen, zijn ze beter aan te pakken. Vakblad voor de Bloemisterij 49: 46-47.

Fransen, J.J., J. Tolsma, M. Marck & A. Breugem, 1993. Tien jaar californische trips in Nederland: waarnemen belangrijk instrument bij bestrijding. Vakblad voor de Bloemisterij 36: 30-33.

Freier, B. & H. Triltsch, 1995. *Harmonia axyridis* (Pallas) - ein interessanter Marienkäfer für den biologischen Pflanzenschutz. Gesunde Pflanzen 47 (7): 269-271.

Fye, R.E. & W.C. McAda, 1972. Laboratory studies on the development, longevity and fecundity of six lepidopterous pests of cotton in Arizona. USDA Technical Bulletin 1454: 73 pp.

Gagné, R.J., 1995. Revision of tetranychid (Acarina) mite predators of the genus *Feltiella* (Diptera: Cecidomyiidae). Annals of the Entomological Society of America 88 (1): 16-30.

Garms, L.M., O.E. Krips, C. Schütte & M. Dicke, 1998. The ability of the predatory mite *Phytoseiulus persimilis* to find a prey colony: effect of host plant species and herbivore induced volatiles. Proceedings of Experimental and Applied Entomology, NEV 9: 67-72.

Gaugler, R. & H.K. Kaya, 1990. Entomopathogenic nematodes in biological control. CRC Press - Boca Raton: 365 pp.

Geest, van der, & Evenhuis, 1991. World Crop Pests 5; Tortricid pests, their biology, natural enemies and control. Elsevier - Amsterdam: 808 pp.

Gerling, D., 1986. Natural enemies of *Bemisia tabaci*, biological characteristics and potential as biological control agents: a review. Agriculture, ecosystems and environment 17: 99-110.

Gerling, D., A.R. Horowitz & J. Baumgaertner, 1986. Autecology of *Bemisia tabaci*. Agriculture, ecosystems and environment 17: 5-19.

Gillespie, D., G. Opit, R. McGregor, M. Johnston, D. Quiring & M. Foisy, 1997. Use of *Cotesia marginiventris* (Cresson) (Hymenoptera: Braconidae) for biological control of cabbage loopers, *Trichoplusia ni* (Lepidoptera: Noctuidae) in greenhouse vegetable crops in British Columbia - final report to the BC Greenhouse Vegetable Research Council Pacific Agriculture Research Centre (Agassiz), Technical report 141: 13 pp.

Gillespie, D., G. Opit, R. McGregor, M. Johnston, D. Quiring & M. Foisy, 1997. Recommendations for the use of the parasitic wasp, *Cotesia marginiventris* (Hymenoptera: Braconidae) for biological control of cabbage looper, *Trichoplusia ni*. Pacific Agri Food Research Centre (Agassiz). Technical report; 142: 2 pp.

Gillespie, D.R. & C.A. Ramey, 1988. Life history and cold storage of *Amblyseius cucumeris* (Acarina: Phytoseiidae). Journal of the Entomological Society of British Columbia 85: 71-76.

Gillespie, D.R. & D.M.J. Quiring, 1994. Rearing the predatory gall midge, *Feltiella acarisuga* (Vallot) (Diptera: Cecidomyiidae). Pacific Agriculture Research Centre (Agrassiz). Technical Report 118: 21 pp.

Gillespie, D.R. & D.M.J. Quiring, 1994. Reproduction and longevity of the predatory mite, *Phytoseiulus persimilis* (Acari: Phytoseiidae) and its prey, *Tetranychus urticae* (Acari: Tetranychidae) on different host plants. Journal of the Entomological Society of British Columbia 91 (December): 3-8.

Gilstrap, F.E. & D.D. Friese, 1985. The predatory potential of *Phytoseiulus persimilis*, *Amblyseius californicus* and *Metaseiulus occidentalis* (Acarina: Phytoseiidae). International Journal of Acarology 11, 3: 163-168.

Goffau, L.J.W. de, 1991. Schade van economische betekenis in de tuinbouw door *Liriomyza huidobrensis*, de nerfmineervlieg (Diptera: Agromyzidae) in 1989 en 1990. Jaarboek Plantenziektenkundige Dienst 1989/1990 - sectie Entomologie: 51-58.

Goffau, L.J.W. de, 1994. Gegroefde lapsnuitkever of taxuskever nader bekeken. Vakblad voor de Bloemisterij 17: 43.

Goot, V. van der, 1989. Zweefvliegen. Sichting Uitgeverij Koninklijke Nederlandse Natuurhistorische Vereniging - Utrecht: 52 pp.

Gratwick, M. (ed.), 1992. Crop pests in the UK. Collected edition of MAFF leaflets. Chapman & Hall - London: 490 pp.

Guldemond, A. & F. Dieleman, 1994. Rode luis blijkt tabaksperzikluis. Groenten + Fruit / Glasgroenten 27: 16-17.

Guldemond, A., 1996. Herkennen en bestrijden van bladluis in chrysant. Vakblad voor de Bloemisterij 13: 32-35.

Guldemond, J.A., 1994. Bladluizen in de sierteelt: Hoe herken en bestrijd je ze? Vakblad voor de Bloemisterij 14: 40-43.

Gunst, J.H. de, 1978. De Nederlandse lieveheersbeestjes (Coleoptera: Coccinellidae). KNNV - Utrecht: 96 pp.

Gutierrez, A.P., J.U. Baumgaertner & K.S. Hagen, 1981. A conceptual model for growth, development, and reproduction in the ladybird beetle, *Hippodamia convergens* (Coleoptera: Coccinellidae). The Canadian Entomologist 113: 21-33.

Hall, R.A., 1982. Control of whitefly, *Trialeurodes vaporariorum* and cotton aphid *Aphis gossypii* in glasshouses by two isolates of the fungus, *Verticillium lecanii*. Annals of applied Biology (1): 1-11.

Halliwell, V, 1995. Biological control of the glasshouse mealybugs, *Pseudococcus* spp and *Planococcus* sp, using the cocconellid predator *Cryptolaemus montrouzieri* - University Bristol: 47 pp.

Harris, K.M., 1982. The aphid midge: a brief history. Antenna 6(4): 286-289.

Hart, A.J., J.S. Bale & J.S. Fenlon, 1997. Development threshold, day-degree requirements and voltinism of the aphid predator *Episyrphus balteatus* (Diptera: Syrphidae). Annals of Applied Biology, 130: 427-437.

Havelka, J., 1980. Effect of temperature on the developmental rate of preimaginal stages of *Aphidoletes aphidimyza* (Diptera, Cecidomyiidae). Entomologia Experimentalis et Applicata 27(1): 83-90.

Helle, W. & M.W. Sabelis, (eds.) 1985. World Crop Pests vol. 1 (A and B); Spider mites. Their biology, natural enemies and control. Elsevier - Amsterdam.

Hendrikse, A., 1980. A method for mass rearing two braconid parasites (*Dacnusa sibirica* and *Opius pallipes*) of the tomato leafminer (*Liriomyza bryoniae*). Mededelingen Faculteit Landbouwwetenschappen Rijksuniversiteit Gent 45/3: 563-571.

Hendrikse, A., R. Zucchi, J.C. van Lenteren & J. Woets, 1980. *Dacnusa sibirica* Telenga and *Opius pallipes* Wesmael (Hym., Braconidae) in the control of the tomato leafminer *Liriomyza bryoniae* Kalt. Bulletin IOBC/WPRS 3(3): 83-98.

Heppner, J.B., J.E. Pena & H. Glenn, 1987. The banana moth, *Opogona sacchari* (Bojer) (Lepidoptera: Tineidae), in Florida. Entomology circular 293. Department of Agriculture and Consumer Services, Division of Plant Industry.

Heungens, A. & E. van Daele, 1980. Schadelijke weekhuidmijten (Tarsonemidae) in de azaleateelt. Mededelingen Faculteit Landbouwwetenschappen Rijksuniversiteit Gent 45/5: 1223-1231.

Hill, D.S., 1987. *Lygocoris pabulinus* (L.) + *Lygus* spp. In: Agricultural insect pests of temperate regions and their control. University Press - Cambridge: 244-245.

Hoddle, M.S., R.G. van Driesche & J.S. Sanderson, 1996. A grower's guide to using biological control for silverleaf whitefly on poinsettias in the northeast United States. University of Massachusetts Cooperative Extension Publication, Floral Facts: 4 pp.

Hoddle, M.S., R.G. van Driesche & J.S. Sanderson, 1998. Biology and use of the whitefly parasitoid *Encarsia formosa*. Annual Review of Entomology 43: 645-669.

Hodek, I. & A. Honek, 1996. Ecology of Coccinellidae. Kluwer Academic Publisher - Dordrecht: 464 pp.

Hodgson, C.J., 1994. The scale family Coccidae: an identification manual to genera. CAB International - Wallingford: 639 pp.

Hoffmann, M.P. & A.C. Frodsham, 1993. Natural enemies of vegetable insect pests. Cornell University - Ithaca: 63 pp.

Hoog, Jr. J. de, 1998. Teelt van kasrozen. Proefstation voor Bloemisterij en Glasgroente - Aalsmeer: 218 pp.

Hoogelander, E., 1996. Chemie deert Floridamot amper. Groenten + Fruit / Glasgroenten 33: 16-17.

Hoogstrate, J, 1987. Dierlijke aantasters. Herkennen en bestrijden van mijten. Vakblad voor de Bloemisterij 40: 45.

Houten, Y.M. van, P.C.J. van Rijn, L.K. Tanigosha & P. van Stratum, 1993. Potential of phytoseiid predators to control western flower thrips in greenhouse crops, in particular during the winter period. Bulletin IOBC/WPRS 16/8: 98-101.

Houten, Y.M. van, P.C.J. van Rijn, L.K. Tanigosha, P. van Stratum & J. Bruin, 1995. Preselection of predatory mites to improve year-round biological control of western flower thrips in greenhouse crops. Entomologia Experimentalis et Applicata 74: 225-234.

Hubert, L., 1991. Snelle aanpak houdt brandnetelwants tegen. Groenten + Fruit / Glasgroenten 23: 18-19.

Hussey, N.W. & N. Scopes (eds.), 1985. Biological pest control: the glasshouse experience. Blandford Press - Dorset: 240 pp.

Hussey, N.W. & W.J. Parr, 1963. Dispersal of the glasshouse red spider mite *Tetranychus urticae* Koch (Acarina: Tetranychidae). Entomologia Experimentalis et Applicata 6: 207-214.

Hussey, N.W., W.H. Read & J.J. Hesling (eds.), 1969. The pests of protected cultivation: the biology and control of glasshouse and mushroom pests. Arnold Publishers, London: 404 pp.

Isenhour, D.J. & K.V. Yeargan, 1981. Effect of temperature on the development of *Orius insidiosus*, with notes on laboratory rearing. Annals of the Entomological Society of America 74 (1): 114-116.

Jacques, R.L. jr, 1988. The potato beetles. The genus *Leptinotarsa* in North America (Coleoptera: Chrysomelidae). Flora & Fauna Handbook no.3. E.J. Brill - Leiden: 144 pp.

Janssen van Bergeijk, K.E., 1995. 3. *Verticillium lecanii*. In: schimmel- en viruspreparaten. IKC Landbouw - Ede: 13-18.

Janssen van Bergeijk, K.E., 1995. Achtergrondinformatie landbouwkundige toepasbaarheid bestrijdingsmiddelen: bacteriepreparaten. IKC Landbouw - Ede: 20 pp.

Jeekel, C.A.W., 1953. (Duizendpootachtigen – Myriapoda.) De millioenpoten (Diplopoda) van Nederland. Wetenschappelijke mededelingen van de koninklijke Nederlandse natuurhistorische vereniging, No.9.

Jeppson. L.R., H.H. Keifer & E.W. Baker, 1975. Mites injurious to economic plants. University of California Press – Berkeley: 614 pp.

Jervis, M. & N. Kidd, 1996. Biological control of green leafhopper on sweet peppers. HDC Project News, August 1996: 6-7.

Jones, M.G., 1942. A description of *Aphis (Doralis) rumicis* and comparison with *Aphis (Doralis) fabae*. Bulletin of Entomological Research 33: 5-20.

Jong, D.J. de & H. Beeke, 1982. Bladrollers in appel- en pereboomgaarden. Mededeling no. 19. Wilhelminadorp: Proefstation voor de fruitteelt: 218 pp.

Kaaij, W. van der & M. Jansen, 1993. Wittevlieg-sluipwespen: *Encarsia formosa* & *Eretmocerus californicus*. Stageverslag Koppert B.V., Agrarische Hogeschool Delft: 32 pp.

Kabuswe, M.L., 1985. Ernährungsbiologie, Produktivität und Populationsdynamik von *Heliothis armigera* (Hübner) (Insecta, Lepidoptera, Noctuidae) im Hinblick auf Bekämpfungsmöglichkeiten in Sambia. Inaugural Dissertation. Bonn: Rheinischen Friedrich Wilhelms Universität: 115 pp.

Karl, E., 1965. Untersuchungen zur Morphologie und Ökologie von Tarsonemiden gärtnerischer Kulturpflanzen II; *Hemitarsonemus latus* (Banks), *Tarsonemus confusus* (Ewing), *Tarsonemus talpae* (Schaarschmidt), *Tarsonemis smithi* (Ewing) und *Tarsonemoides belemnitoides* (Ewing). Biologischer Zentralblatt H3: 331-357.

Keifer, H.H., E.W. Baker, T. Kono, M. Delfinado & W.E. Styer, 1982. An illustrated guide to plant abnormalities caused by eriophyid mites in North America. Agriculture Handbook 573. Agricultural Research Service, USDA Washington: 178 pp.

Kemperman, J., 1987. Bestrijdingsmethoden voor "overige" insekten. Vakblad voor de Bloemisterij 40: 47-50.

Kevan, D.K. McE. & G.D. Sharma, 1964. Observations on the biology of *Hypoaspis aculeifer* (Canestrini, 1884), apparently new to North America (Acarina: Mesostigmata: Laelaptidae) Acarologia 6 fasc. 4: 647-658.

Kiman, Z.B. & K.V. Yeargan, 1985. Development and reproduction of the predator *Orius insidiosus* (Hemiptera: Anthocoridae) reared on diets of selected plant material and arthropod prey. Annals of the Entomological Society of America 78 (4): 464-467.

Klein, E., 1997. Bespuiting treft volwassen dop- en schildluizen nauwelijks. Boomkwekerij 35: 12-13

Kleukers, R., Nieukerken, E. van, Ode, B., Willemse, L., & Wingerden, W. van, 1997. De sprinkhanen en krekels van Nederland. Nederlandse fauna I. Nationaal natuurhistorisch museum, Leiden: 413 pp.

Kono, T. & Papp, C.S., 1977. Handbook of agricultural pests. Aphids, thrips, mites, mites, snails and slugs.California: Dept. Of Food and Agriculture - Sacramento: 205 pp.

Kot, Ya., & T. Plewka, 1974. Biology and ecology of *Trichogramma* spp. Biological agents for plant protection. Moscow: Kolos: 183-200.

Kremer, F.W., 1956. Untersuchungen zur Biologie, Epidemiologie und Bekämpfung von *Bryobia praetiosa* Koch. Pflanzenschutz Nachrichten Bayer, Höfchen Briefe, heft 4: 189-252.

Krieg, A., 1986. *Bacillus thuringiensis*, ein mikrobielles Insektizid: Grundlagen und Anwendung. Acta Phytomedica 10. Paul Parey - Berlin & Hamburg: 191 pp.

Krips, O.E., A. Witul, P.E.L. Willems & M. Dicke, 1998. Intrinsic rate of population increase of the spider mite *Tetranychus urticae* on the ornamental crop gerbera: intraspecific variation in host plant and herbicore. Entomologia Experimentalis et Applicata 89: 159-168.

Kulp, D., M. Fortmann, M. Hommes & H. Plate, 1989. Die räuberische Gallmücke *Aphidoletes aphidimyza* (Rondani) (Diptera: Cecidomyiidae) – Ein bedeutender Blattlausprädator – Nachslagewerk zur Systematik, Verbreitung, Biologie, Zucht und Anwendung. Mitteilungen aus der Biologischen Bundesanstalt für Land- und Forstwirtschaft Berlin-Dahlem, heft 250: 126 pp.

Lacey, L., ed., 1993. Manual of techniques in insect pathology – Biological Techniques Series. Academic Press - San Diego: 40 pp.

Laing, J.E., 1968. Life history and life table of *Phytoseiulus persimilis* Athias-Henriot. Acarologia 10: 578-588.

Laing, J.E., 1969. Life history and life table of *Tetranychus urticae* Koch. Acarologia 11: 32-42.

Lenteren, J.C. van, 19.. Insects, man and environment: who will survive? In: Environmental Concerns. An interdisciplinary exercise. J.A. Hansen (ed.). Elsevier Science Publishing - Londen: 191-210.

Lenteren, J.C. van, P.M. Hulspas-Jordaan & L. Zhao-Hua, 1987. Leaf hairs, *Encarsia formosa* and biological control of whitefly on cucumber. Bulletin IOBC/WPRS 10 (2): 92-96.

Lepesme, P., 1947. Les insectes des palmiers. Paul Chevalier - Paris: 904 pp.

Lewis, T., 1973. Thrips, their biology, ecology and economic importance. Academic Press - London: 349 pp.

Lieten, F., 1998. Geintegreerde aardbeienteelt: ervaringen met lieveheersbeestjes als bladluisbestrijders. Fruitteelt Nieuws 9 okt: 6-8.

Linden, A. van der, 1993. Overwintering of *Liriomyza bryoniae* and *Liriomyza huidobrensis* (Diptera: Agromyzidae) in the Netherlands. Proceedings of Experimental & Applied Entomology, NEV, Amsterdam, 4: 145-150.

Linden, A. van der, 1996. Control of caterpillars in integrated pest management. Bulletin IOBC/WPRS 19 (1): 91-94.

Linden, A. van der, 1996. Rupsen moeten uit gaan kijken. Groenten + Fruit / Glasgroenten 13: 20-21.

Linden A. van der & C. van Achterberg, 1989. Recognition of eggs and larvae of the parasitoids of *Liriomyza* spp. (Diptera: Agromyzidae; Hymenoptera: Braconidae and Eupholidae). Entomologische Berichten 49(9): 138-140.

Lindquist, E.E., M.W. Sabelis & J. Bruin (eds), 1996. World Crop Pests, 6: Eriophyoid mites – their biology, natural enemies and control. Elsevier - Amsterdam: 790 pp.

Lublinkhof, J. & D.E. Foster, 1977. Development and reproductive capacity of *Frankliniella occidentalis* (Thysanoptera: Thripidae) reared at three temperatures. Journal of the Kansas Entomological Society 50 (3): 313-316.

MAFF, 1974. Chrysanthemum leaf miner. Advisory leaflet 550: 4 pp.

Mantel, W.P. & M. van de Vrie, 1988. De Californische trips *Frankliniella occidentalis*, een nieuwe schadelijke tripssoort in de tuinbouw onder glas in Nederland. Entomologische Berichten 48(9): 140-144.

Mantel, W.P., 1986. Handleiding voor het herkennen van enkele in de tuinbouw onder glas voorkomende tripssoorten en hun ontwikkelingsstadia. Instituut voor Planteziektenkundig Onderzoek (IPO) – Wageningen.

Margulis, L. & K.V. Schwartz, 1988. Five Kingdoms. An illustrated guide to the phyla of life on earth. W.H. Freeman and company - San Fransisco: 376 pp.

Markkula, M., K. Tiitanen, M. Hämäläinen & A. Forsberg, 1979. The aphid midge *Aphidoletes aphidimyza* (Diptera, Cecidomyiidae) and its use in biological control of aphids. Annals Entomologica Fennica 45(4): 89-98.

Mayhew, P.J., 1998. Testing the preference-performance hypothesis in phytophagous insects: lessons from chrysanthemum leafminer (Diptera: Agromyzidae). Environmental Entomology 27 (1): 45-52.

McGill, E.I., 1934. On the biology of *Anagrus atomus* (L.) Hal.: an egg parasite of the leaf-hopper *Erythroneura pallidifrons* Edwards. Parasitology 26: 57-63.

McKinlay, R.G., 1992. Vegetable crop pests. Macmillan Press - Basingstoke: 406 pp.

McLeod, D.M., D. Tyrrell & K.P. Carl, 1976. *Entomophthora parvispora* sp. nov., a pathogen of *Thrips tabaci*. Entomophaga 21: 307-312.

McMurtry, J.A., 1977. Some predaceous mites (Phytoseiidae) on citrus in the Mediterranean region. Entomophaga 22 (1): 19-30.

Meer, F.E. van der, J.Th.J. Verhoeven, M.G. Jansen & M. G. Roosjen, 1992. Tabakswittevlieg geeft nieuwe problemen: nu handelen om extra problemen te voorkomen. Vakblad voor de Bloemisterij 49: 48-49.

Meiracker, R.A.F. van den & P.M.J. Ramakers, 1991. Biological control of the western flower thrips *Frankliniella occidentalis*, in sweet pepper, with the anthocorid predator *Orius insidiosus*. Mededelingen Faculteit Landbouwwetenschappen Rijksuniversiteit Gent 56, (2A): 241-249.

Mendel, Z. D. Blumberg & M. Wysoki, 1992. Biological control of four homopterans in Israeli horticulture: achievements and problems. Anglo Israeli Symposium: Non chemical approaches to crop protection in horticulture, Wellesbourne: 32-35.

Merendonk, S. van de & J.C. van Lenteren, 1978. Determination of mortality of greenhouse whitefly *Trialeurodes vaporariorum*, Westwood (Homoptera: Aleyrodidae) eggs, larvae & pupae on four host-plant species: eggplant (*Solanum melongena* L.) cucumber (*Cucumis sativus* L.) tomato (*Lycopersicon esculentum* L.) & paprika (*Capsicum annuum* L.). Mededelingen Faculteit Landbouwwetenschappen Rijksuniversiteit Gent 43/2: 421-429.

Mertens, P., 1993. Biologische bestrijding in boomkwekerij en bij tuinaanleg: Parasitaire aaltjes bestrijden taxuskever. Tuinbouw Visie (17/09/93): 30-32.

Mietkiewski, R. & L.P.S. van der Geest, 1985. Notes on entomophthoraceous fungi infecting insects in the Netherlands. Entomologische Berichten 45: 190-192.

Minkenberg, O.P.J.M., 1990. On seasonal inoculative biological control. Governing *Liriomyza* populations by parasitoids. Thesis. Ponsen & Looijen BV - Wageningen: 230 pp.

Minks, A.K. & P. Harrewijn, 1988. World Crop Pests, vol. 2B; Aphids, their biology, natural enemies and control. Elsevier – Amsterdam: 364 pp.

Moreton, B.D., 1974. New or uncommon plant diseases and pests; *Opogona sacchari*. Plant Pathology 23: 163-164.

Mound, L.A., G.D. Morison, B.R. Pitkin & J.M. Palmer, 1976. Handbooks for the identification of British insects: vol. 1, part 11: Thysanoptera. Royal Entomological Society - London: 79 pp.

Mourikis, P.A. & P. Vassilaina-Alexopoulou, 1981. Data on the biology of the *Opogona sacchari* (Bojer, 1856), a new pest for ornamental plants in Greece. Annals of the Phytopathological Institute Benaki, (N.S.), 13: 59-64.

Murphy, P.W. & M.A. Sardar, 1988. Chapter 22: Resource allocation and utilization contrasts in *Hypoaspis aculeifer* (Can.) and *Alliphis halleri* (G. & R. Can.) (Mesostigmata) with emphasis on food source. In: The Acari 3: 301-311. Eds.: R. Schuster & Murphy, P.W. Chapman Hall - London.

Naumann, I.D., 1994. Systematic and Applied Entomology. An introduction. Melbourne University Press - Carlton: 484 pp.

Nedstam, B. & M. Burman, 1990. The use of nematodes against sciarids in Swedish greenhouses. Bulletin OILB/SROP. XII/5: 147-148.

Nedstam, B., 1985. Development time of *Liriomyza bryoniae* Kalt. (Diptera: Agromyzidae) and two of its natural enemies, *Dacnusa sibirica* Telenga (Hymenoptera: Braconidae) and *Cyrtogaster vulgaris* Walker (Hymenoptera: Pteromalidae) at different constant temperatures. Mededelingen Faculteit Landbouwwetenschappen Rijksuniversiteit Gent 50/2a: 411-417.

Norton, R.A., 1994. Evolutionary aspects of oribatid mite life histories and consequences for the origin of the Astigmata. In: Mites; Ecological and evolutionary analyses of life-history patterns, hst. 5. Ed. M.A. Houck. Chapman & Hall - New York: 99-135

Odermatt, S., 1999. Contribution à la recherche d'un moyen de lutte biologique contre *Nezara viridula* (Linneus) en culture d'aubergine sous abri. Koppert S.A.R.L France/Université d'Avignon et des Pays de Vaucluse: 1-24.

Oetting, R.D. & R.J. Beshear, 1993. Biology of the greenhouse pest *Echinothrips americanus* Morgan (Thysanoptera: Thripidae). Advances in Thysanopterology: Journal of Pure and Applied Zoology 4: 307-315.

Olkowski, W., S. Daar & H. Olkowski, 1991. Common sense pest control: least-toxic solutions for your home, garden, pets and community. Taunton Press - Newtown: 736 pp.

Opit, G., 1996. The life history and management of a new greenhouse pest, *Echinothrips americanus* Morgan. (Thysanoptera: Thripidae): 1-15.

Osborne, L.S. & J. Peña, 1998. More than you want to know about mites and their biological control on ornamentals. University Florida - Apopka: 53-85.

Osborne, L.S. & L.E. Ehler, 1981. Biological control of greenhouse whitefly in California greenhouses. Division of Agricultural Sciences, (University of California), leaflet 21260: 7 pp.

Osborne, L.S., L.E. Ehler & J.R. Nechols, 1985. Biological control of the two-spotted spider mite in greenhouses. Agriculture Experiment Stations, Institute of Food and Agric. Sciences, University Florida, Gainesville; Bulletin (Technical) 853: 40 pp.

Palmer, J.M., L.A. Mound & G.J. du Heaume, 1989. 2. Thysanoptera, CIE Guides to insects of importance to man. CAB International Institute of Entomology - Wallingford: 73 pp.

Parker, R. & U. Gerson, 1994. Dispersal of the broad mite *Polyphagotarsonemus latus* (Banks) (Heterostigmata: Tarsonemidae), by the greenhouse whitefly, *Trialeurodes vaporariorum* (Westwood) (Homoptera: Aleyrodidae). Experimental & Applied Acarology 18: 581-585.

Parrella, M.P., 1984. Effect of temperature on oviposition, feeding and longevity of *Liriomyza trifolii* (Diptera: Agromyzidae). The Canadian Entomologist 116: 85-92.

Parrella, M.P., 1993. IPM with Parrela: a new whitefly species. GrowerTalks 56 (10): 89.

Parrella, M.P. & J.A. Bethke, 1984. Biological studies of *Liriomyza huidobrensis* (Diptera : Agromyzidae) on chrysanthemum, aster, and pea. Journal of Economic Entomology 77: 342-345.

Parrella, M.P., K.L. Robb & J. Bethke, 1983. Influence of selected host plants on the biology of *Liriomyza trifolii* (Diptera: Agromyzidae). Annals of the Entomological Society of America 76: 112-115.

Peeters, I., 1999. Interactie tussen *Podisus maculiventris* (Say) en *Harmonia axyridis* Pallas, twee roofinsecten gebruikt in de biologische bestrijding van schadeverwekkers in kasteelten. Scriptie, Universiteit Gent: 106 pp.

Péricart, J., 1972. *Orius*: 160-190. In: Hémiptères, Anthocoridae, Cimicidae et Microphysidae de l'Ouest-Paléarctique, Faune de l'Europe et du bassin méditerranéen 7. Masson et Cte: Paris: 402 pp.

Piatkowski, J., 1987. Effect of temperature, relative air humidity and kind of food on the development of the predatory mite *Amblyseius mckenziei* sch. et pr. (Acarina: Phytoseiidae). Bulletin EPRS/WPRS X(2): suppl.: 1-6.

Poinar, G.O., 1989. Successful control of pests with entomopathogenic nematodes. Proceedings of the conference: Alternative systems in plant protection. Cesena 3.6.1989: 24 pp.

Poitout, S. & R. Bues, 1982. Les principales noctuelles nuisibles. Phytoma 4: 39-43.

Polk, Ph. 1959... De land-pissebedden (Isopoda, Oniscoidea) van België en Nederland. Koninklijke Nederlandse natuurhistorische Vereniging. Wetenschappelijke mededelingen no. 31, Amsterdam: 12 pp.

Popov, N.A., I.A. Zabudskaya, E.A.S. Shiiko, E.M. Mencher, & V.B. Shevchenko, 1992. Development of glasshouse pests of cucumber under different conditions, in relation to possibilities for their biological control. EPPO Bulletin 22 (3): 529-536.

Porter, J., 1997. The colour identification guide to caterpillars of the British isles (Macrolepidoptera). Viking – London: 275 pp.

Quicke, D., M. Fitton, K. van Achterberg, M. Day, N. Fergusson, J. Noyes & M. Shaw., 1993. Taxonomy and biology of parasitic hymenoptera: short course 28 March - 4 April 1993 at Sheffield University. National History Museum, University Sheffield – London.

Ramakers, P.M.J., 1987. Possibilities for biological control of *Thrips tabaci* lind. (Thysanoptera: Thripidae) in glasshouses. Mededelingen Faculteit Landbouwwetenschappen Rijksuniversiteit Gent 43/2: 463-469.

Rijn, J.F.A.T. van (ed.), 1988. Geïntegreerde bestrijding bij appel. Nederlandse Fruittelers Organisatie (NFO): 45 pp.

Rijn, P.C.J., C. Mollema & G.M. Steenhuis-Broers, 1995. Comparative life history studies of *Frankliniella occidentalis* and *Thrips tabaci* (Thysanoptera: Thripidae) on cucumber. Bulletin of Entomological Research 85: 285-297.

Riudavets, J., 1995. Predators of *Frankliniella occidentalis* (Perg.) and *Thrips tabaci* (Lind.): a review. Biological control of thrips pests. Wageningen Agricultural University Papers 95-1: 43-87.

Robb, K.L., 1989. Analysis of *Frankliniella occidentalis* (Pergande) as a pest of floricultural crops in California greenhouses. PhD dissertation, University of California - Riverside: 57 pp.

Rodriguez-Saona, C. & J.C. Miller, 1999. Temperature-dependent effects on development, mortality, and growth of *Hippodamia convergens* (Coleoptera: Coccinellidae). Environmental Entomology 28 (3): 518-522.

Romeijn, G. 1996. Info: *Duponchelia fovealis* (Zeller). Plantenziektenkundige Dienst Wageningen (04/10/96).

Romeijn, G., 1999. Info: *Opogona sacchari*. Plantenziektenkundige Dienst Wageningen (11/02/99).

Roosjen, M.G. & Th. Duijvestijn, 1989. Nieuwe mineervlieg – Aanpak mineervlieg *Liriomyza huidobrensis*. Plantenziektenkundige Dienst & Consulentschap voor de Tuinbouw: 6 pp.

Roosjen, M., H. Cevat & J. de Goey, 1992. Herkennen van ziekten en plagen. (red.:) Berg, E. van den & K. de Kruijf. Vakblad voor de Bloemisterij Plus 21a. Misset - Doetinchem: 78 pp.

Roosjen, M. & G. Romeijn, 1993. PD wil lastig rupsenduo buiten de deur houden. Vakblad Bloemisterij 45: 36-37.

Rortais, A., 1993. Etude des activités nycthémérales liées a l'alimentation, aux déplacements et a la ponte, et étude du comportement de ponte de *Macrolophus caliginosus* Wagner (Heteroptera; Miridae). INRA Antibes, France; 36 pp.

Sabelis, M.W., 1981. Biological control of two-spotted spider mites using phytoseiid predators. Part I: Modelling the predator-prey interaction at the individual level. Agricultural Research Reports, 910. PUDOC - Wageningen: 242 pp.

Salas, J. & O. Mendoza, 1995. Biology of the sweetpotato whitefly (Homoptera: Aleyrodidae) on tomato. Florida Entomologist 78 (1): 154-160.

Samson, R.A., P.M.J. Ramakers & T.G. Oswald, 1979. *Entomopthora thripidum*, a new fungal pathogen of *Thrips tabaci*. Canadian Journal of Botany 57: 1317-1323.

Sannino, L. & B. Espinosa, 1998. Ciclo biologico di *Mamestra brassicae* e danni alle colture ortive in Campania. Informatore Fitopatologico 5: 59-67.

Sas, J. van, J. Woets & J.C. van Lenteren, 1978. Determination of host-plant quality of gherkin (*Cucumis sativis* L.), melon (*Cucumis melo* L.) and gerbera (*Gerbera jamesonii* Hook) for the greenhouse whitefly, (*Trialeurodes vaporariorum* (Westwood)) (Homoptera: Aleyrodidae). Mededelingen Faculteit Landbouwwetenschappen Rijksuniversiteit Gent 43/2: 409-420.

Schelt, J. van, 1994. The selection and utilisation of parasitoids for aphid control in glasshouses. Proceedings of Experimental & Applied Entomology, N.E.V. Amsterdam 5: 151-157.

Schelt, J. van, J. Klapwijk, M. Letard & C. Aucouturier, 1996. The use of *Macrolophus caliginosus* as a whitefly predator in protected crops. *Bemisia* 1995: Taxonomy, Biology, Damage, Control and Management: 515-521.

Schuler, T., M. Hommes, H. Plate & G. Zimmermann, 1991. *Verticillium lecanii* (Zimmermann) Viégas (Hyphomycetales: Monoliaceae) Geschichte, Systematik, Verbreitung, Biologie und Anwendung im Pflanzenschutz. Mitteilungen aus der Biologischen Bundesanstalt für Land- und Forstwirtschaft, heft 269. Berlin-Dahlem: 154 pp.

Sell, P. & H.L. Kuo-Sell, !987. Leistungsfähigkeit der aphidofagen Gallmücke *Aphidoletes aphidimyza* (Rond.) (Diptera, Cecidomyiidae) in Abhängigkeit von verschiedenen Einflußfaktoren. Journal of Applied Entomology 103: 434-447.

Sen, P., 1931. Preliminary observation on the early stages of *Scatella stagnalis* Fal. (Diptera: Ephydridae). Journal of the Marine Biological Association 16: 847-851.

Sharaf, N.S., 1984. Studies on natural enemies of tetranychid mites infesting eggplant in the Jordan Valley, Journal of Applied Entomology 98: 527-533.

Simpson R.G. & C.C. Burkhardt, 1960. Biology and evaluation of certain predators of *Terioaphis maculata* (Buckton). Journal of Economic Entomology 53(1) 2: 89-94.

Slobbe, M.H.M. van, 1992. Mieren struikelblok bij bestrijding bladluizen. Groenten + Fruit / Glasgroenten 11: 22-23.

Smith, D., G.A.C. Beattie & R. Broadley (eds), 1997. Citrus pests and their natural enemies: Integrated pest management in Australia. DPI - Brisbane: 287 pp.

Smith, I.M., D.G. McNamara, P.R. Scott & M. Holderness, 1991. Quarantine pests for Europe, 2nd edition. CAB International & EPPO:
Cacoecimorpha pronubana: 132-135.
Leptinotarsa decemlineata: 352-357.

Smits, P.H., 1987. Nuclear polyhedrosis virus as biological control agent of *Spodoptera exigua*. Thesis, Pudoc - Wageningen: 127 pp.

Southwood, T.R.E. & D. Leston, 1959. *Liocoris tripustulatis*: 277. In: Land and water bugs of the British Isles. Frederick Warne & Co - London: 436 pp.

Spencer, K.A., 1973. Agromyzidae (Diptera) of economic importance. Series Entomologica, 9. Junk: Den Haag, 418 pp.

Spencer, K.A., 1990. Host specialization in the world Agromyzidae (Diptera). Series Entomologica, 45. Kluwer Academic Publishers – Dordrecht: 643 pp.

Speyer, E.R., 1927. An important parasite of the greenhouse white-fly (*Trialeurodes vaporariorum*, Westwood). Bulletin of Entomological Research 17: 301-308.

Stary, P., 1975. *Aphidius colemani* Viereck: its taxonomy, distribution and host range (Hymenoptera, Aphidiidae). Acta Entomologica Bohemoslovaka 72: 156-163.

Steene, F. van de, J. Vancayzeele & J. van Melckebeke, 1996. Geleide bestrijding van de kooluil: Demonstratieproject Spruitkool. Proeftuinnieuws 7: 37-39.

Steenis, M. van, 1995. Evaluation and application of parasitoids for biological control of *Aphis gossypii* in glasshouse cucumber crops. Proefschrift, Landbouwuniversiteit Wageningen: 217 pp.

Stenseth, C., 1979. Effect of temperature and humidity on the development of *Phytoseiulus persimilis* and its ability to regulate populations of *Tetranychus urticae* (Acarina: Phytoseiidae, Tetranychidae). Entomophaga 24 (3): 311-317.

Stigter, H. & B. Aukema, 1995. Wie is wie bij de bladluizen. Boomkwekerij 13: 10-15.

Stigter, H. & A. van Frankenhuyzen, 1994. Bruine rups beschadigt niet alleen kool. Groenten + Fruit / Fruit 32: 8-9.

Stigter, H.& G. Romeijn, 1997. Onzichtbare beschadiger. Groenten + Fruit / Glasgroenten 14: 6-7.

Sutre, B. & A. Fos, 1997. *Anagrus atomus*, parasitoïde naturel de cicadelles. Essai préliminaire de son efficacité en viticulture. Phytoma – La Défènse des Végétaux 495: 40-44.

Szabo, P., J.C. van Lenteren & P.W.T. Huisman, 1993. Development time, survival and fecundity of *Encarsia formosa* on *Bemisia tabaci* and *Trialeurodes vaporariorum*. Bulletin IOBC/WPRS 16 (2): 173-176.

Takafuji, A. & D.A. Chant, 1976. Comparative studies of two species of predacious phytoseiid mites (Acarina: Phytoseiidae), with special reference to their responses to the density of their prey. Res. Popul. Ecol. 17: 255-310.

Tawfik, M.F.S. & A.M. Ata, 1973. The life history of *Orius albidipennis* (Fieber) (Hemiptera: Anthocoridae). Bulletin Société Entomologique Egypte, 57: 145-151.

Tenhumberg, B. & H.M. Poehling, 1995. Syrphids as natural enemies of cereal aphids in Germany: Aspects of their biology and efficacy in different years and regions. Agriculture, Ecosystems and Environment 52 (1): 39-43.

Tingle, C.C.D. & M.J.W. Copland, 1989. Progeny production and adult longevity of the mealybug parasitoids *Anagyrus pseudococci*, *Leptomastix dactylopii* and *Leptomastidae abnormis* (Hym.: Encyrtidae) in relation to temperature. Entomophaga 34 (2): 111-120.

Tingle, C.C.D., 1985. Biological control of the glasshouse mealybug using parasitic Hymenoptera. PHD-Thesis, Wye College: University of London: 375 pp.

Todd, J.W., 1989. Ecology and behavior of *Nezara viridula*. Annu. Rev. Entomol. 34: 273-292.

Tol, R. van 1995. Onderschat varenrouwmug niet: steeds vaker schade in de kas. Boomkwekerij 25/26: 22-23.

Tommasini, M.G. & G. Nicoli, 1993. Adult activity of four *Orius* species reared on two preys. Bulletin IOBC/WPRS 16(2): 181-184.

Tommasini, M.G. & S. Maini, 1995. *Frankliniella occidentalis* and other thrips harmful to vegetable and ornamental crops in Europe. Biological control of thrips pests. Wageningen Agricultural University Papers 95-1: 1-42.

Tommassini, M.G. & G. Nicoli, 1996. Evaluation of *Orius* spp. as biological control agents of thrips pests. Further experiments on the existence of diapause in *Orius laevigatus*. Bulletin IOBC/WPRS 19(1): 183-186.

Tsai, J.H.& K. Wang, 1996. Development and reproduction of *Bemisia argentifolii* (Homoptera: Aleyrodidae) on five host plants. Environmental Entomology 25 (4): 810-816.

Tsueda, H. & K. Tsuchida, 1998. Differences in spatial distribution and life history parameters of two sympatric whiteflies, the greenhouse whitefly (*Trialeurodes vaporariorum* Westwood) and the silverleaf whitefly (*Bemisia argentifolii* Bellows & Perring), under greenhouse and laboratory conditions. Applied Entomology & Zoology 33 (3): 379-383.

Vacante, V., G. Tropae Garzia, C. Pucci & G.E. Cocuzza, 1995. Notes sur la biologie d'*Orius laevigatus* (Fieber). I. Influence de la photoperiode. Mededelingen Faculteit Landbouwwetenschappen Universiteit Gent 60 (3a): 631-633.

Vacante, V., G.E. Cocuzza, P. de Clercq, M. van de Veire & L. Tirry. Development and survival of *Orius albidipennis* and *Orius laevigatus* (Het.: Anthocoridae) on various diets. Entomophaga 42 (4): 493-498.

Valicente, F.H. & R.J. O'Neil, 1995. Effects of host plants and feeding regimes on selected life history characteristics of *Podisus maculiventris* (Say) (Heteroptera: Pentatomidae). Biological Control 5: 449-461.

Vänninen, I. & H. Koskula 1996. Biology and management of shore flies (*Scatella stagnalis*) in a cucumber seedling crop grown in rockwool. Bulletin IOBC/WPRS 19 (1): 187-190.

Vanparys, L., 1994. Geleide bestrijding: Waarnemingen van de kooluil en de groente-uil. Proeftuinnieuws 14: 42-43.

Veenenbos, J.A.J., 1981. *Opogona sacchari*, a pest risk from imports of ornamental plants of tropical origin. EPPO Bulletin 11 (3): 235-237.

Veenenbos, J.A.J., 1988. EPPO data sheets on quarantaine organisms 154. *Opogona sacchari*. EPPO Bulletin 18: 513-516.

Vehrs, S.L., J.A. Bethke, J. Garcia & M.P. Parrella, 1988. Aphids: common problems on numerous ornamentals. California Florists, january 1988: 1-14.

Verhoeven, J.Th.J., 1996. Impatiens-vlekkenvirus: een nieuwe bedreiging voor de bloemisterij. Vakblad voor de Bloemisterij 35: 22-25.

Verhoeven, J.Th.J., M.G.M. Jansen & M.G. Roosjen, 1992. Tabakswittevlieg brengt nieuwe problemen. Groenten en Fruit – Glasgroenten 46: 10-11.

Verschuuren, G.M.N., H. de Bruin & M.W. Halsema, 1985. Grondslagen van de biologie deel 3, Populaties. 185 pp. Vertaald en bewerkt van "Elements of biological science, William T. Keeton & Carol Hardy McFadden, derde druk. Norton: New York.

Vet, L.E.M. & J.C. van Lenteren, 1981. The parasite-host relationship between *E. formosa* Gah. and *T. vaporariorum*: 10. A comparison of three *Encarsia* spp. and one *Eretmocerus* sp. to estimate their potentialities in controlling whitefly on tomatoes in greenhouses with a low temperature regime. Zeitschrift für angewandte Entomologie 91, 4: 327-348.

Vette, A.W.G.M., 1983. Biologische bestrijding van *Liriomyza bryoniae* op tomaat: een verslag van praktijkproeven met *Opius pallipes*. Stageverslag, vakgroep Entomologie, Landbouwhogeschool, Wageningen: 112 pp.

Vierbergen, B., 1997. Amerikaanse trips: nieuwe belager in paprika. Groenten + Fruit / Glasgroenten 24: 12-13.

Vierbergen, G., 1993. Een schadelijke trips voor de Nederlandse sierteelt onder glas: *Echinothrips americanus* Morgan. Diagnostische circulaire, PD Entomologie: 3 pp.

Vrie, M. van de, 1977. *Spodoptera exigua* (Lepidoptera: Noctuidae) in sierteeltgewassen Gewasbescherming 8(2): 67-70.

Vrie, M. van de, P. Wilders & J.H.L. Kemperman, 1988. Bestrijding moeizaam: mijten richten veel schade in gewassen aan. Vakblad voor de Bloemisterij 27: 48, 49, 51.

Wahab, W.A., 1985. Observations on the biology and behaviour of *Aphelinus abdominalis* Dalm. (Hym., Aphelinidae), a parasite of aphids. Zeitschrift für angewandte Entomologie 100: 290-296.

Westerman, P.R., L.J.W. Goffau & N.W.F. Steeghs, 1993. Het is niet alles varenrouwmug wat er vliegt. Vakblad voor de Bloemisterij 28: 38-39.

Wiedenmann, R.N. & R.J. O'Neil, 1991. Searching behavior and time budgets of the predator *Podisus maculiventris*. Entomologia Experimentalis et Applicata 60: 83-93.

Wilder, P., 1987. Bodemdieren onder de loep genomen. Vakblad voor de Bloemisterij 40: 51-53.

Wilding, N. & G. Latteur, 1987. The entomophthorales-problems relative to their mass production and their utilisation. Mededelingen Faculteit Landbouwwetenschappen Rijksuniversiteit Gent 52/2a: 159-164.

Willemse, C., 1971. De in Nederland voorkomende oorwormen (Dermaptera). Wetenschappelijke mededelingen van de koninklijke Nederlandse natuurhistorische vereniging, nr. 4: 16 pp.

Woets, J., 1997. Genoeg middelen voor bestrijding dopluis. Groenten + Fruit / Hard- en zachtfruit 7: 8-9.

Woets, J. & H. Stigter, 1995. De nieuwste plaag heet anjerbladroller. Groenten + Fruit / Vollegrondsgroenten 8: 12-13.

Woets, J. & J.C.v. Lenteren, 1976. The parasite-host relationship between *Encarsia formosa* (Hymenoptra: Aphelinidae) and *Trialeurodes vaporariorum* (Homoptera: Aleyrodidae). VI: The influence of the host plant on greenhouse whitefly and its parasite *Encarsia formosa*. Bulletin OILB/SROP 4: 151-164.

Wouts, M.W., 1980. The biology and life cycle of a New Zealand population of *Heterorhabditis heliothidis*. Nematologica 25: 191-202.

Wright, E.M. & Chambers, R.J., 1994. The biology of the predatory mite *Hypoaspis miles* (Acari: Laelapidae), a potential biological control agent of *Bradysia paupera* (Dipt.: Sciaridae). Entomophaga 39 (2): 225-235.

Wyckhuys, K., 2000. Interactie tussen de eiparasitoïd *Trissolcus basalis* (Woll.) en de roofwants *Podisus maculiventris* (Say), twee natuurlijke vijanden van de fytofage schildwants *Nezara viridula*. Universiteit Gent.

Zhang, Z-Q, 1995. Mites of glasshouses and nurseries: identfication, biology and control. CAB International Institute of Entomology - London: 95 pp.

Zöllner, U. & H.M. Poehling, 1994. Influence of different aphid species on the efficiency of gall midge larvae (*Aphidoletes aphidimyza*) (Rond.) (Diptera, Cecidomyiidae). Mededelingen Faculteit Landbouwwetenschappen Universiteit Gent, 59/2a: 281-286.

Index of Latin and English names

Name	Page (bold=detailed description)
7-spot ladybird – *Coccinella septempunctata*	9, 19, 149, 152
Acheta domestica – house cricket	17
Aculops lycopersici – tomato russet mite	20, 39, **44-45**
Adalia bipunctata	19, 144, **150-152**, 168-169
Adalia decempunctata	19, 149-150, 152
Agromyzidae – leaf miners	18, **112-114**, 235, 241, 276
Agrotis – cutworms	19, 171, 173, 193
Aleyrodes lonicera – strawberry whitefly	16, 55, 56, **66**
Aleyrodes proletella – cabbage whitefly	16, 55, 56, **66**
Aleyrodidae – whiteflies	16, **56-58**, 276
Alloxysta	18, 163, **164**, 167
Amblyseius barkeri = *Neoseiulus barkeri*	20, 47, 84, 102
Amblyseius californicus = *Neoseiulus californicus*	20, 21, 27, **31-33**, 37, 38, 47, 48, 84, 102
Amblyseius cucumeris = *Neoseiulus cucumeris*	20, 21, 27, 47-48, 84, 96, **97-99**, 100, 102, 109-110
Amblyseius degenerans = *Iphiseius degenerans*	20, 21, 27, 47-48, 84, 96, **100**, 102, 109-110
american serpentine leaf miner – *Liriomyza trifolii*	18, 111-114, **116-117**, 120-122, 126
Anagrus atomus	18, 144, 248, **262**
ants – Formicidae	18, 263, **265**
Aphelinus abdominalis	18, 144, 159, **161-162**, 163-164, 166, 168-169
Aphidius colemani	18, 144, **159-161**, 166, 168-169
Aphidius ervi	18, 144, 159, **160-161**, 166, 168-169
Aphidius matricariae	18, 159, 166
Aphidius	18, 125, 159, 163-164
Aphidoidea – aphids	16, 129, **130-134**, 276
Aphidoletes aphidimyza	18, 144, **145-147**, 168-169, 240
aphids – Aphidoidea	16, 129, **130-134**, 276
Aphis fabae – black bean aphid	16, 140-141
Aphis gossypii – cotton aphid	16, 130-132, **136-137**, 140, 159-160
Aphis nasturtii – buckthorn aphid	16, 132, 140, 142
armoured scales – *Diaspididae*	17, 205-206, **218-219**
Aspidiotus nerii – oleander scale	17, 206, 218-219
Auchenorrhyncha – leafhoppers	17, **258-259**, 276
Aulacorthum circumflexum – lily aphid	16, 140, 143
Aulacorthum solani – glasshouse potato aphid	16, 130-131, **138-139**, 140-141, 159-160, 169
Autographa gamma – silver-Y moth	19, 171-173, 175, 177, **182**, 193, 203
azalea moth – *Caloptilia azaleella*	18, 197
Bacillus thuringiensis	11, 14, 171-172, 180, 198, **199-200**, 203, 227, 234
banana moth – *Opogona sacchari*	19, 171-173, **188-189**, 195
beet armyworm – *Spodoptera exigua*	19, 171, 173, 175, **180-181**, 192
beetles – Coleoptera	19, **226**
Bemisia tabaci – tobacco whitefly	16, 55, 56, 58, **61-65**, 67, 71
bird cherry-oat aphid – *Rhopalosiphum padi*	16, 130, 140, 143, 159, 168-169
black bean aphid – *Aphis fabae*	16, 140-141
black vine weevil – *Otiorhynches sulcatus*	19, 225-226, **228-229**, 234
Boisduval scale – *Diaspis boisduvalii*	17, 218-219
Brachycaudus helichrysi – leaf curling plum aphid	16, 140, 142
Bradysia	18, 237, 240
Brevipalpus	20, 42-43
broad mite – *Polyphagotarsonemus latus*	20, 39, **48-51**
brown lacewings – Hemerobiidae	19, 21, 27, 154, **155-156**
brown scale – *Parthenolecanium corni*	16, 206, 213-214, **217**
brown soft scale – *Coccus hesperidum*	16, 206, 213, **215**
Bryobia	20, 39, **40-41**
buckthorn aphid – *Aphis nasturtii*	16, 132, 140, 142
butterflies and moths – Lepidoptera	19, **173-175**, 276
cabbage leafroller – *Clepsis spectrana*	19, 171, 173, **185-186**, 193
cabbage moth – *Mamestra brassicae*	19, 171-174, **179**, 192, 203
cabbage whitefly – *Aleyrodes proletella*	16, 55, 56, **66**
Cacoecimorpha pronubana – carnation leafroller	18, 171, 173, **189-190**, 196
Caloptilia azaleella – azalea leaf miner	18, 194
carnation leafroller – *Cacoecimorpha pronubana*	18, 171, 173, **189-190**, 196
Cecidomyiidae – gall midges	18, 240
Chromatomyia syngenesiae – chrysanthemum leaf miner	18, 111-114, **119**, 120, 241
chrysanthemum leaf miner – *Chromatomyia syngenesiae*	18, 111-114, **119**, 120, 241
Chrysodeixis chalcites – tomato looper	19, 171-173, 175, **177**, 191, 203
Chrysoperla carnea	19, 84, 144, 168-169, 262
Chrysopidae – green lacewings	19, **154-155**, 156
citrus mealy bug – *Planococcus citri*	16, 206-209, **210-211**, 224

Clepsis spectrana – cabbage leafroller	19, 171, 173, **185-186**, 193
click beetles – Elateridae	19, 225-226, **230**, 234
Coccidae – soft scales	16, 205-206, **213-214**
Coccinella septempunctata – 7-spot ladybird	9, 19, 149, 152
Coccinellidae – ladybirds	19, **148-150**
Coccotrypes dactyliperda – palm seedborer	19, 225-226, **231**, 234
Coccus hesperidum – brown soft scale	16, 206, 213, **215**
Coenosia	18, 241
Coleoptera – beetles	19, **226**
Collembola – springtails	14, 263, **264**
Colorado beetle - *Leptinotarsa decemlineata*	19, 225-226, **227**
common earwig – *Forficula auricularia*	17, **266**
common nettle bug – *Liocoris tripustulatis*	16, 248-250, **252-257**
cotton aphid – *Aphis gossypii*	16, 130-132, **136-137**, 140, 159-160
crickets – Gryllidae	17, 263, **267**
Cryptolaemus montrouzieri	19, 153, 205, 207, 220, **221-222**, 224
currant lettuce aphid – *Nasonovia ribis-nigri*	16, 140, 142
cutworms – *Agrotis*	19, 171, 173, 193
cyclamen mite – *Tarsonemus pallidus*	20, 39, 48, **50-52**
Dacnusa sibirica	18, 111, 121-123, **124-127**, 128
Dendrocerus carpenteri	18, 163, **164**, 167
Dermaptera – earwigs	17, 263, **266**
Diaspididae – armoured scales	17, 205-206, **218-219**
Diaspis boisduvalii – boisduval scale	17, 218-219
Diglyphus isaea	18, 111, 121, **122-123**, 125, 127-128
Diplopoda – millipedes	14, 263, **269**
Diptera – flies	18, 236-237
Drosophila	18, 240
Drosophilidae – fruit flies	18, 235, 240
Duponchelia fovealis	19, 171-173, **186-187**, 195
dust lice – Psocoptera	17, 263, **268**
earwigs – Dermaptera	17, 263, **266**
Echinothrips americanus	17, 83, 85, 87, **92-93**, 95, 110, 171, 173
Egyptian cotton leafworm – *Spodoptera littoralis*	19, 171-173, **183**, 194
Elateridae – click beetles	19, 225-226, **230**, 234
Empoasca vitis – grape leafhopper	17, 248, 258-259, **260**, 262
Encarsia formosa	18, 55, 67, **68-71**, 74, 81-82
Entomophthorales	14, 236, **239**
Episyrphus balteatus – marmalade hoverfly	18, 144, 156, **157-158**, 168-169, 241
Eretmocerus eremicus	18, 55, 67, **72-73**, 81-82
Eretmocerus mundus	18, 55, 67, 74, 75, 81-82
Eriophyidae – gall mites	20, 39, **44**
European plant bug – *Lygus rugulipennis*	16, 248-251, **254-257**
false spider mites – Tenuipalpidae	20, 39, **42-43**
Feltiella acarisuga	18, 21, 27, **34-35**, 37, 38, 240
flies – Diptera	18, 236-237
Forficula auricularia – common earwig	17, **266**
Formicidae – ants	18, 263, **265**
Frankliniella occidentalis – western flower thrips	17, 83, 84-87, **89-91**, 94, 97, 99
fruit flies – Drosophilidae	18, 235, 240
fruit spider mite – *Panonychus ulmi*	20, 22
fungus gnats – Sciaridae	18, 235-236, **237-238**, 240, 276
gall midges – Cecidomyiidae	18, 240
gall mites – Eriophyidae	20, 39, **44**
garden symphylan – *Scutigerella immaculata*	12, 14, 263, **270-271**
Gastropoda – slugs and snails	15, 263, **272-273**
Geolaelaps aculeifer = *Hypoaspis aculeifer*	20, 47, 84, 96, **101**, 102, 109-110, 235, 242, 244, **245-246**, 247
Geolaelaps miles = *Hypoaspis miles*	20, 47, 84, 96, **101**, 102, 109-110, 235, 242, **244-245**, 247
glasshouse potato aphid – *Aulacorthum solani*	16, 130-131, **138-139**, 140-141, 159-160, 169
glasshouse whitefly – *Trialeurodes vaporariorum*	16, 55-58, **59-61**, 65, 70, 71
grain aphid – *Sitobion avenae*	16, 130, 140, 143, 160, 162
grape leafhopper – *Empoasca vitis*	17, 248, 258-259, **260**, 262
green capsid bug – *Lygocoris pabulinus*	16, 248-250, **251-252**, 257
green lacewings – Chrysopidae	19, **154-155**, 156
Gryllidae – crickets	17, 263, **267**
Harmonia axyridis – multicoloured asiatic lieveheersbeestje	19, 152
Helicoverpa armigera – tomato fruitworm	19, 171, 173, **184**, 194
Heliothrips haemorhoidales	83, 85, 92
Hemerobiidae – brown lacewings	19, 21, 27, 154, **155-156**
hemispherical scale – *Saissetia coffeae*	16, 206, 213-214, **216**
Hercinothrips femoralis	83, 85, 92
Heteroptera – true bugs	16, **249-250**

Heterorhabditis bacteriophora	12, 14, 225, **232-233**, 234
house cricket – *Acheta domestica*	17, 267
hover flies – Syrphidae	18, **156-157**, 241
Hypoaspis aculeifer = *Geolaelaps aculeifer*	20, 47, 84, 96, **101**, 102, 109-110, 235, 242, 244, **245-246**, 247
Hypoaspis miles = *Geolaelaps miles*	20, 47, 84, 96, **101**, 102, 109-110, 235, 242, **244-245**, 247
Iphiseius degenerans = *Amblyseius degenerans*	20, 21, 27, 47-48, 84, 96, **100**, 102, 109-110
Isopoda – woodlice	15, 263, **274**
Lacanobia oleracea – tomato moth	19, 171-173, 175, **178**, 191, 203
ladybirds – Coccinellidae	19, **148-150**
leaf curling plum aphid – *Brachycaudus helichrysi*	16, 140, 142
leaf miners – Agromyzidae	18, **112-114**, 235, 241, 276
leafhoppers – Auchenorrhyncha	17, **258-259**, 276
Lepidoptera – butterflies and moths	19, **173-175**, 276
Leptinotarsa decemlineata – Colorado beetle	19, 225-226, **227**
Leptomastix dactylopii	18, 205, 220, **223**, 224
lily aphid – *Aulacorthum circumflexum*	16, 140, 143
Liocoris tripustulatis – common nettle bug	16, 248-250, **252-257**
Liriomyza bryoniae – tomato leaf miner	18, 111-114, **115-116**, 117, 120, 122, 124, 126
Liriomyza huidobrensis – pea leaf miner	18, 111-114, **118**, 120, 122
Liriomyza trifolii – american serpentine leaf miner	18, 111-114, **116-117**, 120-122, 126
Liriomyza	18, 241
long-tailed mealy bug – *Pseudococcus longispinus*	16, 208, **212**
Lycoriella	18, 237
Lygocoris pabulinus – green capsid bug	16, 248-250, **251-252**, 257
Lygus rugulipennis – European plant bug	16, 248-251, **254-257**
Macrolophus caliginosus	16, 21, 27, 37-38, 55, 67, **76-78**, 81-82, 84, 111, 121, 172, 198, 203, 249, 262
Macrosiphum euphorbiae – potato aphid	16, 130, **137-138**, 140-141, 159, 162, 169
Macrosiphum rosae – rose aphid	16, 140, 143
Mamestra brassicae – cabbage moth	19, 171-174, **179**, 192, 203
marmalade hoverfly – *Episyrphus balteatus*	18, 144, 156, **157-158**, 168-169, 241
mealy bugs – Pseudococcidae	16, 205-206, **207-209**
millipedes – Diplopoda	14, 263, **269**
moss mites – Oribatidae	20, 39, **53**
moth flies – Psychodidae	18, 235, 241
multicoloured asiatic ladybird – *Harmonia axyridis*	19, 152
muscid flies – Muscidae	18, 241
Muscidae – muscid flies	18, 241
Myzus ascalonicus – shallot aphid	16, 140-141
Myzus persicae var. *nicotianae* – tobacco aphid	16, 130, 132, **135-136**, 140, 159-160
Myzus persicae var. *persicae* – peach potato aphid	16, 130, 132, **135-136**, 140, 159-160
Nasonovia ribis-nigri – currant lettuce aphid	16, 140, 142
Neoseiulus barkeri = *Amblyseius barkeri*	20, 47, 84, 102
Neoseiulus californicus = *Amblyseius californicus*	20, 21, 27, **31-33**, 37, 38, 47, 48, 84, 102
Neoseiulus cucumeris = *Amblyseius cucumeris*	20, 21, 27, 47-48, 84, 96, **97-99**, 100, 102, 109-110
Nezara viridula – southern geen stink bug	16, **255-257**
obscure mealy bug – *Pseudococcus affinis*	16, 208, **211-212**
oleander scale – *Aspidiotus nerii*	17, 206, 218-219
onion thrips – *Thrips tabaci*	17, 83-85, 87, **88**, 89-92, 94
Opius pallipes	18, 111, 121-123, **124-127**, 140
Opogona sacchari – banana moth	19, 171-173, **188-189**, 195
Oribatida – moss mites	20, 39, **53**
Orius insidiosus	103
Orius laevigatus	16, 84, 96, 103-104, **105-107**, 109-110
Orius laticollis	103
Orius majusculus	16, 84, 96, 103-104, **105-107**, 109-110, 169
Orius minutus	103
Orius niger	103
Orius vicinus	103
Orius	16, **103-105**, 109-110, 129, 172, 198, 249
Otiorhynches sulcatus – black vine weevil	19, 225-226, **228-229**, 234
palm seedborer – *Coccotrypes dactyliperda*	19, 225-226, **231**, 234
palm thrips – *Parthenothrips dracaenae*	17, 84, 93, 95
Panonychus ulmi – fruit spider mite	20, 22
Parthenolecanium corni – brown scale	16, 206, 213-214, **217**
Parthenothrips dracaena – palm thrips	17, 84, 93, 95
pea leaf miner – *Liriomyza huidobrensis*	18, 111-114, **118**, 120, 122
peach potato aphid – *Myzus persicae* var. *persicae*	16, 130, 132, **135-136**, 140, 159-160
Philaenus spumarius	17, 248, 258-259, **261**
Phytoseiulus persimilis	20-21, 27, **28-30**, 37-38, 98, 102
Pinnaspis spp.	17, 206, 218-219
Planococcus citri – citrus mealy bug	16, 206-209, **210-211**, 224
Polyphagotarsonemus latus – broad mite	20, 39, **48-51**

potato aphid – *Macrosiphum euphorbiae*	16, 130, **137-138**, 140-141, 159, 162, 169
Praon volucre	18, 167
Pseudococcidae – mealy bugs	16, 205-206, **207-209**
Pseudococcus affinis – obscure mealy bug	16, 208, **211-212**
Pseudococcus longispinus – long-tailed mealybug	16, 208, **212**
Pseudococcus	16, 206-208, 224
Psocoptera – dust lice	17, 263, **268**
Psychodidae – moth flies	18, 235, 241
Rhodobium porosum – yellow rose aphid	16, 140, 142
Rhopalosiphum padi – bird cherry oat aphid	16, 130, 140, 143, 159, 168-169
rose aphid – *Macrosiphum rosae*	16, 140, 143
rose thrips – *Thrips fuscipennis*	17, 83-85, 87, **91**, 94
Saissetia coffeae – hemispherical scale	16, 206, 213-214, **216**
Scatella stagnalis	18, **239-240**
Sciara	18, 237
Sciaridae – fungus gnats	18, 235-236, **237-238**, 240, 276
Scutigerella immaculata – garden symphylan	12, 14, 263, **270-271**
Scymnus	19, 153
shallot aphid – *Myzus ascalonicus*	16, 140-141
shore flies – Ephydridae	18, 236, **239**
silver-Y moth – *Autographa gamma*	19, 171-173, 175, 177, **182**, 193, 203
Sitobion avenae – grain aphid	16, 130, 140, 143, 160, 162
slugs and snails – Gastropoda	15, 263, **272-273**
soft scales – Coccidae	16, 205-206, **213-214**
southern green stink bug – *Nezara viridula*	16, **255-257**
spider mites – Tetranychidae	21, 22
Spodoptera exigua – beet armyworm	19, 171, 173, 175, **180-181**, 192
Spodoptera littoralis – Egyptian cotton leafworm	19, 171, 173, **183**, 194
springtails – Collembola	14, 263, **264**
Steinernema feltiae	12, 14, 235, 242, **243**, 247
Steneotarsonemus	20, 48
Stethorus punctillum	19, 21, 27, **36**, 153
storage mites – *Tyrophagus*	20, 39, **46-47**, 101
strawberry whitefly – *Aleyrodes lonicera*	16, 55, 56, **66**
Syrphidae – hover flies	18, **157-157**, 241
tarsonemid mites – Tarsonemidae	20, 39, **48**
Tarsonemidae – tarsonemidae	20, 39, **48**
Tarsonemus pallidus – cyclamen mite	20, 39, 48, **50-52**
Tenuipalpidae – false spider mites	20, 39, **42-43**
Tenuipalpus	20, 42-43
Tetranychidae – spider mites	20, 21, 22
Tetranychus urticae – two-spotted spider mite	20, 22, **23-26**, 37, 38
thrips – Thysanoptera	17, **85-87**, 276
Thrips fuscipennis – rose thrips	17, 83-85, 87, **91**, 94
Thrips major	84, 91, 94
Thrips palmi	17, 84, 93, 95
Thrips tabaci – onion thrips	17, 83-85, 87, **88**, 89-92, 94
Thysanoptera – thrips	17, **85-87**, 276
tobacco aphid – *Myzus persicae* var. *nicotianae*	16, 130, 132, **135-136**, 140, 159-160
tobacco whitefly – *Bemisia tabaci*	16, 55, 56, 58, **61-65**, 67, 71
tomato fruitworm – *Helicoverpa armigera*	19, 171, 173, **184**, 194
tomato leaf miner – *Liriomyza bryoniae*	18, 111-114, **115-116**, 117, 120, 122, 124, 126
tomato looper – *Chrysodeixis chalcites*	19, 171-173, 175, **177**, 191, 203
tomato moth – *Lacanobia oleracea*	19, 171-173, 175, **178**, 191, 203
tomato russet mite – *Aculops lycopersici*	20, 39, **44-45**
Trialeurodes vaporariorum – glasshouse whitefly	16, 55-58, **59-61**, 65, 70, 71
Trichogramma brassicae	18, 172, 198, **201-202**, 203
true bugs – Heteroptera	16, **249-250**
two-spotted spider mite – *Tetranychus urticae*	20, 22, **23-26**, 37, 38
Tyrophagus – storage mites	20, 39, **46-47**, 101
Verticillium lecanii	14, 55, 67, **79-80**, 81-82, 84, 96, **108**, 109-110, 144, **165**
western flower thrips – *Frankliniella occidentalis*	17, 83, 84-87, **89-91**, 94, 97, 99
whiteflies – Aleyrodidae	16, **56-58**, 276
woodlice – Isopoda	13, 15, 263, **274**
yellow rose aphid – *Rhodobium porosum*	16, 140, 142

Knowing and recognizing **Photography**

Photography

Where there is more than one figure under one number the sequence on the page is:

A   D
B   E
C   F

Bio-Bee, Kibbutz Sde Eliyahu, Israel: 2.6, 2.10, 6.26, 7.63, 7.64, 9.31, 9.32, 9.33, 9.34, 9.35

R. Bühl, Stuttgart, Germany: 2.15, 4.27, 5.15, 5.25, 5.34, 7.78, 10.2, 10.6

Dutch National Information Service: 5.1, 5.6, 5.8

Elsevier, Doetinchem, Netherlands: 3.10B, 3.17B, 3.17C, 3.22B, 8.25, 8.42, 9.3, 9.4, 9.15, 9.16, 9.28, 9.29, 13.12

Koppert B.V., Berkel en Rodenrijs, Netherlands: 2.1, 2.2, 2.5, 2.7, 2.9, 2.11, 2.14, 2.16, 2.22, 3.4, 4.3, 4.9A, 4.9B, 4.15, 4.19, 4.25, 4.29, 4.34, 4.44, 4.45, 4.46, 4.47, 5.3, 5.4, 5.5, 5.7, 5.10, 5.13, 5.24, 5.28, 5.31, 6.5, 6.9, 6.12, 6.13, 6.18, 6.21, 6.22, 6.23, 6.24, 6.25, 6.27, 6.30, 6.31, 7.2, 7.3, 7.6A, 7.6B, 7.6C, 7.8, 7.10, 7.12A, 7.12B, 7.12C, 7.13B, 7.15A, 7.15B, 7.17, 7.23, 7.30, 7.34, 7.39A, 7.43, 7.46, 7.61, 7.68, 7.76, 7.91, 8.3, 8.6, 8.8, 8.12, 8.14, 8.16, 8.18A, 8.20A, 8.22B, 8.22C, 8.23B, 8.23C, 8.31, 8.37, 8.38, 8.39, 8.40, 8.49, 8.51, 8.55, 9.2, 9.6, 9.9, 10.4, 10.7, 10.11, 10.12, 11.1, 11.4, 11.5, 11.7, 11.8, 11.12, 12.2, 12.3, 12.7, 12.10B, 12.14, 12.15, 12.16B, 13.1, 13.2, 13.7B, 13.8, ch. 14 sticky plates

Koppert B.V., J. van Schelt: 2.3, 2.4, 2.8, 2.12, 2.13, 2.19, 2.20, 2.21, 2.23, 2.24, 2.25, 2.26, 2.27, 2.28, 2.30, 3.3, 3.7B, 3.13, 3.14, 3.15, 3.17A, 3.20, 3.21, 4.1, 4.2, 4.5, 4.6, 4.7, 4.8, 4.10, 4.11, 4.12, 4.13, 4.14, 4.16, 4.17, 4.18, 4.28, 4.32, 4.33, 4.36, 4.36a, 4.36b, 4.37, 4.38, 4.40, 4.41, 4.43, 5.2, 5.9, 5.11, 5.12, 5.14, 5.17, 5.18, 5.21, 5.22, 5.26, 5.27, 5.29, 5.30, 5.32, 5.33, 6.1, 6.6, 6.7, 6.10, 6.11, 6.15, 6.17, 7.1, 7.4, 7.9, 7.11, 7.12D, 7.13A 7.14, 7.15C, 7.16, 7.18, 7.19, 7.20, 7.22, 7.24, 7.25, 7.26, 7.27, 7.28, 7.29, 7.31, 7.32, 7.36, 7.38, 7.39B, 7.40, 7.42, 7.44, 7.48, 7.49, 7.50, 7.51, 7.53, 7.54, 7.55, 7.57, 7.58, 7.62, 7.65, 7.67, 7.69, 7.71, 7.72, 7.73, 7.74, 7.79, 7.80, 7.81, 7.82, 7.83, 7.84, 7.85, 7.89, 7.90, 8.19B, 8.32, 8.50, 8.61, 8.62, 9.1, 9.5, 9.7, 9.8, 9.10, 9.11, 9.12, 9.13, 9.20, 9.21, 9.25, 10.10, 11.2, 11.3, 11.6, 11.9, 11.11, 12.15A, 12.17, 12,20, 12.21, 12.25, 12.26, 12.27, 12.28, 12.29, 13.3, 13.4, 13.5, 13.10, 13.11, 13.13, 13.14, 13.15, 14.1

Koppert B.V., S. Mulder: 3.16, 5.23, 6.2, 6.8

Koppert Biological Systems, Spain, M. Fajardo: 3.10A

KVL, Denmark: 7.37, 7.88

LFP Stuttgart, Germany: 5.16, 5.35, 6.19, 7.59, 9.27, 9.30

B. Mans, Naaldwijk, Netherlands: 2.17, 2.18, 3.12, 4.22, 4.23, 4.24, 4.30, 4.31, 4.39, 4.42, 5.19, 5.20, 7.35, 7.41, 7.45, 7.70, 7.77, 8.1, 8.5, 8.15, 8.17A, 8.17B, 8.17C, 8.18B, 8.20B, 8.20C, 8.27, 8.30, 8.33, 8.34, 12.1, 12.6, 12.8, 12.9, 12.11, 12.12, 12.13, 13.7A

Novo Nordisk, Bagsvaerd, Denmark: 8.53, 8.54

Plant Pathology Service, Wageningen, Netherlands, A. van Frankenhuyzen: 2.29, 3.2, 3.8, 3.24, 7.56, 7.60, 7.66, 8.9, 8.41, 9.19, 10.9, 13.6, 13.16

Plant Pathology Service, Wageningen, M. Jansen: 9.26

Plant Pathology Service, Wageningen, W. van Lienden: 3.23

Plant Pathology Service, Wageningen, H. Stigter: 3.1, 3.5, 3.6, 3.7A, 3.11, 3.22A, 6.3, 6.14, 8.7, 8.10, 8.21.a, 8.21.b, 8.22A, 8.23A, 8.24, 8.26, 8.43, 8.44, 8.45, 8.48, 9.14, 9.24, 10.1, 10.5, 13.9

Plant Pathology Service, Wageningen, G. Vierbergen: 3.9

Plant and Environmental Practice Research Unit, Naaldwijk: 4.9C, 7.6D, 7.6E, 12.10A

Research Institute for Green Space, Alterra, Wageningen, A. van Frankenhuyzen: 4.4, 4.20, 4.21, 7.33, 7.75, 8.2, 8.4, 8.11, 8.13, 8.17D, 8.18C, 8.19A, 8.19C, 8.19D, 8.19E, 8.21, 8.28, 8.29, 8.35, 8.36, 8.46, 8.47, 9.17, 9.18, 9.22, 9.23, 10.3, 10.8, 12.5, 12.22, 12.23, 12.24, 12.30, 12.31, 12.32

SRVP, Nord pas de Calais, France: 7.21

W. Stepman, BCP Ltd.: 6.4, 6.16, 6.20

Texas A & M, USA, D. Gouge: 11.10, 12.4, 12.18, 12.19

University of Wageningen, J. van Lenteren: 6.28, 6.29